Network analysis and practice

A. K. WALTON
Department of Physics, Sheffield University

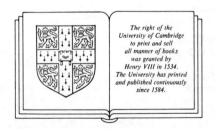

CAMBRIDGE UNIVERSITY PRESS
CAMBRIDGE
LONDON NEW YORK NEW ROCHELLE
MELBOURNE SYDNEY

Published by the Press Syndicate of the University of Cambridge
The Pitt Building, Trumpington Street, Cambridge CB2 1RP
32 East 57th Street, New York, NY 10022, USA
10 Stamford Road, Oakleigh, Melbourne 3166, Australia

© Cambridge University Press 1987

First published 1987

Printed in Great Britain at University Press, Cambridge

British Library Cataloguing in Publication Data
Walton, A.K.
Network analysis and practice.
1. Electronic network analysis
I. Title
621.319'2   TK454.2

Library of Congress Cataloguing in Publication Data
Walton, A. K. (Alan Keith), 1933–
Network analysis and practice.
Includes index.
1. Electric network analysis. I. Title.
TK454.2.W32 1987   621.319'2   87-4255

ISBN 0 521 26459 6 hard covers
ISBN 0 521 31903 X paperback

CONTENTS

Preface ix

1 Electric charge, field and potential
1.1 Electric charge 1
1.2 The inverse square law 2
1.3 Force, electric field and potential 6

2 Electric current, resistance and electromotive force
2.1 Electrical conduction, electric current and current density 11
2.2 Ohm's law and electrical resistance 14
2.3 Resistors and nonlinear circuit elements 18
2.4 Electromotive force 21
2.5 Internal resistance, sources and matching 25

3 Direct-current networks
3.1 Kirchhoff's laws 30
3.2 Resistors in series and parallel 33
3.3 Generality of analysis by Kirchhoff's laws 36
3.4 Mesh current analysis 37
3.5 Node-pair potential analysis 39
3.6 The superposition and reciprocity theorems 41
3.7 The Thévenin and Norton theorems 45
3.8 Measurement of direct current, potential difference and resistance 54
3.9 The Wheatstone bridge 59
3.10 Load-line analysis 62

4 Capacitance, inductance and electrical transients
4.1 Capacitance and capacitors 65
4.2 Inductance and inductors 71
4.3 Transient responses of C–R and L–R circuits to a step e.m.f. 80
4.4 Basic four-terminal C–R networks 84
4.5 Transient response of an L–C–R circuit to a step e.m.f. 89

5 Introduction to the steady-state responses of networks to sinusoidal sources
5.1 Sinusoidal sources and definitions 97
5.2 Responses of purely resistive, purely capacitive and purely inductive circuits to sinusoidal e.m.f.s 100
5.3 Sinusoidal response through differential equation solution 102
5.4 Steady-state sinusoidal response from phasor diagram 105
5.5 Steady-state sinusoidal response through complex representation 107
5.6 Series resonant circuit 110
5.7 Parallel resonant circuits 114
5.8 Power dissipation associated with sinusoidal current 119
5.9 Sinusoidal sources in nonlinear circuits 124

6 Transformers in networks
6.1 Mutual inductance 127
6.2 Transformers 131
6.3 Reflected impedance and matching by transformers 135
6.4 Critical coupling of resonant circuits 139

7 Alternating-current instruments and bridges
7.1 Alternating-current meters 144
7.2 Measurement of impedance by a.c. meters 148
7.3 Measurement of impedance by the Wheatstone form of a.c. bridge 151
7.4 A.c. bridges for determining inductance 155
7.5 The Schering bridge for determining capacitance 157
7.6 The Heydweiller bridge for determining mutual inductance 158

7.7	A.c. bridges for determining the frequency of a source	160
7.8	Transformer ratio-arm bridges	161

8 Attenuators and single-section filters
8.1	Attenuators	167
8.2	Simple single-section filters	173
8.3	Wien, bridged-T and twin-T rejection filters	179
8.4	Phase-shift networks	186

9 Multiple-section filters and transmission lines
9.1	Ladder filters	189
9.2	Constant-k filters	193
9.3	m-Derived filters	202
9.4	Asymmetric sections	209
9.5	Transmission lines	212

10 Signal analysis of nonlinear and active networks
10.1	Two-terminal nonlinear networks	220
10.2	Four-terminal nonlinear networks	223
10.3	Small-signal equivalent circuits and analysis	230
10.4	Feedback	235
10.5	Operational amplifiers	240
10.6	Nyquist's criterion and oscillators	244
10.7	Amplifier instability and Bode diagrams	249

11 Fourier and Laplace transform techniques
11.1	Fourier analysis of periodic nonsinusoidal signals	252
11.2	Fourier analysis of pulses	256
11.3	The Laplace transform	259
11.4	Commonly required Laplace transforms	261
11.5	Inverse Laplace transforms	265
11.6	Network analysis by Laplace transformation	269
11.7	Pole-zero plots in the complex s-plane	275

12 Filter synthesis
12.1	Introduction	278
12.2	Butterworth filters	279
12.3	Chebyshev filters	282
12.4	Synthesis of high-pass filters	285
12.5	Band filter synthesis	286

Mathematical background appendices
1 Harmonic functions — 289
2 Exponential functions — 291
3 Phasors and complex representation — 293
4 Linear differential equations with constant coefficients — 296

Problems — 302
Answers — 313
Solutions — 318
Index — 331

PREFACE

Where to begin constitutes a difficulty in expounding most subjects. For completeness' sake, the present treatment of the analysis of electrical networks begins by establishing from first principles those basic electrical concepts such as current, potential and electromotive force in terms of which analysis is executed. In covering these basic concepts in the first two chapters it is, of course, recognised that some students will already be thoroughly conversant with them, some will merely need to 'brush up' on them and others will prefer to acquire them through studying more-detailed physical texts.

Network analysis begins in earnest in chapter 3 where network laws and theorems, such as Kirchhoff's laws and Thévenin's theorem, are introduced in the easy context of direct-current networks. Following descriptions of the physical nature of capacitance and inductance, traditional methods of deducing transient and sinusoidal steady-state responses are developed. These encompass the solution of linear differential equations and the application of phasor and complex algebraic methods. Consideration of the powerful Fourier and Laplace transform techniques is delayed until towards the end of the book, by which stage it is hoped that any reader will have acquired considerable mathematical and physical insight regarding the signal responses of circuits. Overall, the intention is that the book will take a student from 'scratch' to a level of competence in network analysis that is broadly commensurate with a graduate in Electrical or Electronic Engineering, or one in Physics if specialising somewhat in electrical aspects.

A concerted attempt has been made throughout the text to relate the analysis to as great a variety of practical circuits and situations as possible. The reader is strongly recommended to put theory to the test by building real circuits, observing their response and discovering whether there is accordance with design expectation. Fortunately, the construction and

x   *Preface*

testing of practical circuits can be quick and quite inexpensive. As a further aid to assessing whether the contents have been grasped, a collection of over 50 relatively straightforward problems has been incorporated together with worked solutions to many and answers to all of them.

Assistance with preparation of the manuscript in the Physics Department at Sheffield is gratefully acknowledged. During a difficult staffing period several individuals have been involved, but thanks are especially due to Mrs S. Stapleton, Mrs E. Lycett and Mrs J. Hedge for most of the typing and to Mrs K. J. Batty for all the line-drawings.

Physics Department, University of Sheffield, 1986   A. K. WALTON

1

Electric charge, field and potential

1.1 Electric charge

Before embarking on a study of electrical networks, it is important to understand thoroughly such basic electrical concepts as charge, field, potential, electromotive force, current and resistance. The first few sections of this book are concerned with developing these concepts in a logical sequence.

Phenomena due to static electricity have been observed since very early times. According to Aristotle, Thales of Miletus (624–547 B.C.) was acquainted with the force of attraction between rubbed pieces of amber and light objects. However, it was William Gilbert (1540–1603) who introduced the word electricity from the Greek word $\eta\lambda\varepsilon\kappa\tau\rho o\nu$ (electron) signifying amber, when investigating this same phenomenon. Many other materials have been found to exhibit similar properties. For example, if a piece of ebonite is rubbed with fur, then on separation they attract each other. However, an ebonite rod rubbed with fur repels another ebonite rod similarly treated. Also, a glass rod rubbed with silk repels another glass rod rubbed with silk. Yet an ebonite rod that has been rubbed with fur attracts a glass rod that has been rubbed with silk.

All of these observations can be explained by stating that two kinds of electricity exist, that on glass rubbed with silk being called *positive* electricity and that on ebonite rubbed with fur *negative* electricity. In addition, *like* kinds of electricity *repel*, *unlike* kinds *attract*. Actually, all substances taken in pairs become oppositely electrified when rubbed together.

Initially it was thought that positive and negative electricity were weightless fluids, a preponderance of one or the other determining the overall sign of electrification. There was also a single-fluid theory in which a normal amount corresponded to zero electrification, an excess to electrification of one sign and a deficiency to electrification of the other sign. It is now known that electrification comes about as a result of the transfer of tiny subatomic particles known as *electrons*. Each electron has a definite mass and carries a definite amount of the entity that produces the electric forces. For descriptive convenience this entity is called *charge*. Note that the charge carried by an electron is of the kind which has been called negative. Consequently, positive or negative charging corresponds to the subtraction or addition of electrons respectively. Electrical charge is also carried by certain other subatomic particles. All matter is formed from atoms and each atom comprises a minute but relatively massive central nucleus surrounded by electrons. Electrical neutrality is preserved through the nucleus containing protons equal in number to the surrounding electrons with each proton carrying positive charge of electronic magnitude. Protons are over one thousand times heavier than electrons and the remainder of the nucleus is made up of similarly massive neutrons that are electrically neutral. Particles called positrons, identical to electrons except that they carry charge of opposite sign, exist transiently in connection with nuclear reactions. Such reactions are brought about by deliberate nuclear bombardment or arise naturally as a result of cosmic ray showers incident from space. Although there are many other subatomic particles both charged and uncharged, from the electrical point of view the important point to emerge is that the electronic charge appears to be the smallest charge magnitude that can be physically separated.

Of great importance to network analysis and physics generally is the law of *conservation of charge*. Experiments clearly show that the algebraic sum of charges is constant in an isolated system. This is true both on the macroscopic scale and at the fundamental particle level. When ebonite is rubbed with fur, upon separation the ebonite is found to carry a negative charge equal in magnitude to the positive charge carried by the fur. An interesting example of charge conservation at the fundamental particle level is provided by positron–electron annihilation. When these two particles come into close proximity, they may simply disappear, converting all their mass into energy according to the Einstein relation, energy equals mass times the velocity of light squared. The energy appears as two oppositely directed γ-rays. Significantly, the net charge is zero before and after the event; charge is conserved whereas mass is not, being convertible into energy.

1.2 The inverse square law

Priestley (1773–1804) asserted that the law of variation of force F between charged bodies situated a distance r apart is

1.2 The inverse square law

$$F \propto q_1 q_2 / r^2 \tag{1.1}$$

where q_1 and q_2 are the charges on the bodies. In arriving at this law he was influenced by the similar gravitational law of force between two masses. It was already well known from the theory of gravitation that a mass inside a uniform spherical shell of matter experiences no net force due to the material of the shell. Priestley observed that the charges on a hollow conductor exert no force on a small charged body placed inside.

Direct verification of the inverse square law in electrostatics to any degree of precision is difficult. The law is sometimes known alternatively as Coulomb's law because, in 1785, Coulomb used a torsion balance to roughly establish its truth by direct measurement. In his experiments Coulomb examined the force between like charges on two gilt pith balls. One of the balls was fixed and the other mounted on the end of a light rod suspended by a fine silver wire. The repulsive force between the charges created a balancing twist in the suspension wire. To halve the distance between the balls it was necessary to twist the wire four times as much. Further, on removing the fixed ball and sharing its charge with another identical uncharged ball (so that its charge was halved by symmetry) and then replacing the ball, it was found that the twist required to give the same separation as previously was halved. For clarity, the essential features of Coulomb's experiments are illustrated in figure 1.1. The work of Coulomb suffered from the charges not being point but distributed over spherical surfaces, charge leaking away and charges being induced on the case and components of the balance which causes interfering forces. A case was of course necessary to exclude draughts.

Confidence in the *precise* nature of the inverse square law stems from *indirect* experiments similar to those of Priestley but in which absence of an electric force on a charge inside a hollow spherical conductor carrying a uniform distribution of charge is tested. Consider the force that would act on unit positive charge at a point P within a uniformly positively charged spherical conductor if the force between point charges varied inversely as the nth power of their separation. To this end, construct an elementary cone of solid angle $d\omega$ with its vertex at P as shown in figure 1.2. Let the cone intersect the surface in areas dS_1 and dS_2 ($<dS_1$) and let O be the centre of the sphere. Clearly from the geometry

$$dS_1 \cos \alpha = r_1^2 \, d\omega$$
$$dS_2 \cos \alpha = r_2^2 \, d\omega$$

where r_1 and r_2 ($<r_1$) are respectively the distances of dS_1 and dS_2 from P. Thus with surface charge density σ over the sphere, the net force on unit charge at P due to the charges residing on elements dS_1 and dS_2 obeys

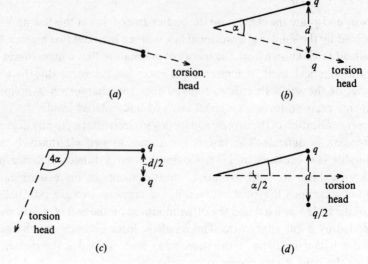

1.1 Illustration of Coulomb's experimental verification of the inverse square law using a torsion balance. (a) With uncharged balls, suspension is untwisted, (b) with charge q on each ball, suspension twist is α, (c) separation is halved by rotating torsion head until twist is 4α and (d) half the original charge on the fixed ball gives the original separation when the torsion head is rotated to halve the twist in the suspension compared with (b).

$$dF \propto \frac{\sigma\, dS_1}{r_1^n} - \frac{\sigma\, dS_2}{r_2^n} = \frac{\sigma\, d\omega}{\cos\alpha}(r_1^{2-n} - r_2^{2-n}) \tag{1.2}$$

From this equation, if the force law is precisely inverse square, that is, if $n=2$, $dF=0$ and, since the whole charge on the spherical surface can be handled by similar conical constructions, the resultant force on unit charge at P is exactly zero. On the other hand, if $n<2$, since $r_2<r_1$ for all conical constructions, $dF>0$ always and, on summing over all cones, symmetry considerations reveal that a resultant force will exist acting radially outwards from the centre. Similarly if $n>2$, $dF<0$ and a resultant force will act radially inwards towards the centre.

Cavendish and at a later date Maxwell supported a sphere A inside a second sphere B so that the two spheres were insulated from each other except when connection was made between them by a hinged wire as indicated in figure 1.3(a). Sphere B was first positively charged, then A was connected to B by means of the hinged wire following which the hinged connection was broken again. From the above discussion A would be positively charged by this sequence if $n>2$, negatively charged if $n<2$ and uncharged if $n=2$. This is because unless $n=2$, a radial force is exerted on

1.2 The inverse square law

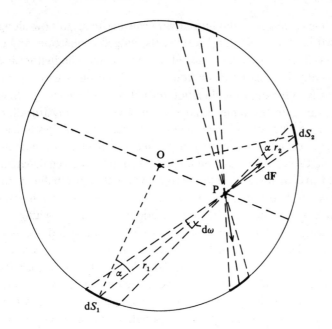

1.2 Elementary conical construction to find the force on unit positive charge at a point P inside a uniformly positively charged spherical conductor assuming an inverse dependence of the force between charges on the nth power of their separation.

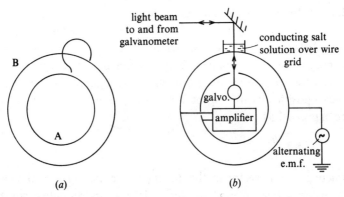

1.3 Indirect experimental verification of the inverse square law of force between point charges; (*a*) basic arrangement of early static charge experiments and (*b*) later equivalent arrangement using an alternating supply.

the mobile electrons in the hinged wire causing them to flow along it as discussed in section 2.1. On breaking the hinged connection, any charge that has flowed becomes trapped on the inner conductor and can subsequently be measured. Of course, the force on negatively charged electrons is in the opposite direction to that on positive charge. Should n be less than 2, for example, electrons would be forced inwards along the wire giving A a negative charge on interconnecting the spheres. Since A and B are spherical and concentric, it is possible to calculate the magnitude of charge that A would acquire on interconnection for a given departure of n from 2. Cavendish searched for charge on A with a pith ball electroscope but failed to detect any. From his limiting sensitivity, that is minimum detectable charge, he was able to conclude that n must be within 1% of 2. Later attempts by Faraday with a gold-leaf electroscope and Maxwell with a quadrant electrometer also failed to detect any charge on A. Maxwell was able to conclude that n could not differ from 2 by more than 0.003%. Modern versions of this experiment permanently connect the two spheres but charge and discharge the outer from an alternating source. Alternating charge flow between the spheres is sought with electronic amplifying equipment mounted inside the inner sphere. Figure 1.3(b) shows the experimental arrangement of Plimpton and Lawton (1936), who found that $n = 2$ to within 1 part in 10^9. More recently it has been shown that $n = 2$ to 1 part in 10^{16}.

1.3 Force, electric field and potential

It has been established that the force between point charges q_1 and q_2 a distance r apart is proportional to $q_1 q_2 / r^2$ and acts along the line joining them being repulsive if the charges have like signs and attractive otherwise. Consequently, if one charge is regarded as being situated at vector position **r** with respect to the other, the *vector force* acting on it may be expressed as

$$\mathbf{F} = \frac{q_1 q_2}{4\pi \varepsilon r^2} \hat{\mathbf{r}} \tag{1.3}$$

where $\hat{\mathbf{r}}$ is unit vector in the direction of **r**, the signs of the charges are incorporated in q_1 and q_2 and ε is a proportionality constant. The magnitude of the force depends on the medium in which the charges are situated and so ε is a characteristic parameter of the medium called the *permittivity*. Note that it is customary to refer to the special value of ε relevant to a vacuum as the *permittivity of free space* and to denote it by ε_0. Clearly, the permittivity of a medium can alternatively be described in terms of the dimensionless ratio $\varepsilon/\varepsilon_0$ which is known as the *dielectric*

1.3 Force, electric field and potential

constant. The factor 4π is introduced into equation (1.3) so that it will be present on the whole in expressions for quantities where spherical symmetry is involved but not where the symmetry is plane. Its introduction in this way is said to *rationalise* the equations but it also affects the system of units to be used. Appropriate units for rationalised equations are known as *rationalised units*.

In the system of units known as *Système International*, or *SI*, the *metre*, *kilogram* and *second*, respectively denoted by m, kg and s, are adopted as the fundamental units of length, mass and time. Of the four physical quantities involved in the force equation (1.3), namely force, distance, charge and permittivity, the dimensions of the first two are directly expressible in terms of mass, length and time. The fundamental unit of force, known as the *newton* and denoted by N, is precisely that which will give a mass of one kilogram an acceleration of one metre per second squared. In order to describe electrical (and magnetic) quantities, a fourth basic dimension must be chosen. On the SI system, the fourth basic dimension is effectively that of charge (strictly current, see section 2.1) leaving permittivity to be expressed in terms of charge, mass, length and time through equation (1.3). The fundamental SI unit of charge is chosen to be the practical *coulomb* unit (written C). Fixing the unit of charge fixes the magnitude of ε_0, for the force between unit charges situated unit distance apart is a definite magnitude that can be measured. It turns out that ε_0 is very close to $10^{-9}/36\pi\,\mathrm{C^2\,m^{-2}\,N^{-1}}$ or $10^{-9}/36\pi\,\mathrm{F\,m^{-1}}$ (see later, sections 4.1 and 4.2) in the rationalised SI system of units. Incidentally, the electronic charge magnitude turns out to be close to $1.6 \times 10^{-19}\,\mathrm{C}$.

Although it is possible to analyse electrical situations directly in terms of the force law between point charges, introduction of the electric field and potential concepts is extremely helpful in many instances. The *electric field* vector **E** at any point is defined such that the force on an infinitesimal test charge δq at that point is

$$\mathbf{F} = \delta q \mathbf{E} \tag{1.4}$$

It follows immediately that the electric field at vector position **r** with respect to a point charge q is

$$\mathbf{E} = \frac{q}{4\pi\varepsilon r^2}\hat{\mathbf{r}} \tag{1.5}$$

The point of specifying an infinitesimal test charge in defining electric field is that the test charge must not disturb the charge causing the field. If the charge producing the field is fixed in position then the size of the test charge will not matter, but if movable charge is producing the field then a finite test charge will displace the main charge and modify the field. In the field concept a charge is considered to produce something in the space round it called an electric field which then interacts with any other charge present to exert a force on it. The field due to a set or distribution of point charges is the vector sum of the fields due to the individual charges.

The *potential difference* $V_B - V_A$ between two points A and B in an electric field is defined as the work done in conveying unit positive charge from point A to point B. Thus the idea of potential is concerned with the direction in which electric charge moves or tends to move. To obtain a quantitative expression for potential difference, consider any path between points A and B. Let the electric field be **E** at an element d**l** of this path as indicated in figure 1.4(*a*). If the angle between **E** and d**l** is θ then the work expended in moving unit positive charge through d**l** towards B is $-E \cos \theta \, dl$. Notice that if $\theta < 90°$, the work done is negative, that is, work is supplied by the field. Writing dV for the potential difference between the end and beginning of the element d**l**.

$$dV = -E \cos \theta \, dl \qquad (1.6)$$

and integrating over the path between A and B

$$\int_A^B dV = V_B - V_A = -\int_A^B E \cos \theta \, dl \qquad (1.7)$$

It is important to appreciate that the work expended in taking charge between two points in an electric field is independent of the path. If this were not so, differing paths such as ACB and ADB in figure 1.4(*b*) would correspond to differing amounts of work. Now, traversing a path between two points in a field with a charge but in opposite directions involves the

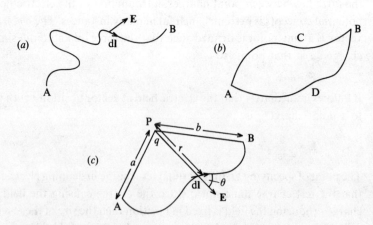

1.4 Work done on unit positive charge over paths between points A and B.

1.3 Force, electric field and potential

opposite sign of work since each element is reversed in sign. Thus, if and only if the work depended on the path between two points, taking charge round a closed path such as ACBDA in figure 1.4(b) could lead to a negative amount of work being expended. It would be possible to generate energy just by taking charge round a closed path in a field due to static charge, which task would not necessitate the expenditure of any other energy. This would violate the law of conservation of energy and therefore the work involved in taking charge between points in an electric field must be independent of the path. In turn, this means that the electric potential difference between points is single valued.

In the special case of a path along the field, in accordance with equation (1.6), $dV = -E\, dl$. This shows that an electric field is directed from a point of higher towards a point of lower potential and that positive charge moves or tends to move to a place of lower potential.

The potential difference between two points A and B distance a and b respectively from a point charge q is readily found. With reference to figure 1.4(c)

$$dV = -\frac{q}{4\pi\varepsilon r^2}\cos\theta\, dl = -\frac{q\, dr}{4\pi\varepsilon r^2}$$

Hence

$$\int_A^B dV = -\frac{q}{4\pi\varepsilon}\int_a^b \frac{dr}{r^2}$$

and

$$V_B - V_A = \frac{q}{4\pi\varepsilon}\left(\frac{1}{b} - \frac{1}{a}\right) \tag{1.8}$$

As expected, the potential difference is independent of path between A and B; it depends only on the positions of A and B. Since any electric field can be represented as a vector summation of fields due to point charges, equation (1.8) actually confirms that potential difference is quite generally independent of path as claimed earlier. Path irrelevance is often neatly expressed through stating that the work expended in taking unit positive charge round a closed path is zero, that is,

$$\oint E \cos\theta\, dl = 0 \tag{1.9}$$

The physical origin of this important feature is the fact that the electric field due to a point charge is *central*, that is, radial. Any path can be formed from radial and tangential elements, only radial elements contributing to the work done in taking a charge over it. Over a closed path radial inwards and outwards work cancels while the potential difference between two points is simply determined by the difference in radial distance.

Strictly only potential *difference* exists. However, by arbitrarily assigning zero potential to some point, it is possible to talk about the *potential* of any point, by which is meant the potential difference between that point and where it has arbitrarily been assigned as zero. Whenever the term potential is used in this way, the origin of potential should be specified. A common placing of the origin of potential is at infinity in view of its theoretical convenience. With regard to the electric field due to a point charge, this means putting $V_A = 0$ when $a = \infty$ in equation (1.8) which leads to $V_B = q/4\pi\varepsilon b$ and the potential at a distance r from a point charge q being

$$V = q/4\pi\varepsilon r \tag{1.10}$$

In the case of a distribution of point charges, the potential will be just the scalar sum of the potentials due to the individual charges.

Another common choice of origin of potential of considerable practical convenience is the potential of the *earth*. This is because the earth, being a good electrical conductor, is at a relatively uniform potential (as may be appreciated from the contents of chapter 2) and possessing an enormous capacitance does not change its potential significantly when charge is transferred to it (see section 4.1).

Basic SI units of electric field and potential are deduced as follows. The SI unit of work or energy is that of force times distance so that it is the newton metre which has been termed the *joule* and is often written as J. Since work done equals potential difference times charge, the SI unit of potential difference or potential is the joule/coulomb which is precisely the practical unit known as the *volt* or V. It follows immediately, for example from equation (1.6), that the SI unit of electric field is the *volt/metre* or $V\,m^{-1}$.

2

Electric current, resistance and electromotive force

2.1 Electrical conduction, electric current and current density

When a *steady* electric field is maintained in a medium, some *continuous* flow of charge always occurs. Just how a steady field can be maintained is discussed in section 2.4. The flow of charge arises because the field exerts a force on mobile charged particles in the medium causing them to acquire a drift velocity. Under a steady field, the flow of charge is relatively very large in metals, intermediate in magnitude in semiconductors and electrolytic solutions but negligibly small for most purposes in gases and certain other solids and liquids which are termed *insulators*. The phenomenon of charge flow is described as electrical *conduction*, the rate of flow of charge being referred to as the electric *current*. A charge dQ flowing through a surface, such as a cross-section of a solid wire for example, in an infinitesimal time dt constitutes an electric current

$$I = dQ/dt \tag{2.1}$$

The mobile charge can be positive or negative. In electrolytes, both positive and negative ions, that is, atoms or groups of atoms with a deficit or surplus of electrons, comprise the mobile charged particles. In metals, it is the outer electrons of the atoms that are free to move and provide the mobile charge. The outer valence electrons of semiconductors also provide mobile charge but only when excited, for example thermally, across an energy gap to essentially free conduction states. Interestingly and importantly, the residual vacancies created in the valence states through excitation of valence electrons appear positively charged and are also mobile by virtue of other valence electrons moving to occupy the vacancies thereby forming new vacancies. The positively charged vacancies in the valence states clearly carry a charge of electronic magnitude and are known as positive holes. Direct experimental observation of the mechanical momentum change associated with reversing a current in a coil confirms that electrons are the current carriers in metals. Similar electromechanical experiments verify that ions carry the current in electrolytes. Confirmation of the existence of electron and hole carriers of current in semiconductors comes from Hall-effect experiments in which a magnetic field deflects a current to a given side of a specimen, for it is found that the resultant charging of that side can be positive or negative.

Whatever the nature of the mobile charged particles, positive charge flows in the direction of the field while negative charge flows in the opposite direction because of the opposite force. In terms of charge transfer, both are equivalent to the flow of positive charge along the field and *positive current* is taken to correspond to *positive charge flow*. Note that the contributions to the electric current by positive and negative ions in electrolytes and by positive holes and electrons in semiconductors are additive.

As mentioned in section 1.3, in the SI system of units it is actually electric current which is taken as the fourth basic dimension rather than charge. All electric and magnetic quantities can then be described in terms of mass, length, time and current. The practical unit of current known as the *ampere* or *amp* and denoted by A is chosen to be the fundamental unit of current. This makes the fundamental unit of charge the ampere second, that is, the coulomb as stated in section 1.3. Choice of the practical unit as the fundamental unit of the fourth basic dimension leads to a highly convenient system of electromagnetic units.

Because charge is a scalar quantity, electric current defined according to equation (2.1) is also scalar. It is described by its magnitude and sign. Some confusion may arise because current is due to charge flow which has direction. However, current is a flux concept; if steady it is the quantity of positive charge crossing a given surface each second. To help clarify the situation, consider the radial flow of charge through a conducting medium between two coaxial cylindrical electrodes. The current is the total rate of flow of charge between the electrodes but charge flows in all cross-sectional directions as illustrated in the cross-sectional diagram of figure 2.1(*a*). If current was to be a vector, what would be its direction in this case? If local currents were defined so as to have the local direction of charge flow, the total current would be the vector sum which would be zero!

Turning to the flow of charge round a single conducting plane loop path, notice that, although charge flows in all directions in the plane, in equilibrium the current or rate of flow of charge through any cross-section of the path is the same. Of course, the charge does flow round the path in a particular sense, clockwise as illustrated in figure 2.1(*b*), or anticlockwise. The sense of positive current is taken to be that of positive charge flow.

The problems of charge flowing in a local direction and the rate of flow of

2.1 Conduction, current and current density

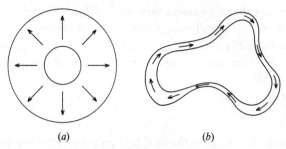

2.1 (a) Radial flow of charge between concentric electrodes and (b) flow of charge round a plane loop path.

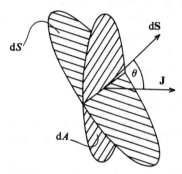

2.2 Current density at a surface element.

charge varying over a surface are countered by introducing the concept of *current density*. Current density is a vector quantity, usually denoted by **J**, that describes both the magnitude of the local flow and its direction. The magnitude of the current density at a point is the current dI through an elementary area dA normal to the flow of charge at that point divided by the area dA. That is

$$dI = J\, dA \tag{2.2}$$

Now consider any surface S and let the current density at an element dS of it be **J**. It is customary to represent a surface element by a vector d**S** where dS indicates the area and the vector direction that of the surface normal. In general, **J** will not be normal to the surface element, as illustrated in figure 2.2. The current dI through d**S** will be that through the area of d**S** projected normal to **J**, that is through dA in the figure. Since the angle of projection is the angle θ between **J** and d**S**,

$$dI = J\, dA = J\, dS \cos\theta$$

or in vector notation

$$dI = \mathbf{J} \cdot d\mathbf{S} \tag{2.3}$$

14   *Electric current, resistance and e.m.f.*

Particularly notice that the current through a surface element is the *scalar* product of the current density and surface element vectors. Also observe that from equation (2.2) or (2.3), the fundamental unit of current density is the ampere per square metre (A m^{-2}). Integrating over the surface S, the total current is

$$I = \int_S \mathbf{J} \cdot d\mathbf{S} \tag{2.4}$$

While the current density gives the local rate of flow of charge in magnitude and direction, the current gives the total rate of flow of charge through a given surface. In the example of radial flow of charge between cylindrical electrodes, the current density is everywhere radial, its magnitude depending on radial distance, while the current through any concentric cylindrical surface is the same. In the example of a single circuit loop, the current density changes magnitude and direction round the loop, but the current through any section of the loop is the same.

2.2   Ohm's law and electrical resistance

Usually, but not always, the current density at a point in a medium is proportional to the local electric field. It is therefore convenient to write

$$\mathbf{J} = \sigma \mathbf{E} \tag{2.5}$$

the parameter σ introduced in this way being characteristic of the medium and known as the electrical *conductivity*.

Theoretical support for proportionality between current density and field is readily obtained for conduction in a solid on consideration of the conduction mechanism. In the case of a metal, the outer electrons are free to move and because of thermal energy are in rapid random motion in the absence of an applied electric field. The motion is jerky because the electrons frequently encounter lattice atoms or impurities and suffer *scattering* collisions. Clearly the random thermal motion averages to zero over any macroscopic volume and does not produce any net current density. On application of an electric field $\mathbf{E}$, each outer electron experiences an accelerating force $-e\mathbf{E}$, where e is the magnitude of the electronic charge. Because the mobile outer electrons are subject to forces not only due to the applied electric field but also the remainder of the atomic structure, they actually respond to the applied field as though they possess an effective mass m^* which differs somewhat from the normal vacuum mass. The outer electrons consequently undergo an acceleration $-e\mathbf{E}/m^*$ over the time T between scattering collisions to reach an additional imparted velocity of $-eT\mathbf{E}/m^*$. Assuming that the scattering collisions randomise the motion, as far as net transport is concerned it is as

2.2 Ohm's law and electrical resistance

if the field imparted motion were removed at each collision and the electron accelerated from rest again afterwards. Evidently, if the time T between collisions were always the same, the field would impart an average drift velocity of

$$\mathbf{v} = -(eT/2m^*)\mathbf{E}$$

In fact, there is a probability distribution of free times which causes the mean drift velocity to be given by

$$\mathbf{v} = -(e\tau/m^*)\mathbf{E} \tag{2.6}$$

where τ is the *mean free time*. Alternatively expressed in terms of *mobility* μ, which is the magnitude of the mean drift velocity of charge carriers per unit electric field strength,

$$\mathbf{v} = -\mu\mathbf{E} \tag{2.7}$$

where

$$\mu = e\tau/m^* \tag{2.8}$$

Now all the mobile electrons within a perpendicular distance v of an area dA situated normal to the flow will pass through dA each second. Thus the current through dA will be the mobile charge within volume $v\,dA$. If n is the density of mobile electrons, this amounts to $nv\,dA(-e)$ and it follows from equation (2.2) that the current density is

$$\mathbf{J} = -ne\mathbf{v} \tag{2.9}$$

Combining equations (2.7), (2.8) and (2.9)

$$\mathbf{J} = ne\mu\mathbf{E} = (ne^2\tau/m^*)\mathbf{E} \tag{2.10}$$

It will be observed that the electrical conductivity is given by

$$\sigma = ne\mu = ne^2\tau/m^* \tag{2.11}$$

and that provided τ and m^* are independent of the field E, then the current density J is proportional to E as already asserted. In fact, only at extraordinarily high field strengths do τ and m^* become dependent on E since only then does the additional energy acquired by the current carriers in the field become appreciable compared with the thermal energy.

Proportionality between the current density J and electric field E is actually the common property of a conducting medium upon which *Ohm's law* depends. As long ago as 1826, Ohm discovered that a metallic conductor maintained at a fixed temperature passes an electric current directly proportional to the potential difference applied across it. The connection between $J \propto E$ and Ohm's law is easily shown for a rectangular slab of uniformly conducting material in an insulating environment with two opposite end faces a and b arranged to be differing equipotentials. With reference to figure 2.3, the current density $\mathbf{J}$ in the slab is everywhere

16 *Electric current, resistance and e.m.f.*

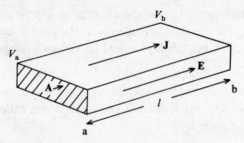

2.3 A rectangular conducting slab with an electric field applied along its length.

the same magnitude and parallel to the length **l**. Thus in terms of current density, the current is

$$I = \int_A \mathbf{J} \cdot d\mathbf{A} = \int_A J \, dA = J \int_A dA = JA \tag{2.12}$$

where **A** is the cross-sectional area. However because

$$\mathbf{E} = \mathbf{J}/\sigma \tag{2.13}$$

the electric field **E** is also everywhere the same magnitude and parallel to the length **l**. Consequently the potential difference between the ends of the slab may be expressed as

$$\int_a^b dV = V_b - V_a = -\int_0^l \mathbf{E} \cdot d\mathbf{l} = -El$$

The negative sign here merely means that end b is at lower potential than end a and in terms of the potential difference $V = V_a - V_b$

$$V = El \tag{2.14}$$

Combining equations (2.12), (2.13) and (2.14)

$$I = (\sigma A/l)V \tag{2.15}$$

demonstrating that if $J \propto E$ so that σ is independent of E and hence of V, $I \propto V$, that is to say, Ohm's law applies. Equivalence between $J \propto E$ and $I \propto V$ for more general geometries follows from consideration of the relations

$$I = \int_S \mathbf{J} \cdot d\mathbf{S}$$

and

$$V_b - V_a = -\int_0^l \mathbf{E} \cdot d\mathbf{l}$$

In the development of electrical theory it is conventional to introduce a

2.2 Ohm's law and electrical resistance

quantity R known as the electrical *resistance* by writing

$$I = V/R \tag{2.16}$$

irrespective of whether Ohm's law applies. The greater the resistance, the less the current for a given potential difference. Defining resistivity ρ as the reciprocal of conductivity σ and conductance G as the reciprocal of resistance R, it follows from equations (2.15) and (2.16) that

$$R = 1/G = l/\sigma A = \rho l/A \tag{2.17}$$

According to the definition implicit in equation (2.16), the fundamental SI unit of resistance is the volt/ampere which is precisely that practical unit which has been called the *ohm* and is usually denoted by Ω. The SI units of conductance, conductivity and resistivity are correspondingly from equation (2.17) the reciprocal ohm known as the *siemen* and denoted by S, the siemen/metre written $S\,m^{-1}$ and the ohm metre written $\Omega\,m$.

For the conductivity to be constant it is necessary for the temperature to be maintained constant. In metals the conductivity falls slowly with increasing temperature as a result of τ decreasing. In semiconductors σ can rise very rapidly with temperature through the rise in the number of mobile carriers of charge taking part in conduction swamping any variation in τ. At some critical low temperature, usually below 10 K, the conductivity of many metals and alloys suddenly jumps to an enormous value. This striking phenomenon is known as superconductivity. Many other physical conditions besides temperature affect the conductivity, for example, the presence of a magnetic field. Indeed, such dependences are often used in conjunction with electronic equipment to detect the conditions.

Although it has already been indicated at the beginning of this chapter that the range of conductivities exhibited by different materials is very high, just how enormous the range is even at room temperature and under other normal conditions is worthy of comment. Altogether it extends to around 27 orders of magnitude! At the high end, the conductivities of metallic elements mostly lie between those of copper and mercury which amount to 5.8×10^7 and $10^6\,S\,m^{-1}$ respectively. Conductivities of metallic alloys occupy the lower part of this range, those of brass, manganin, stainless steel and nichrome, for example, being 1.6×10^7, 2.4×10^6, 1.8×10^6 and $9 \times 10^5\,S\,m^{-1}$ respectively. The conductivities of semiconductors are not only much lower but cover a much wider range. For instance, pure indium antimonide, germanium and silicon have conductivities of 1.8×10^4, 2.1 and $4.3 \times 10^{-4}\,S\,m^{-1}$ respectively. The equilibrium conductivities of insulators are difficult to establish but vary even more. As a guide those of ivory, shellac, diamond, paraffin, pyrex glass, polythene and polytetrafluorethylene (PTFE) are of the orders of 10^{-6}, 10^{-7}, 10^{-10}, 10^{-11}, 10^{-12}, 10^{-14}–10^{-18} and $10^{-19}\,\mathrm{S\,m^{-1}}$ respectively.

2.3 Resistors and nonlinear circuit elements

An electrical component which has been deliberately fabricated so as to exhibit a certain resistance between its two terminals is known as a *resistor*. The purpose of resistors in electrical networks is to control the current. For a resistor to be ideal, it is clear that Ohm's law must apply so that the resistance is independent of the current. Whenever Ohm's law is applicable to an electrical component to a sufficient approximation, the component is said to be *linear*.

Inevitably, the behaviour of all conducting elements deviates appreciably from Ohm's law at sufficiently high currents due to the heating effect of the current itself which raises the temperature and changes the resistance. In accordance with the definition of potential difference introduced in section 1.3, a charge Q in dropping through potential difference V loses potential energy QV. The kinetic energy gained in dropping through the potential difference is given up to the lattice through the scattering mechanism and reappears as additional lattice vibration, that is, heat energy. The rate of expenditure of energy when a current I flows, called the electrical *power*, is the rate of production of heat and is given by

$$P = V\,\mathrm{d}Q/\mathrm{d}t = VI = RI^2 = V^2/R \qquad (2.18)$$

From this equation the SI unit of electrical power is the V A or V C s^{-1} or J s^{-1} which is named the *watt* and is normally denoted by W. Resistors should be physically large enough to render the resistance change, corresponding to the rise in temperature caused by the current, negligible. It helps of course to make the resistors from a material having a small temperature dependence of conductivity.

Practical fixed resistors are manufactured in several ways. *Wirewound* resistors use manganin, constantan or nickel–chrome alloy wire, all of which have appreciable resistivity with a low temperature coefficient so that the resistors are of *high stability*. Manganin is particularly useful in this respect having an exceedingly tiny temperature coefficient of resistance and low thermal e.m.f. (see following section) against copper. On the other hand, constantan and nickel–chrome resist corrosion well. Satisfactory resistors for general purposes other than those requiring a high-power rating are made from a baked *composition* rod of graphite filler and resin, or *carbon* deposited on a ceramic rod, or *metal film* formed on some insulating substrate. Such resistors, although of lower stability, exhibit much smaller inductance and capacitance (see chapter 4) than wirewound resistors, especially so in the case of metal film resistors, and are therefore particularly

2.3 Resistors and nonlinear circuit elements

suitable for alternating-current circuits (see chapter 5 *et seq.*) which are a common feature of electronic instrumentation. Resistors are formed in integrated circuits by producing doped regions of the required resistivity within the semiconducting chip.

Individual fixed resistors are normally produced in a *preferred range* comprising powers of 10 times 10, 12, 15, 18, 22, 27, 33, 39, 47, 56, 68 and 82 Ω. Values and tolerances are indicated by a colour code consisting of four bands marked on the body. The virtue of this code is that the value of the resistor can be read from any angle when mounted on a circuit board. In the code, black, brown, red, orange, yellow, green, blue, violet, grey and white respectively represent the digits 0–9. Reading inwards from the end, the first two bands indicate the first two significant digits of the magnitude of the resistance and the third band the power of ten multiplier. The fourth band, if present, designates the resistance tolerance, silver for 10%, gold for 5% and pink for 2%. If it is absent the tolerance is only 20%. Production of the preferred range is in a number of power ratings, e.g. $\frac{1}{2}$ W. Figure 2.4 presents the *circuit symbols* commonly used to depict fixed resistors. The first labelled (*a*) is that recommended by the Institute of Electrical and Electronic Engineers and is convenient for engineering drawing, but many authors, circuit designers and teachers still prefer the second but older form labelled (*b*) which is easier to draw freehand, for example on a blackboard, and is certainly more evocative of the concept of resistance. The straight line extensions are of course just meant to represent the highly conducting leads by which circuit connections can be implemented. Note that it is customary to draw open circles whenever leads terminate without connection.

Variable resistors are made with wirewound or graphite elements and have stepped or continuously variable contacts. The stepped varieties are operated by removing plugs or rotating decade dials to change resistance. Continuously variable types either have a linear sliding contact or a contact that moves over a circular arc by rotating a shaft. When the fixed end and variable contacts are all utilised, the circuit symbol is as shown in figure 2.4(*c*). In this case the device is often termed a *potentiometer* since it can be

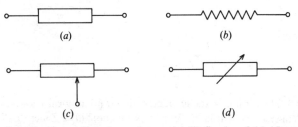

2.4 Circuit symbols for resistors; (*a*), (*b*) fixed and (*c*), (*d*) variable.

used to divide potential. When connected as a two-terminal resistor of adjustable magnitude, the circuit diagram can be reduced to that of figure 2.4(d). The older alternative variable resistor symbols have the box replaced by the zig-zag of figure 2.4(b).

A conducting element to which Ohm's law does not apply is described as *nonlinear*. One simple example is the filament of an electric lamp which is designed to rise dramatically in temperature when current passes through it. The behaviour of nonlinear devices is usually represented in graphical plots of current versus potential difference. Such plots are known as *static characteristics* or just *characteristics*, the term static meaning that data for the plot has been collected in the steady state. The form of the characteristic for a lamp filament is presented in figure 2.5(b) and its nonlinear nature should be noted compared with the linear characteristic for an ideal resistor shown in figure 2.5(a). Because the magnitude of the current through resistors and lamp filaments is independent of the sign of the potential difference, their characteristics are symmetric about the origin. That of the

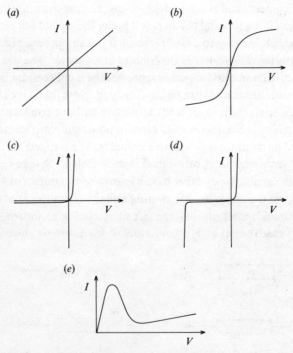

2.5 The form of static characteristic for (a) an ideal resistor, (b) a tungsten filament lamp, (c) a P–N junction, (d) a Zener diode and (e) a Gunn effect device.

lamp shows that its resistance V/I rises rapidly as the heating effect of an increasing current grows. An enormous number of devices have been invented over the years with nonlinear characteristics and corresponding important applications. Three further illustrative examples will be considered. The static characteristic of a P–N junction diode is shown in figure 2.5(c). It is seen to be highly nonlinear and asymmetric about the origin; a low resistance is presented to current flow in one direction but a very high resistance to current flow in the other direction. This behaviour has wide application in the field of electronics including conversion of alternating to direct current and achievement of radio and television transmission and reception. A Zener diode exhibits the characteristic shown in figure 2.5(d). The reverse characteristic shows the useful feature of a sudden steep rise in current at a critical potential difference. This feature can be used in the control of potential difference, such systems being known as voltage stabilisers. A Gunn effect device exhibits the form of characteristic shown in figure 2.5(e). Above a certain potential difference, electrons acquire sufficient kinetic energy between scattering collisions to transfer to different states of much lower mobility. The current then falls with increasing potential difference until all the conduction electrons are in the lower mobility state when the current begins to rise again with increasing potential difference. In a suitable circuit a Gunn device can generate microwaves.

Fascinating as a discussion of nonlinear devices is, further comment on this topic is out of place early in a text concentrating on linear electrical networks and their analysis, apart from one important aspect. Even when the characteristic is nonlinear, small enough changes ΔV in the potential difference from some standing level V produce proportional changes ΔI in the current given by $\Delta V/(\mathrm{d}V/\mathrm{d}I)$. Appropriately $\mathrm{d}V/\mathrm{d}I$ is known as the *differential*, *incremental* or *small-signal* resistance. For small enough signals about a standing level, Ohm's law applies, and such behaviour is clearly within the scope of this text and will be returned to later. In the meantime, note that the differential resistance is the reciprocal of the slope of the characteristic and that it can be positive or negative. For a fuller discussion of nonlinear devices the reader is referred to an electronics textbook.

2.4 Electromotive force

For electric current to flow steadily, a closed circuit is necessary. If current flows along an open-circuit path then charge builds up at the extremities, increasingly opposing the flow until it stops. Consider, for example, an uncharged conducting body B insulated from its surroundings that is suddenly introduced into a region where an electric field exists due to another positively-charged body A. The initial situation is as depicted in figure 2.6(a). However, this situation only occurs instantaneously since the electric field immediately causes the mobile charges in B to flow so that the end near A becomes negatively charged and the other end positively charged. The charging process continues until the electric field in B is reduced to zero by the charge distribution on it and no more current flows, as depicted in figure 2.6(b). Notice that current has only flowed until a new equilibrium state has been reached with B an equipotential.

Even in a closed circuit, current will only flow steadily if some agency does the work involved in driving it round. Such an agency cannot be electrostatic because, as has been seen in chapter 1, the nature of an electric field due to static charge is such that $\oint \mathbf{E} \cdot \mathbf{dl} = 0$, that is, the work done on a charge by an electrostatic field round a closed path is zero. The agency must be capable of continuously transforming some other form of energy into electrical energy. An agency of this kind is appropriately known as an *electromotive force* or *e.m.f.*, the abbreviated form being extremely popular on account of the frequency of its usage. A device which delivers an e.m.f. is correspondingly referred to as a *source* of e.m.f. In sources of steady e.m.f. called batteries, the other form of energy that is converted is chemical. In thermocouples, steady e.m.f.s are produced by the transformation of thermal energy.

Whatever its source, a pure e.m.f. acts between two points in an electrical circuit so as to maintain a potential difference that is independent of the current delivered. A steady or *direct* e.m.f. is one which also maintains the same potential difference over all time. Good batteries provide a steady e.m.f. to a good approximation.

To help understand the nature of electromotive force, consider an

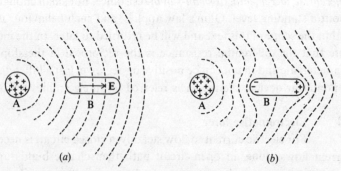

2.6 Effect of the sudden introduction of a conductor B into the electric field due to a charged body A; (a) instantaneous and (b) final equilibrium situation. The dashed contours indicate equipotentials.

2.4 Electromotive force

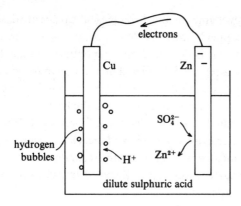

2.7 A primary battery connected to an external conducting path.

electrical battery of the type invented by Volta (1745–1827) comprising one copper and one zinc electrode dipping into dilute sulphuric acid as depicted in figure 2.7. Note that diluting sulphuric acid with water causes a degree of dissociation into positive and negative ions that can be represented by

$$H_2SO_4 \to 2H^+ + SO_4^{2-}$$

The important point is that zinc shows a marked tendency to go into this solution from the solid state in the form of Zn^{2+} ions. Consequently the immersed zinc electrode acquires a substantial negative charge of electronic origin. Entry of zinc ions into solution proceeds until the chemical force causing it is balanced by mutual attraction of the ions to the charge left behind on the electrode. Because copper only exhibits a slight tendency to enter the solution as copper ions, there is clearly a potential difference between the zinc and copper electrodes of this system on open circuit. When the electrodes are connected through some external conducting path, current flows between them, hydrogen ions in the solution migrating in the field towards and collecting at the copper electrode. However, for every two electrons that leave the zinc electrode for the external circuit, an additional zinc ion goes into solution. Also each electron that arrives at the copper electrode via the external circuit is neutralised by a hydrogen ion, neutral hydrogen so produced being liberated as gas bubbles. Most importantly, it will be appreciated that, irrespective of the current flowing, the chemical process maintains the electrical charge on the electrodes and hence the potential difference between them constant. One snag with this simple battery is that hydrogen released at the copper electrode creates a back e.m.f. which opposes the main e.m.f. due to the entry of zinc ions into solution. In practical batteries this problem is overcome by adding a depolarising agent that just oxidises the hydrogen to form water.

Batteries in which irreversible chemical changes take place are described as *primary*. In the zinc–copper battery just described, zinc is consumed from the zinc electrode as current is delivered and the battery ultimately fails unless the zinc electrode is replaced. Depleted materials can be renewed in different kinds of battery described as *secondary* by forcing a current through the cell in a reverse direction. In *charging* the battery in this way, the chemical reaction that led to the discharge is reversed. The most important secondary batteries are lead–acid and nickel–cadmium–alkaline types. Charging is implemented by connecting the secondary cell to a larger e.m.f. to force a reverse current through it. Rectified transformed mains is often the origin of the larger e.m.f.

In circuit diagrams, the presence of a direct e.m.f. is indicated by a two-bar symbol as shown at the left of figure 2.8(*a*) while leads or other conducting paths of negligible resistance are denoted by continuous lines, also as shown in figure 2.8(*a*). The longer of the two bars in an e.m.f. symbol denotes the terminal that is at higher potential, that is, the terminal that is positive with respect to the other. In this text, the symbol $\mathscr{E}$ will be adopted to denote the magnitude of an e.m.f., that is, the potential difference between its terminals. Within an e.m.f. that is connected to an external circuit, current flows up the potential gradient as emphasised in figure 2.8(*a*) and $\mathscr{E}$ is the work done per unit charge by the other form of energy. When charge Q passes through an e.m.f. $\mathscr{E}$, work

$$W = \mathscr{E} Q \qquad (2.19)$$

is done by the e.m.f. on the charge. When an e.m.f. $\mathscr{E}$ delivers current I to an external circuit, the e.m.f. works at rate

$$P = \mathscr{E} I \qquad (2.20)$$

precisely balancing the rate of dissipation of energy (recall equation (2.18))

$$P = VI$$

by the rest of the circuit, since $V = \mathscr{E}$. The SI unit of e.m.f. is evidently that of potential difference, namely the volt. However, although the unit is the

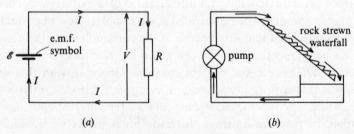

2.8 Analogy between simple electric and water circuits.

same, it is vital to distinguish between the nature of e.m.f. and that of potential difference across a resistor. As already pointed out, in the former case $\mathscr{E}$ is independent of the current while in the latter case potential difference and current are proportional. In particular, the potential difference across a resistor reverses sign with reversal of current through it, but an e.m.f. acts in the same sense independent of current through it. Also the potential difference vanishes across a resistor when the current through it reaches zero, whereas $\mathscr{E}$ remains unchanged and finite at zero current.

A convention will be adopted in this book of indicating the sense of a denoted potential difference by an arrow as in figure 2.8(a). The denoted potential difference will be the potential of the arrowed end minus that of the other end. Applying equation (2.16) to the simple circuit of figure 2.8(a).

$$I = V/R = \mathscr{E}/R \tag{2.21}$$

It may be instructive to draw an analogy between the electrical circuit of figure 2.8(a) and the water circuit depicted in figure 2.8(b) where a pump returns water from the bottom to the top of a waterfall. The water corresponds to charge and its flow to current. Water flowing down the fall gains kinetic energy at the expense of potential energy but this is dissipated at the various rocks of the fall and water arrives at the bottom with negligible kinetic energy. The rocks correspond to electron scattering centres and the fall corresponds to the resistor. Expenditure of electrical or mechanical energy in the pump raises water back to the top of the fall so that the pump is analogous to the e.m.f.

2.5 Internal resistance, sources and matching

All sources of e.m.f. exhibit some *internal resistance*. For example, in the case of the copper–zinc battery considered in the previous section, the electrolyte and even the electrodes present resistance to current. There is also resistance at the contacts between the electrolyte and metal electrodes, the insulating effect of hydrogen gas adjacent to the copper electrode being especially significant in this respect. A source of e.m.f. is therefore really equivalent to some e.m.f. $\mathscr{E}$ in *series* with some resistance r, so that when it is connected to an external circuit, as shown in figure 2.9(a) for example, the potential difference V_t across its terminals is related to the current I delivered by

$$V_t = \mathscr{E} - rI \tag{2.22}$$

In these circumstances $V_t < \mathscr{E}$. Only when $I = 0$, that is, when the source is *open circuit*, does $V_t = \mathscr{E}$. Note in passing that an external circuit connected to a source is generally described as a *load* if current is delivered to it. Further, if a resistor happens to serve as the load, it is described as a *load*

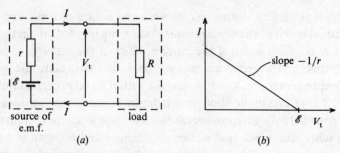

2.9 (a) Source of e.m.f. connected to a load and (b) relation between the terminal current and potential difference.

resistor. Equation (2.22) only applies in practice with fixed values of $\mathscr{E}$ and r over a finite range of I. Beyond the range, $\mathscr{E}$ and r depend on I. When the external circuit is another larger source of e.m.f. connected to drive a reverse current through the source under consideration, equation (2.22) applies but with I negative so that $V_t > \mathscr{E}$.

The dependence of I on V_t can be determined experimentally for any given source by connecting a variable load resistor across it and measuring corresponding values of I and V_t. Rearranging equation (2.22) shows that the dependence will obey

$$I = -\frac{V_t}{r} + \frac{\mathscr{E}}{r} \tag{2.23}$$

In particular, over the range where $\mathscr{E}$ and r are constant, the dependence will be linear and a plot of I versus V_t will give a straight line graph as shown in figure 2.9(b). Such a graph reveals the magnitude of $\mathscr{E}$ as the intercept on the V_t axis while the magnitude of r is given by the slope which is $-1/r$. Since $V_t = \mathscr{E}$ when $I = 0$, $\mathscr{E}$ can of course be determined directly if measurements of V_t can be made under sufficiently open-circuit conditions. With respect to determining the source resistance r, if the resistance of the load resistor is R

$$V_t = RI \tag{2.24}$$

so that

$$I = \frac{\mathscr{E}}{R + r} \tag{2.25}$$

and

$$V_t = \left(\frac{R}{R + r}\right)\mathscr{E} \tag{2.26}$$

It follows that r can be conveniently determined as the value of load resistance R which halves the potential difference V_t from its open-circuit value, for when $R = \infty$, $V_t = \mathscr{E}$, whereas when $R = r$, $V_t = \mathscr{E}/2$. In situations where $\mathscr{E}$ and r are not constant, note that it is possible to define an

2.5 Internal resistance, sources and matching

incremental or differential internal resistance $-dV_t/dI$ (cf. end of section 2.3).

Equation (2.25) shows that the internal resistance r of a source limits the maximum current that can be drawn to the *short-circuit* ($R=0$) value of $\mathscr{E}/r$. For a source of e.m.f. to be capable of supplying a large current, r must be small. In the case of a battery, this means that the areas of the electrodes in contact with the electrolyte must be large and the strength of the electrolyte such that its resistance is low. Although the internal resistance of a dry battery might be as high as $\sim 1\,\Omega$ that of a lead–acid accumulator might be only $\sim 0.01\,\Omega$. Connection of a shorting wire across the terminals of a 2 V accumulator would cause a current ~ 200 A to flow and generate power dissipation of ~ 400 W within the accumulator which would harm it. The connecting wire might also vaporise explosively.

Sources of e.m.f. having very low internal resistance are often described as *constant-voltage sources* because, in accordance with equation (2.22), the terminal potential difference (or voltage) V_t only varies very slowly with current drawn, that is to say, V_t is almost constant as the load varies over a wide range. Important sources also exist which, to a good approximation, deliver the same current whatever the load and these are appropriately referred to as *constant-current sources*. Inserting a very large resistance r' in series with a source of very high e.m.f. $\mathscr{E}$ is one way of obtaining a constant-current source, for when connected to an external resistance R the current is $\mathscr{E}/(R+r'+r)$ which is finite and almost independent of R. A Van de Graaff generator provides an interesting example of an intrinsic constant-current source. Electrical charge is sprayed from one terminal onto a fast-moving belt which conveys it to another insulated terminal at a constant rate thereby delivering a constant current. The terminals become oppositely charged, raising the potential difference V across them until, with a resistance R between them, the current $I = V/R$ balances the flow of charge dQ/dt supplied by the belt. Maximum potential difference is produced when R is a maximum, that is, when no resistor is deliberately connected between the terminals and R corresponds to charge leakage. Notice that in this particular source it is mechanical energy that is converted to electrical energy.

In fact, any direct source can be represented by some e.m.f. $\mathscr{E}$ in series with some resistance r *or* by a related constant-current source in *parallel* with the *same* resistance. If r is small, it often may be neglected in the former representation which is therefore more appropriate. If, on the other hand, r is large, it often may be neglected in the latter representation which becomes more appropriate. To prove the equivalence of the two representations, consider the circuits of figure 2.10. For the circuit of figure 2.10(a)

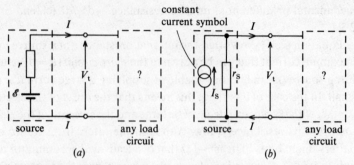

2.10 Equivalent representations of direct sources; (a) e.m.f. and series resistance and (b) constant current and parallel resistance.

$$V_t = \mathscr{E} - rI$$

while for that of figure 2.10(b) (notice the use of a heavy dot to denote a junction of conductors)

$$V_t = r_S(I_S - I)$$

since of the current I_S, I does not pass through r_S. Here we are anticipating Kirchhoff's laws of section 3.1. It is apparent that both circuit representations of the source develop the same *form* of relationship between V_t and I and the relationship becomes identical if

$$r_S = r, \quad I_S = \mathscr{E}/r \tag{2.27}$$

Thus a direct source may be represented by some e.m.f. $\mathscr{E}$ in series with some resistance r *or* by a constant-current source $\mathscr{E}/r$ in parallel with resistance r.

As the resistance R of a load resistor connected across a source of e.m.f. falls, the current through it rises according to equation (2.25) but the potential difference across it falls according to equation (2.26). Consider the power $P = V_t I$ developed in the load resistor. When $R \to \infty$, $I \to 0$ while $V \to \mathscr{E}$ so that $P \to 0$. When $R \to 0$, $I \to \mathscr{E}/r$ while $V_t \to 0$ so that $P \to 0$ again. As R varies between these limits the power is finite and is given by

$$P = RI^2 = R\left(\frac{\mathscr{E}}{R+r}\right)^2 \tag{2.28}$$

Maximum power dissipation occurs when $dP/dR = 0$, that is, when

$$\frac{dP}{dR} = \mathscr{E}^2 \left[\frac{(R+r)^2 - 2R(R+r)}{(R+r)^4} \right] = 0$$

or

$$R = \pm r \tag{2.29}$$

Although the negative solution has significance in certain signal circuits in electronics where negative differential resistance can arise, in direct circuits

2.5 Internal resistance, sources and matching

only the positive solution is physically admissible. Making the resistance of a load resistor equal to the internal resistance of a source of e.m.f. to achieve maximum power transfer from the source to load is referred to as *matching*.

Any component which can supply power continuously to a load is said to be *active* while components that can only consume or store energy are said to be *passive*.

3

Direct-current networks

3.1 Kirchhoff's laws

An electrical network is just a system of interconnected electrical components. For reference purposes, closed paths in networks are called *meshes* while junctions of paths are often called *nodes*. In these terms, the simple network of figure 2.8(a) considered in chapter 2 can be described as comprising a single mesh and containing no nodes. As mentioned in chapter 2, it is current practice to denote a node in a circuit diagram by a pronounced dot (see figure 2.10(b), for example). Absence of such a dot where paths cross over implies that the paths do not join there. This contemporary convention contrasts sharply with previous practice where any crossover was assumed to be a node unless indicated otherwise by a small semicircular jump in one path at the crossover.

Direct-current networks only include sources of direct e.m.f. or direct current and resistive loads. Except in section 3.10, attention will be confined in this chapter to *linear* direct-current networks in which the loads are resistors to which Ohm's law applies. Currents and potential differences in direct-current networks obey Kirchhoff's laws and, although these laws have already been invoked in certain simple situations in chapter 2, they will now be carefully established and stated.

Because charge is conserved, if $\sum I$ represents the *net* current out of any region, the charge Q in that region satisfies

$$\sum I = -dQ/dt \tag{3.1}$$

In the steady state, nothing changes with time and in particular $dQ/dt = 0$ so that $\sum I = 0$ also. Considering a small portion of a single-mesh network, this means that if current I flows into it, current I also flows out of it. Thus, in the steady state, the current is the same all the way round a single isolated mesh as indicated in figure 2.8(a). The implication of $dQ/dt = 0$ at a node is that, in

3.1 Kirchhoff's laws

the steady state, the sum of the currents entering a node equals the sum of currents leaving it. Regarding current leaving a node as negative current entering it, this last aspect can be alternatively expressed by saying that the algebraic sum of the steady currents entering a node is zero, that is,

$$\sum_i I_i = 0 \qquad (3.2)$$

Equation (3.2) is a statement of *Kirchhoff's first law* and it is worth stressing that it is only strictly applicable in the steady state. Under nonsteady conditions, charging is possible. Fortunately, however, in any reasonably conducting medium any accumulated charge disappears so quickly (actually, as may become apparent from the theory developed in sections 4.1 and 4.3, in times of the order ε/σ where ε and σ are the permittivity and conductivity respectively) that, provided the frequency of the current fluctuations is not extremely high, Kirchhoff's first law still holds to a good approximation. This is especially true of nodes formed by joining good conductors like copper for which the time constant of decay of charge is around 2×10^{-19} s.

Kirchhoff's second law states that the algebraic sum of the e.m.f.s acting in any mesh is equal to the algebraic sum of the products of the resistances of the various parts of the mesh and the currents in them. Mathematically this may be expressed by the equation

$$\sum_j \mathscr{E}_j = \sum_k R_k I_k \qquad (3.3)$$

where the subscript j identifies a particular e.m.f. in a mesh and the subscript k a particular part of that mesh. The second law embodied in equation (3.3) is just a certain way of expressing the law of conservation of energy in terms of resistance. Considering a single mesh for simplicity and multiplying both sides of equation (3.3) by the current, the left-hand side becomes the electrical power delivered by the e.m.f.s and the right-hand side the power dissipated by the remainder of the circuit. Put another way, equation (3.3) states that the potential at any point in a mesh is unique. In circumnavigating a mesh, the net potential rise experienced through the e.m.f.s precisely balances the net potential drop encountered through the resistors. In the water circuit analogy of figure 2.8(*b*), the potential energy at the top of the waterfall, for example, is regained as the water circulates round once.

Kirchhoff's current law expressed by equation (3.2) is not of course relevant to networks comprising only a single mesh. In the trivially simple example of the single-mesh network shown in figure 3.1(*a*), applying

32 Direct-current networks

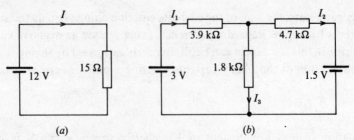

3.1 Two simple direct-current networks.

Kirchhoff's voltage law expressed by equation (3.3) to find the current gives

$$12 = 15I$$

where I is in amperes so that

$$I = (12/15)\,\text{A} = 0.8\,\text{A}$$

Now consider the network of figure 3.1(b) involving two coupled meshes. Let the currents in the three *branches* be I_1, I_2 and I_3 in mA units as shown. Note that when resistances are of the kΩ order, as is common, especially in electronic circuits, it is convenient to work in mA rather than A units. According to Kirchhoff's current law

$$I_1 = I_2 + I_3 \tag{3.4}$$

while Kirchhoff's voltage law applied to the left and right meshes in turn yields

$$3 = 3.9 I_1 + 1.8 I_3 \tag{3.5}$$

and

$$1.5 = -4.7 I_2 + 1.8 I_3 \tag{3.6}$$

respectively. The three unknown currents I_1, I_2 and I_3 are now found by simultaneous solution of equations (3.4)–(3.6) which can be rewritten

$$I_1 - I_2 - I_3 = 0$$
$$3.9 I_1 + 1.8 I_3 - 3 = 0$$
$$-4.7 I_2 + 1.8 I_3 - 1.5 = 0$$

The determinant form of solution is

$$\frac{I_1}{\begin{vmatrix} -1 & -1 & 0 \\ 0 & 1.8 & -3 \\ -4.7 & 1.8 & -1.5 \end{vmatrix}} = \frac{-I_2}{\begin{vmatrix} 1 & -1 & 0 \\ 3.9 & 1.8 & -3 \\ 0 & 1.8 & -1.5 \end{vmatrix}} = \frac{I_3}{\begin{vmatrix} 1 & -1 & 0 \\ 3.9 & 0 & -3 \\ 0 & -4.7 & -1.5 \end{vmatrix}}$$

$$= \frac{-1}{\begin{vmatrix} 1 & -1 & -1 \\ 3.9 & 0 & 1.8 \\ 0 & -4.7 & 1.8 \end{vmatrix}}$$

3.2 Resistors in series and parallel

and evaluating the determinants gives

$$I_1 = 0.497 \text{ mA}, \quad I_2 = -0.093 \text{ mA}, \quad I_3 = 0.590 \text{ mA}$$

The negative value obtained for I_2 simply means that the current in that branch is in the opposite direction to that assigned to I_2. With sufficient physical insight or luck, I_2 would have been assigned the opposite direction to that in figure 3.1(b), when it would have been found to be positive, but this is of no consequence.

Clearly, as the number of meshes in networks increases, they rapidly become more cumbersome to solve by straightforward application of Kirchhoff's laws and it is important to find ways of simplifying the analysis of such networks.

3.2 Resistors in series and parallel

Networks can often be reduced and their analysis thereby simplified by replacing several resistors connected in series or parallel by a single equivalent resistor. Consider first three resistors connected in *series* as shown in figure 3.2(a). The potential difference V across the series combination equals the sum of the potential differences V_1, V_2 and V_3 across the individual resistors, that is,

$$V = V_1 + V_2 + V_3$$

while the current I through the combination is common to the individual resistors. Application of equation (2.16) gives the equivalent resistance of the series combination as

$$R = \frac{V}{I} = \frac{R_1 I + R_2 I + R_3 I}{I} = R_1 + R_2 + R_3$$

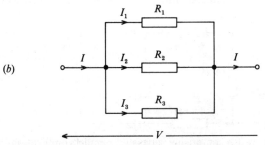

3.2 Resistors connected (a) in series and (b) in parallel.

In general then, the equivalent resistance of a set of series resistances R_i is

$$R = \sum_i R_i \qquad (3.7)$$

When three resistors are connected in parallel as in figure 3.2(b), the same potential difference exists across the combination as across the individual resistors while the current I through the combination divides among the individual resistors so that

$$I = I_1 + I_2 + I_3$$

Applying equation (2.16) again, the resistance of the combination is

$$R = \frac{V}{I} = \frac{V}{I_1 + I_2 + I_3} = \frac{V}{V/R_1 + V/R_2 + V/R_3} = \frac{1}{1/R_1 + 1/R_2 + 1/R_3}$$

Extension of this argument to any number of resistors shows that the equivalent resistance R of parallel resistances R_i is given by

$$\frac{1}{R} = \sum_i \frac{1}{R_i} \qquad (3.8)$$

It is also useful in network analysis to know how potential difference divides across series resistances and how current divides through parallel resistances. From the theory of this section, the potential difference V_i across a resistance R_i of a series combination is the fraction $R_i/\sum_i R_i$ of the potential difference across the combination. In the particular case of two series resistances R_1 and R_2, the fraction of the total potential difference that appears across R_1 is $R_1/(R_1 + R_2)$. Again from the theory of this section, the current I_i through resistance R_i of a parallel combination is the fraction $[R_i \sum_i (1/R_i)]^{-1}$ of the current through the combination. In the case of two parallel resistances R_1 and R_2, the fraction of the total current that passes through R_1 is $R_2/(R_1 + R_2)$.

As an example of the convenience of the formulae for the equivalent resistances of series and parallel combinations of resistances, suppose that it is desired to find the current delivered by the 6 V e.m.f. in the circuit of figure 3.3. The equivalent resistance connected between the terminals of the e.m.f.

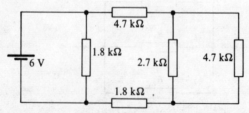

3.3 A network comprising series and parallel combinations of resistors.

3.2 Resistors in series and parallel

is $1.8\,\text{k}\Omega\,\|\,(4.7\,\text{k}\Omega+1.8\,\text{k}\Omega+2.7\,\text{k}\Omega\,\|\,4.7\,\text{k}\Omega)$ where $\|$ denotes 'in parallel with'. This resistance is approximately $1.8\,\text{k}\Omega\,\|\,(6.5\,\text{k}\Omega+1.7\,\text{k}\Omega)=1.8\,\text{k}\Omega\,\|\,8.2\,\text{k}\Omega\approx1.48\,\text{k}\Omega$ so that the current delivered amounts to about $6\,\text{V}/1.48\,\text{k}\Omega\approx4.05\,\text{mA}$. For comparison, to solve this problem by the straightforward application of Kirchhoff's laws, the voltage law would have to be used in three independent meshes and the current law at two suitable nodes. This would still leave five simultaneous equations requiring solution to find the current delivered by the e.m.f.

A transformation that is sometimes useful in reducing networks is the *star–delta* transformation shown in figure 3.4. For these two networks to be equivalent, the resistance must be the same between any pair of terminals in each. Thus

$$R_A + R_B = \frac{R_{AB}(R_{BC} + R_{CA})}{R_{AB} + R_{BC} + R_{CA}}$$

$$R_B + R_C = \frac{R_{BC}(R_{CA} + R_{AB})}{R_{AB} + R_{BC} + R_{CA}}$$

$$R_C + R_A = \frac{R_{CA}(R_{AB} + R_{BC})}{R_{AB} + R_{BC} + R_{CA}}$$

Subtracting the second relation from the first and adding the last gives

$$2R_A = [R_{AB}(R_{BC} + R_{CA}) - R_{BC}(R_{CA} + R_{AB}) + R_{CA}(R_{AB} + R_{BC})]$$
$$/(R_{AB} + R_{BC} + R_{CA})$$

which reduces to

$$R_A = \frac{R_{AB} R_{CA}}{R_{AB} + R_{BC} + R_{CA}} \qquad (3.9)$$

Similar cyclic relations follow for R_B and R_C. The inverse relations can be shown to be

$$R_{AB} = (R_A R_B + R_B R_C + R_C R_A)/R_C \quad \text{etc.} \qquad (3.10)$$

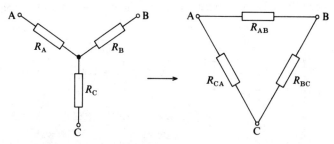

3.4 Star–delta transformation.

3.3 Generality of analysis by Kirchhoff's laws

Application of Kirchhoff's current and voltage laws to the simple networks of figure 3.1 led to their complete solution. However, the question naturally occurs, does the application of these two laws always lead to a complete solution of a network? It is helpful in answering this question to let the number of branches, nodes and meshes in a network be denoted by b, n and m respectively. Certainly, if the currents in all the branches of a network can be found, then all potential differences in the network can be obtained via relation (2.16) and it is completely solved. Thus b independent equations in the branch currents must be generated by Kirchhoff's laws in order to solve a network. Kirchhoff's current law applied to the n nodes of a network yields n equations of course, but one of these equations is not independent so that $(n-1)$ independent equations are obtained in this way. To understand this, consider the formation of a network branch by branch. From a given starting node, the first branch only produces one further node and one equation on application of Kirchhoff's current law because only one current exists – it is the same at both nodes (see figure 3.5(a)). Addition of another branch so as to create a further node (see, for example, figure 3.5(b)) leads to another equation, which is independent since it involves the new branch current. If a branch is added between two existing nodes (see, for example, figure 3.5(c)), it does not lead to another independent equation in the branch currents because it only changes the currents in the equations already existing from the application of Kirchhoff's current law. Kirchhoff's voltage law applied to the m independent meshes of the network under consideration provides another m independent equations in the branch currents so that altogether $(n-1+m)$ independent equations are obtained through assiduous application of the two laws of Kirchhoff. Incidentally, note that the network of figure 3.5(c) only contains two independent meshes. The outer mesh is just the sum of the two inner meshes and is therefore not independent.

Now every time a branch is added to a network, either a new node or new independent mesh is created (see figures 3.5(d)–(g)). From this argument the number of branches in a network is just the number of nodes, apart from

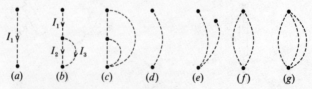

(a) (b) (c) (d) (e) (f) (g)

3.5 Branches, nodes and meshes.

3.4 Mesh current analysis

one in particular, plus the number of independent meshes, that is,

$$b = (n-1) + m \tag{3.11}$$

It follows that application of Kirchhoff's two laws to a network produces just the number of independent equations in the branch currents that are required for complete solution. In the network of figure 3.1(b), for example, $n=2$, $m=2$, so that there is $n-1=1$ equation generated by Kirchhoff's current law and $m=2$ independent equations generated by Kirchhoff's voltage law giving altogether $b=3$ independent equations from which the $b=3$ branch currents can be obtained, as was shown in section 3.1.

3.4 Mesh current analysis

In a more elegant approach to network analysis, the unknown branch currents are replaced by unknown *mesh currents*. This procedure reduces the number of unknowns at the outset, thereby simplifying the solution. Currents in the branches are thought of as arising from combinations of mesh currents, each of which circulates completely round the mesh in which it exists. Such mesh currents have the valuable attribute of automatically satisfying Kirchhoff's current law, for each mesh current that flows into a particular node also flows out of it. For example, in the network illustrated in figure 3.6(a), mesh currents I_A and I_B flow into and

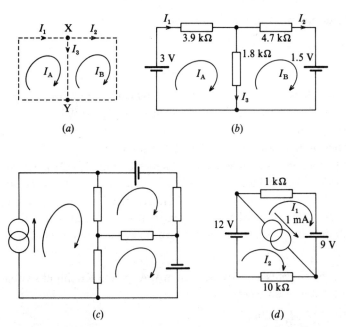

3.6 Networks used to illustrate mesh current analysis in the text.

out of node X and node Y. Working with mesh currents, the $(n-1)$ independent equations generated by applying Kirchhoff's current law at the nodes become redundant. There remain just m independent equations generated by applying Kirchhoff's voltage law in each independent mesh, precisely the number required to find each independent mesh current. Whereas $b=(n-1)+m$ independent equations require solution when working with branch currents, only m require solution when working with mesh currents. Branch currents are found by combining appropriate mesh currents and, if required, potential differences follow with the aid of relation (2.16) as before. For the particular network depicted in figure 3.6(a), $I_1 = I_A$, $I_2 = I_B$ and $I_3 = I_A - I_B$ which maintains $I_1 - I_2 - I_3 = 0$.

To illustrate the mesh current method of network analysis, consider its application to the circuit of figure 3.1(b) which has already been analysed by the branch current method in section 3.1. For easy reference the circuit is redrawn in figure 3.6(b). If I_A and I_B represent the two mesh currents in mA units, application of Kirchhoff's voltage law yields

$$\left. \begin{array}{l} 3 = 3.9 I_A + 1.8(I_A - I_B) \\ 1.5 = -4.7 I_B - 1.8(I_B - I_A) \end{array} \right\} \tag{3.12}$$

or on rearranging

$$5.7 I_A - 1.8 I_B - 3 = 0$$
$$-1.8 I_A + 6.5 I_B + 1.5 = 0$$

In determinant form the solution is

$$\frac{I_A}{\begin{vmatrix} -1.8 & -3 \\ 6.5 & 1.5 \end{vmatrix}} = \frac{-I_B}{\begin{vmatrix} 5.7 & -3 \\ -1.8 & 1.5 \end{vmatrix}} = \frac{1}{\begin{vmatrix} 5.7 & -1.8 \\ -1.8 & 6.5 \end{vmatrix}}$$

from which $I_A = 0.497$ mA and $I_B = -0.093$ mA. The branch currents are therefore $I_1 = 0.497$ mA, $I_2 = -0.093$ mA and $I_3 = 0.590$ mA as obtained before. In this particular instance, the mesh method has reduced the number of equations to be solved from three to two.

When a constant current supply is included in a single mesh, as in the network of figure 3.6(c), it defines the current in that mesh. To solve the network, Kirchhoff's voltage law must then be applied to the other independent meshes. The constant current supply in the network of figure 3.6(d) defines the difference in the two inner mesh currents, that is,

$$I_2 - I_1 = 1 \tag{3.13}$$

where the currents are expressed in mA. Applying Kirchhoff's voltage law to the outer mesh yields

$$21 = I_1 + 10 I_2 \tag{3.14}$$

so that on combining the two equations, $I_1 = 1$ mA and $I_2 = 2$ mA.

3.5 Node-pair potential analysis

In another efficient method of analysing networks, the unknown quantities are taken to be any complete set of independent potential differences between pairs of nodes. This procedure again reduces the number of initial unknowns and therefore simplifies the solution. The easiest way of obtaining a complete set of independent node-pair potential differences is to take the potential of one node as a reference, for then the potentials of all the other nodes with respect to it constitute the required set. Quite clearly there are $(n-1)$ independent node-pair potential differences. To take a particular example, in the circuit of figure 3.7(a), node O could act as the reference node and the potentials of nodes A and B with respect to it could serve as the independent unknowns and be labelled V_A and V_B. The

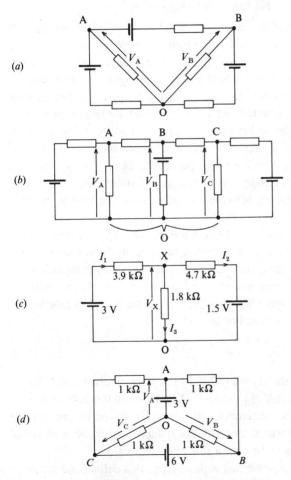

3.7 Networks used to illustrate node-pair analysis in the text.

potential difference between nodes A and B in figure 3.7(a) is not independent of V_A or V_B; indeed it is expressible as $V_A - V_B$. Where branches of zero, or in practice negligible, resistance connect nodes, as for nodes O in figure 3.7(b), separate labelling is superfluous, for they are at the same potential, and it is usual to take their common potential as the reference.

Node-pair potentials have the virtue of automatically satisfying Kirchhoff's voltage law. Round any mesh starting from node A, say, the path proceeds through a closed sequence of nodes such as P, X, F, etc., back to A. The potential differences across successive branches can be written in terms of reference node O node-pair potentials V_A, V_P, V_X, V_F etc. as $(V_P - V_A)$, $(V_X - V_P)$, $(V_F - V_X)$ through to $(V_A - V_?)$ and no matter what the potentials

$$(V_P - V_A) + (V_X - V_P) + (V_F - V_X) + \cdots + (V_A - V_?) = 0$$

which is an expression of Kirchhoff's voltage law. Working with the $(n-1)$ independent node-pair potentials, the m equations generated by applying Kirchhoff's voltage law round the m independent meshes become redundant. There remain just $(n-1)$ independent equations generated by applying Kirchhoff's current law at $(n-1)$ nodes, precisely the number required to find the complete set of independent node-pair potentials. Whereas $b = (n-1) + m$ independent equations require solution when working with branch currents as the unknowns, only $(n-1)$ require solution when working with node-pair potentials. The potential difference across any branch is either one of the set of independent node-pair potentials or the difference between two of them, while branch currents follow with the aid of relation (2.16).

Consider now the solution of the circuit of figure 3.1(b) by the method of node-pair potential analysis. This circuit has already been solved by the branch and mesh current methods of analysis so that ready comparison can be made. The circuit is redrawn in figure 3.7(c) for ease of reference with the single unknown potential difference between nodes X and O labelled V_X. Applying Kirchhoff's current law to node X

$$\frac{V_X}{1.8} + \frac{V_X - 3}{3.9} + \frac{V_X - 1.5}{4.7} = 0 \tag{3.15}$$

from which it is found that $V_X = 1.062$ V. Making use of relation (2.16), $I_1 = (3 - 1.062)/3.9 = 0.497$ mA, $I_2 = (1.062 - 1.5)/4.7 = -0.093$ mA and $I_3 = 1.062/1.8 = 0.590$ mA in agreement with before. It should be appreciated that the node-pair potential method has reduced the number of initial equations and unknowns to one in this particular case.

Where a branch comprises just a pure e.m.f., this defines the potential difference across it. Thus, for example, the solution of the network of figure

3.6 Superposition and reciprocity theorems

3.7(d) follows from

$$V_A = 3 \tag{3.16}$$

$$V_B - V_C = 6 \tag{3.17}$$

and the application of Kirchhoff's current law at nodes O and A which gives

$$\frac{V_B}{1} + \frac{V_C}{1} + \frac{V_B - V_A}{1} + \frac{V_C - V_A}{1} = 0 \tag{3.18}$$

From equations (3.16)–(3.18), $V_A = 3$ V, $V_B = 4.5$ V and $V_C = -1.5$ V.

Whether mesh current or node-pair potential analysis is more efficient in the case of a particular network depends on the number of independent meshes m compared with the number of independent node pairs $(n-1)$. If $m < (n-1)$ then mesh current analysis generates less initial unknowns and corresponding equations for simultaneous solution and is therefore less laborious. On the other hand, if $(n-1) < m$ then node-pair potential analysis provides the easier approach with less initial unknowns and equations. When $(n-1) = m$ the two methods are similar in complexity. The more parallel in form that a circuit is, the greater m will be compared with $(n-1)$ and the more likely it is that node-pair analysis will be more appropriate. All of the networks analysed in this and the preceding section have either the same number of independent node pairs or one less than the number of independent meshes. Figure 3.8 shows an example of a network that contains fewer independent meshes than independent node pairs. One other aspect that influences which is the shorter method of analysis is whether it is a branch current or potential difference that is ultimately required.

3.6 The superposition and reciprocity theorems

Consider a general direct-current network having n independent meshes numbered $1, 2, 3, \ldots, j, \ldots, n$ for identification and let the mesh currents in them in a particular sense, say clockwise, be correspondingly designated $I_1, I_2, I_3, \ldots, I_j, \ldots, I_n$ as illustrated in figure 3.9. It turns out that a particularly convenient scheme for denoting the elements of the

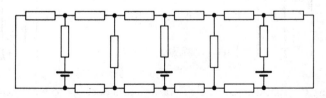

3.8 An example of a network with less independent meshes than independent node pairs.

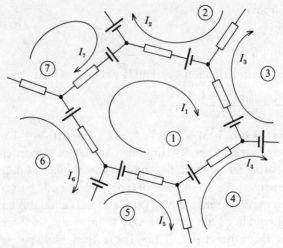

3.9 General direct-current network.

network is to let the total e.m.f. and total resistance acting in mesh j, the former in the same sense as the labelled mesh current, be represented by $\mathscr{E}_j$ and R_{jj} respectively. Resistance that is common to meshes i and j is appropriately represented by $R_{ij} = R_{ji}$. Applying Kirchhoff's voltage law to the network

$$\left.\begin{aligned}\mathscr{E}_1 &= R_{11}I_1 - R_{12}I_2 - R_{13}I_3 - \cdots - R_{1n}I_n\\ \mathscr{E}_2 &= -R_{21}I_1 + R_{22}I_2 - R_{23}I_3 - \cdots - R_{2n}I_n\\ \mathscr{E}_3 &= -R_{31}I_1 - R_{32}I_2 + R_{33}I_3 - \cdots - R_{3n}I_n\\ \vdots&\;\;\vdots\qquad\quad\vdots\qquad\quad\vdots\qquad\qquad\;\;\vdots\\ \mathscr{E}_n &= -R_{n1}I_1 - R_{n2}I_2 - R_{n3}I_3 - \cdots + R_{nn}I_n\end{aligned}\right\} \qquad (3.19)$$

from which it is apparent that the notation adopted is consistent with standard matrix notation, and in matrix language the independent equations in the mesh currents can be expressed as

$$\begin{pmatrix}\mathscr{E}_1\\ \mathscr{E}_2\\ \mathscr{E}_3\\ \vdots\\ \mathscr{E}_n\end{pmatrix} = \begin{pmatrix}R_{11}, & -R_{12}, & -R_{13}, & \ldots, & -R_{1n}\\ -R_{21}, & R_{22}, & -R_{23}, & \ldots, & -R_{2n}\\ -R_{31}, & -R_{32}, & R_{33}, & \ldots, & -R_{3n}\\ \vdots & \vdots & \vdots & & \vdots\\ -R_{n1}, & -R_{n2}, & -R_{n3}, & \ldots, & R_{nn}\end{pmatrix}\begin{pmatrix}I_1\\ I_2\\ I_3\\ \vdots\\ I_n\end{pmatrix}$$

(3.20)

Knowing the significance of the notation, it is possible to write down the elements of the resistance transformation matrix, if so desired, without recourse to Kirchhoff's laws. In this connection, however, note that all

3.6 Superposition and reciprocity theorems

diagonal elements are positive but that all off-diagonal elements are negative.

Multiplying both sides of equation (3.20) by the inverse of the resistance matrix, the elements of which have the dimension of conductance and are conveniently written as G_{ij}, it immediately follows that

$$\begin{pmatrix} I_1 \\ I_2 \\ I_3 \\ \vdots \\ I_n \end{pmatrix} = \begin{pmatrix} G_{11}, & G_{12}, & G_{13}, & \ldots, & G_{1n} \\ G_{21}, & G_{22}, & G_{23}, & \ldots, & G_{2n} \\ G_{31}, & G_{32}, & G_{33}, & \ldots, & G_{3n} \\ \vdots & \vdots & \vdots & & \vdots \\ G_{n1}, & G_{n2}, & G_{n3}, & \ldots, & G_{nn} \end{pmatrix} \begin{pmatrix} \mathscr{E}_1 \\ \mathscr{E}_2 \\ \mathscr{E}_3 \\ \vdots \\ \mathscr{E}_n \end{pmatrix}$$

(3.21)

or in fully written-out form

$$\left. \begin{array}{l} I_1 = G_{11}\mathscr{E}_1 + G_{12}\mathscr{E}_2 + G_{13}\mathscr{E}_3 + \cdots + G_{1n}\mathscr{E}_n \\ I_2 = G_{21}\mathscr{E}_1 + G_{22}\mathscr{E}_2 + G_{23}\mathscr{E}_3 + \cdots + G_{2n}\mathscr{E}_n \\ I_3 = G_{31}\mathscr{E}_1 + G_{32}\mathscr{E}_2 + G_{33}\mathscr{E}_3 + \cdots + G_{3n}\mathscr{E}_n \\ \vdots \quad \vdots \quad \vdots \quad \vdots \quad \vdots \\ I_n = G_{n1}\mathscr{E}_1 + G_{n2}\mathscr{E}_2 + G_{n3}\mathscr{E}_3 + \cdots + G_{nn}\mathscr{E}_n \end{array} \right\}$$

(3.22)

The elements G_{ij} are of course only functions of the elements of the resistance transformation matrix and just as $R_{ij} = R_{ji}$, $G_{ij} = G_{ji}$.

In the simple case of just two coupled meshes all of the foregoing is easy to verify. According to Kirchhoff's voltage law,

$$\mathscr{E}_1 = R_{11}I_1 - R_{12}I_2$$
$$\mathscr{E}_2 = -R_{21}I_1 + R_{22}I_2$$

Multiplying the first equation by R_{22} and the second by R_{12} and adding gives

$$I_1 = (R_{22}\mathscr{E}_1 + R_{12}\mathscr{E}_2)/(R_{11}R_{22} - R_{12}R_{21})$$

Similarly, multiplying the first equation by R_{21} and the second by R_{11} and adding gives

$$I_2 = (R_{21}\mathscr{E}_1 + R_{11}\mathscr{E}_2)/(R_{11}R_{22} - R_{12}R_{21})$$

Evidently, in the case of just two coupled meshes, equations (3.22) apply with $\mathscr{E}_{k \neq 1,2} = 0$ and

$$G_{11} = R_{22}/\Delta; \quad G_{12} = G_{21} = R_{12}/\Delta; \quad G_{22} = R_{11}/\Delta$$

where

$$\Delta = R_{11}R_{22} - R_{12}R_{21}$$

Equations (3.22) show that mesh currents and hence branch currents *superpose*. That is to say, they show that the total mesh current in a mesh of a network due to various e.m.f.s acting throughout the network equals the sum of the mesh currents that would exist in that mesh due to each e.m.f. acting alone. The *superposition theorem* just stated often eases network analysis when more than one source is present. Its validity depends on the linearity of equations (3.19) and hence on the applicability of Ohm's law to the passive elements of the network. Of course, in considering the effect of one source of e.m.f. at a time, the other sources must be replaced by their internal resistances; only the other e.m.f.s must be set to zero. The theorem also applies to current sources since any current source is equivalent to some source of e.m.f. In applying the superposition theorem, when considering the effect of one source at a time, each other constant-current source must be replaced by an open circuit and each other pure e.m.f. by a short circuit.

As an example of the application of the superposition theorem, suppose it is wished to find the current I in the circuit of figure 3.10(a). Considering first the current due to the 24 V e.m.f., the 2 mA source is replaced by an open circuit. Hence the 24 V e.m.f. contributes $[24/(3.3+4.7)10^3]$ A = 3 mA to I. The contribution of the 2 mA source to I is found by replacing the 24 V e.m.f. by a short circuit so that it is $[-2 \times 3.3/(3.3+4.7)10^3]$ A = -0.825 mA. It immediately follows that $I = 2.175$ mA.

Another theorem that is occasionally helpful is the *reciprocity theorem*. This states that if an e.m.f. introduced into one branch of a passive network causes a current I to flow in a second branch then the same e.m.f. introduced

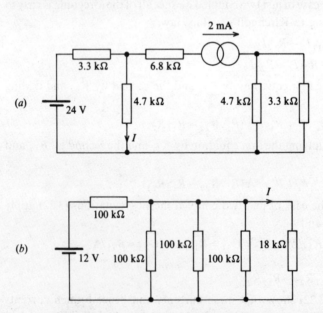

3.10 Circuits to which (a) the superposition and (b) the reciprocity theorems are applied in the text to find the branch current I.

3.7 Thévenin and Norton theorems

into that second branch causes the same current I to flow in the first branch. The reciprocity theorem follows once again from equations (3.22). Let an e.m.f. $\mathscr{E}$ be introduced into the branch that is common to meshes l and m so that $\mathscr{E}_l = \mathscr{E}$ and $\mathscr{E}_m = -\mathscr{E}$. The current caused in the branch common to meshes p and q is

$$I_q - I_p = (G_{ql}\mathscr{E} - G_{qm}\mathscr{E}) - (G_{pl}\mathscr{E} - G_{pm}\mathscr{E}) = (G_{ql} + G_{pm} - G_{qm} - G_{pl})\mathscr{E}$$

If instead e.m.f. $\mathscr{E}$ is introduced into the branch common to meshes p and q so that $\mathscr{E}_p = \mathscr{E}$ and $\mathscr{E}_q = -\mathscr{E}$, the current caused in the branch common to meshes l and m is

$$I_m - I_l = (G_{mp}\mathscr{E} - G_{mq}\mathscr{E}) - (G_{lp}\mathscr{E} - G_{lq}\mathscr{E}) = (G_{mp} + G_{lq} - G_{mq} - G_{lp})\mathscr{E}$$

But $G_{ij} = G_{ji}$, hence

$$(I_q - I_p) = (I_m - I_l)$$

and the reciprocity theorem is proved.

As an illustration of the usefulness of the reciprocity theorem, consider the calculation of the current I in the circuit of figure 3.10(b). The direct method involves calculating the resistance of the three 100 kΩ resistances in parallel with 18 kΩ which is awkward. Thus the current through the 12 V e.m.f. is $12/[100 + (\frac{1}{100} + \frac{1}{100} + \frac{1}{100} + \frac{1}{18})^{-1}] \times 10^3$ A $= (12/111.69)$ mA. The potential difference across the 18 kΩ resistor is therefore $12(1 - 100/111.69)$ V $= 1.256$ V and the current through it 69.8 μA. It is much easier to calculate this current using the reciprocity theorem. To do this the 12 V e.m.f. is moved to the 18 kΩ branch and the current calculated in the 100 kΩ branch from which it has been removed. Thus $I = \frac{1}{4}[12/(18+25)10^3]$ A $= 69.8$ μA. Of course, the reciprocity method is much easier in this contrived example because of the particular resistor values.

3.7 The Thévenin and Norton theorems

The importance to network analysis of the pair of powerful theorems to be introduced in this section cannot be stressed too much. Beyond analysis they give valuable insight into the operation of circuits. They are especially useful in the field of electronics because they apply not only to linear direct-current networks as discussed here but also to linear networks containing signal sources.

In the context of direct-current networks, *Thévenin's theorem* states that any complicated linear network containing sources and resistances, as far as any two terminals of it are concerned, may be replaced by an *equivalent circuit* comprising just a pure e.m.f. in series with a resistance. The e.m.f. is the potential difference between the two terminals of the original network when no load is connected, that is, when the terminals are left externally open-circuit. The series resistance is the resistance between the two

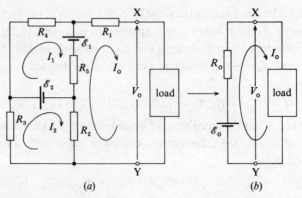

3.11 Illustration of Thévenin's theorem; (a) original network and (b) equivalent network.

terminals of the original network when all sources are replaced by their internal resistances.

To establish the validity of Thévenin's theorem, consider first the particular network shown to the left of the terminals X and Y in figure 3.11(a). The aspect of interest here is the behaviour of the network as far as the terminals X and Y are concerned for any external load whatsoever connected between them. Since the complete circuit with load connected has three independent node pairs and three independent meshes it is equally amenable to node-pair potential or mesh current analysis. Choosing the latter, application of Kirchhoff's voltage law to the independent meshes indicated gives

$$\mathscr{E}_1 - (R_1 + R_2 + R_5)I_o + R_5 I_1 + R_2 I_2 = V_o$$

$$\mathscr{E}_2 - \mathscr{E}_1 + R_5 I_o - (R_4 + R_5)I_1 = 0$$

$$\mathscr{E}_2 - R_2 I_o + (R_2 + R_3)I_2 = 0$$

Using the last two equations to substitute for I_1 and I_2 in the first equation in terms of I_o leads to

$$V_o = \mathscr{E}_1 - (R_1 + R_2 + R_5)I_o + \frac{R_5}{R_4 + R_5}(\mathscr{E}_2 - \mathscr{E}_1 + R_5 I_o)$$

$$+ \frac{R_2}{R_2 + R_3}(R_2 I_o - \mathscr{E}_2)$$

or on regrouping

$$V_o = \mathscr{E}_1 - \frac{R_2}{R_2 + R_3}\mathscr{E}_2 + \frac{R_5}{R_4 + R_5}(\mathscr{E}_2 - \mathscr{E}_1) - \left(R_1 + \frac{R_2 R_3}{R_2 + R_3} + \frac{R_4 R_5}{R_4 + R_5}\right) \quad (3.23)$$

It is most instructive to write this equation in the form

$$V_o = \mathscr{E}_o - R_o I_o \quad (3.24)$$

3.7 Théverin and Norton theorems

where

$$\mathcal{E}_o = \mathcal{E}_1 - \frac{R_2}{R_2 + R_3}\mathcal{E}_2 + \frac{R_5}{R_4 + R_5}(\mathcal{E}_2 - \mathcal{E}_1) \tag{3.25}$$

$$R_o = R_1 + \frac{R_2 R_3}{R_2 + R_3} + \frac{R_4 R_5}{R_4 + R_5} \tag{3.26}$$

The reason is that the relationship between the terminal quantities I_o and V_o of the circuit of figure 3.11(b) is just the same as that represented by equation (3.24). Thus, no matter what the load, the simple circuit to the left of the terminals X and Y in figure 3.11(b) is equivalent to the complicated circuit to the left of the terminals X and Y in figure 3.11(a), provided $\mathcal{E}_o$ and R_o are given by equations (3.25) and (3.26). Inspection of figure 3.11(a) reveals that R_o as given by equation (3.26) amounts to the resistance between terminals X and Y of the circuit to the left of X and Y with the e.m.f.s replaced by their internal resistances, that is, by short circuits. Further inspection of figure 3.11(a) also shows that $\mathcal{E}_o$ as given by equation (3.25) amounts to the potential difference between the terminals X and Y with the load removed, that is, with the circuit to the left of X and Y externally open-circuit. With the load removed, the potential difference between X and Y in figure 3.11(a) is $\mathcal{E}_1$ plus the fraction $R_5/(R_4 + R_5)$ of the e.m.f. $(\mathcal{E}_2 - \mathcal{E}_1)$ acting in the mesh carrying current I_1 minus the fraction $R_2/(R_2 + R_3)$ of the e.m.f. $\mathcal{E}_2$ acting in the mesh carrying current I_2. Apparently Thévenin's theorem is true for the particular network to the left of terminals X and Y in figure 3.11(a). Careful study of the way in which the terms arise in equation (3.23) will show that whatever is added to the network of figure 3.11(a), the form of equation (3.24) will apply. Moreover, $\mathcal{E}_o$ and R_o will always have the values specified by Thévenin's theorem so that Thévenin's theorem is true for all direct-current networks. The reader is advised at this point to try various modest additions to the network of figure 3.11(a) and, by similar analysis to the foregoing, check that Thévenin's theorem still holds.

An alternative way of demonstrating the truth of Thévenin's theorem is the following. Consider any network A connected, as shown in figure 3.12(a), through terminals X and Y of it to another network B comprising an e.m.f. $\mathcal{E}$ in series with a resistance R. Of course, according to Thévenin's theorem, the $\mathcal{E}, R$ combination represents all other possible networks that can be connected to X and Y. The relationship between the current I_o and potential difference V_o delivered to network B is

$$V_o = \mathcal{E} + RI_o \tag{3.27}$$

Now suppose that an e.m.f. $\mathcal{E}'$ is inserted in series between A and B as indicated in figure 3.12(b) so as to reduce the current to zero. In these circumstances it is clear that the potential difference $(V_o)_{I_o=0}$ between X and

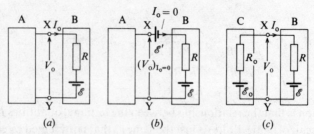

3.12 Thévenin's theorem; (a) original network A connected to network B, (b) e.m.f. $\mathscr{E}'$ inserted to reduce current delivered to B to zero and (c) equivalent network C connected to network B.

Y is given by

$$(V_o)_{I_o=0} = \mathscr{E} + \mathscr{E}' \tag{3.28}$$

However, according to the superposition theorem, the additional e.m.f. $\mathscr{E}'$ acting on its own must deliver that current which reduces the total to zero. Hence

$$0 = I_o - \frac{\mathscr{E}'}{R + R_o} \tag{3.29}$$

where R_o is the resistance of network A between terminals X and Y with all sources replaced by their internal resistance. Eliminating $\mathscr{E}'$ between equations (3.28) and (3.29),

$$(V_o)_{I_o=0} = \mathscr{E} + (R + R_o)I_o \tag{3.30}$$

and making use of equation (3.27)

$$(V_o)_{I_o=0} = V_o + R_o I_o$$

or

$$V_o = (V_o)_{I_o=0} - R_o I_o \tag{3.31}$$

For comparison, the simple equivalent circuit C of figure 3.12(c) delivers current I_o at potential difference V_o to B such that

$$V_o = \mathscr{E}_o - R_o I_o \tag{3.32}$$

Evidently network C is equivalent to network A provided R_o is the resistance between terminals X and Y of the original network when sources of e.m.f. are replaced by their internal resistance and $\mathscr{E}_o$ is the open-circuit potential difference of the original network between terminals X and Y, which is a restatement of Thévenin's theorem. Notice in passing that the short-circuit current $(I_o)_{V_o=0}$ delivered by network C between X and Y is $\mathscr{E}_o/R_o$ so that R_o can be obtained alternatively as

$$R_o = (V_o)_{I_o=0}/(I_o)_{V_o=0} \tag{3.33}$$

That is, R_o is the ratio of the open-circuit voltage across to the short-circuit current between the terminals of the original network.

3.7 Thévenin and Norton theorems

A source of e.m.f. being representable as a pure e.m.f. in series with an internal resistance is one trivial example of the applicability of Thévenin's theorem. Whenever a network is not too complicated, the theorem may be applied immediately to the whole network, as far as two terminals of it are concerned, to reduce it to a simple Thévenin equivalent. In more complicated networks, where immediate application of the theorem to the whole network can be awkward, application to successive parts of it often proves effective. Thus a part of the network to which Thévenin's theorem is easily applied is selected and the remainder of the network regarded as a load. Having simplified one part, this is now combined with a further convenient portion of the complete network and Thévenin's theorem applied again. Repetition of the process soon reduces the whole network to the Thévenin equivalent. Sometimes, it will only be of interest to Thévenin reduce part of a network. The relevance of Thévenin's theorem to matching a load to a network or one part of a network to another should be obvious.

One situation in which the application of Thévenin's theorem is helpful is the design of a potential divider circuit formed by connecting two resistors across an e.m.f. The purpose of such a circuit is to provide a source of smaller e.m.f. across one of the resistors. With reference to the divider depicted in figure 3.13(a), the open-circuit potential difference between X and Y is $[R_1/(R_1 + R_2)]\mathscr{E}$ while the resistance between X and Y with sources replaced by their internal resistance is $R_1 \parallel R_2$. The Thévenin equivalent circuit as far as the terminals X and Y are concerned is therefore as shown in figure 3.13(b). This equivalent shows that to obtain a given output e.m.f. from a given starting e.m.f. $\mathscr{E}$, the ratio R_1/R_2 is fixed. If R_1 and R_2 are made large, the current drain on the source providing e.m.f. $\mathscr{E}$ is small but the penalty is that the 'internal' resistance $R_1 \parallel R_2$ of the divided source of e.m.f. is high. This means that not much current can be supplied to a load by the subsidiary divided source without the potential difference at the terminals

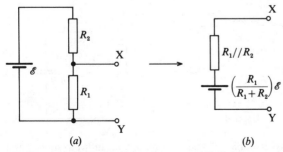

3.13 Thévenin equivalent of a potential-divider network.

falling appreciably. Conversely, making R_1 and R_2 small makes the series resistance of the divided source desirably small but excessively drains the primary source of e.m.f. $\mathscr{E}$. Incidentally, the Thévenin series resistance is often described as the *output resistance*. Clearly, a compromise must be struck in the design of a potential divider. The circuit works best when the ratio R_1/R_2 can be small for then R_1 can be made small to procure a low output resistance while R_2 can be made large to reduce the current drain on $\mathscr{E}$.

Use of Thévenin's theorem to find I_2 in the circuit of figure 3.1(b), which has been solved by various other methods already in this chapter, is illustrated in figure 3.14(a). First, the Thévenin equivalent of the 3 V e.m.f. and 1.8 kΩ, 3.9 kΩ potential divider is found. Current I_2 then follows from Kirchhoff's voltage law in the equivalent series circuit and

$$I_2 = \left(\frac{1.8 \times 3}{5.7} - 1.5\right) \bigg/ (4.7 + 3.9 \| 1.8) 10^3 \text{ A}$$
$$= -0.553/5.932 \text{ mA} = -0.093 \text{ mA}$$

in agreement with before. Current I_1 could be found similarly of course.

The way in which Thévenin's theorem can be applied to the circuit of figure 3.6(d) to find the current I_1 is as follows. With reference to figure 3.14(b) where the circuit is redrawn for easy reference, the branch carrying current I_1 is treated as the load and the Thévenin equivalent of the remainder is found. Replacing the 1 mA current source by an open circuit

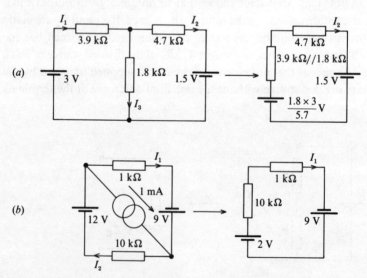

3.14 Two examples of Thévenin reduction; in (a) to find I_2 and in (b) to find I_1.

3.7 Thévenin and Norton theorems

and the 12 V e.m.f. by a short circuit reveals that the Thévenin series resistance is 10 kΩ. With the branch carrying current I_1 open-circuited, the potential difference across the current generator is $12\,\text{V} - (10\,\text{k}\Omega \times 1\,\text{mA}) = 2\,\text{V}$ because the current source forces 1 mA through the 10 kΩ resistor. The Thévenin equivalent is consequently as shown at the right of figure 3.14(b) and $I_1 = 11\,\text{V}/11\,\text{k}\Omega = 1\,\text{mA}$ in agreement with before.

An interesting network which particularly demonstrates the analytic power of Thévenin's theorem is the so-called R–$2R$ digital-to-analogue converter. The essential circuit of an eight-bit version is shown in figure 3.15(a) although the output would normally be followed by an operational amplifier to suitably scale up the output and prevent undue loading by following circuitry. Bits of a binary number from the least to most significant bit (L.S.B. to M.S.B.) are fed into the terminals A to H. A unity or zero bit is entered at each of the eight terminals as appropriate by connecting respectively to the high or low side of the e.m.f. $\mathscr{E}$ using a switching arrangement not shown. Finding the effect on the output of such a binary input is eased by the superposition theorem. It is the sum of the effects of each unity bit acting separately with all the other bits at zero. Consider for example the effect on the output of the entry 00000100 which means that terminal C is connected through the e.m.f. $\mathscr{E}$ to the common line while all other inputs are shorted to the common line as shown in figure 3.15(b). Thévenin reduction of the circuit to the left of node c gives the equivalent circuit shown in figure 3.15(c). Successive reduction of the $2R$–$2R$ potential divider sections eventually gives the Thévenin equivalent of figure 3.15(d). A little thought will show that a one at the nth most significant bit gives an output e.m.f. of $\mathscr{E}/2^n$ so that the superposed effects of ones occurring at various bits, depending on the binary input, will give an e.m.f. which is an analogue representation of the digital input.

The equivalence, discussed in section 2.5, between a source of e.m.f. and a source of current means that there is a current-source version of Thévenin's theorem. This current-source version is known as *Norton's theorem*. In the context of direct-current networks, Norton's theorem states that any complicated network containing sources and resistances, as far as any two terminals of it are concerned, may be replaced by an equivalent circuit comprising just a constant-current source in parallel with a resistance. The constant-current source is that current which flows between the terminals of the original network when they are externally short-circuited. The parallel resistance is the resistance between the two terminals of the original network with all sources replaced by their internal resistances.

Figure 3.16 shows the Thévenin and Norton direct equivalent circuits side by side, each connected to a load. The short-circuit current and open-

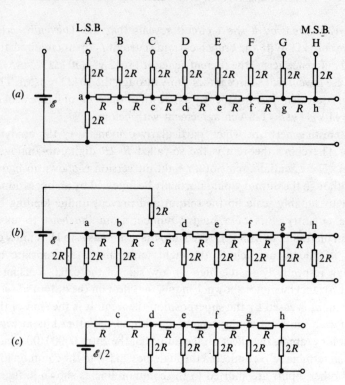

3.15 Thévenin reduction of R–$2R$ digital-to-analogue conversion network; (a) basic eight-bit network without digital input, (b) eight-bit network with binary input 00000100, (c) the same, Thévenin reduced up to point c, and (d) the same completely Thévenin reduced with respect to the output terminals.

circuit potential difference are respectively $\mathscr{E}_o/R_o$ and $\mathscr{E}_o$ for the Thévenin equivalent and I_N and $R_N I_N$ for the Norton equivalent. When the Thévenin and Norton equivalents represent the same original network

$$I_N = \mathscr{E}_o/R_o \tag{3.34}$$

and

$$\mathscr{E}_o = R_N I_N \tag{3.35}$$

so that

$$R_N = R_o \tag{3.36}$$

3.7 Thévenin and Norton theorems

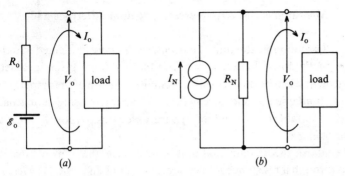

3.16 (a) Thévenin and (b) Norton equivalent direct network connected to a load.

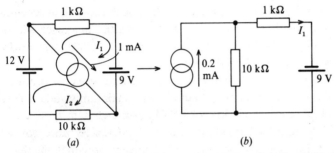

3.17 Norton reduction of a network; (a) the original and (b) the reduced version.

As an example of the application of the Norton equivalent circuit to network analysis, suppose once more that it is required to find the current I_1 in the 1 kΩ resistor of figure 3.6(d). The Norton reduction of the network, which is reproduced in figure 3.17(a) for easy reference, proceeds as follows. First the branch comprising the 12 V e.m.f. in series with the 10 kΩ resistor is transformed into its Norton equivalent of a 1.2 mA current source in parallel with a 10 kΩ resistor according to equations (3.34) and (3.36). Combining the 1.2 mA and 1.0 mA current sources leads to the equivalent circuit shown in figure 3.17(b) in which the 0.2 mA source contributes $(10/11) \times 0.2$ mA and the 9 V e.m.f. $[9/(10+1) \text{mA}] = 9/11$ mA to I_1. By superposition $I_1 = 1$ mA as before. As an alternative, the circuit of figure 3.17(b) can be further reduced to just a Norton equivalent feeding the 1 kΩ resistor only. In this case $R_N = 10$ kΩ again but I_N, found as the current delivered to a short across the 1 kΩ, is 1.1 mA because 9 V appears across the 10 kΩ resistor causing 0.9 mA to flow through it. The current I_1 is now $(10/11) \times 1.1$ mA $= 1$ mA again.

3.8 Measurement of direct current, potential difference and resistance

The fundamental unit of current, the ampere, is defined in terms of the magnetic force between two conductors in a given geometrical arrangement when both carry this current. Realisation of the ampere occurs through fundamental measurement of such force using a mechanical balance of suitable design and the apparatus for this purpose is known as a *current balance*.

To establish the fundamental unit of resistance, the ohm, a current I is passed through a resistance R to create a potential difference $V = RI$ across it. The same current also passes through coils connected in series to create a magnetic field in which conducting discs rotate. An e.m.f. is generated between the axis and rim of a disc, given by MnI where M is the mutual inductance defined in section 6.1 and n the rate of rotation in revolutions/second, and this is balanced against the potential difference across the resistance R by adjusting the frequency of rotation. At balance, $R = Mn$ and R follows since M can be calculated.

A standard current passed through a standard resistor creates a standard of potential difference. National laboratories maintain standard resistors and standard sources of e.m.f. of the Weston cadmium type, the latter calibrated in terms of standard potential difference. More convenient instruments for everyday measurement of current, potential difference and resistance are calibrated in terms of secondary standards derived from the national primary standards. Thermoelectric e.m.f.s interfere with all direct measurements but current reversal leads to their elimination since only potential difference due to current reverses with it.

The easiest way to measure current is through the series insertion of an *ammeter*. Such instruments are normally of the moving-coil type in which a coil mounted on bearings carries the current so that it experiences a deflecting force in the field of a permanent magnet. A steady deflection arises when the restoring force provided by a spring balances the magnetic force. In *galvanometers*, which are more sensitive, the moving coil is suspended by a long fine wire and the magnetic deflecting force is balanced by twist created in the suspension. Often, the deflection is magnified by an optical lever arrangement in which a beam of light is reflected from a small mirror carried by the suspension to give a large displacement at sufficient distance, usually 1 m. Ammeters exhibit sensitivities as high as full-scale deflection (f.s.d.) for only 50 μA current while galvanometers exist which produce up to 1 mm deflection of a light beam at 1 m distance for as little as 1 nA current. When it is only required to know the current in a circuit with an ammeter included, the resistance of the ammeter is unimportant.

3.8 Measurement of direct current, p.d. and resistance

However, to measure current in a circuit by inserting an ammeter in series with it, the resistance of the ammeter must be small enough to avoid changing the current appreciably. Unfortunately greater sensitivity is achieved when the meter coil has more turns which inevitably means more resistance and an ammeter reading 50 μA f.s.d. usually has a resistance of around 2 kΩ. Shunting a meter with a much lower resistance, that is, placing a much lower resistance in parallel with it, bypasses most of the current to create a current measuring system of lower sensitivity. At the same time, the resistance of the measuring system is lowered. Shunting a 50 μA/2 kΩ meter with 0.1 Ω converts it to a 1 A/0.1 Ω meter.

The potential difference between any two points may be measured by connecting a *voltmeter* between them. A common form of voltmeter comprises just an ammeter in series with a known resistance R. When connected across a potential difference V, a current $V/(R+r)$ will flow through the ammeter where r is the resistance of the ammeter. Thus the ammeter scale can be calibrated in volts. For example, a 50 μA/2 kΩ ammeter connected in series with a 180 kΩ resistor and an 18 kΩ resistor constitutes a 10 V voltmeter.

When attempting to measure potential difference by means of a voltmeter, the circuit to which it is connected is inevitably loaded. In accordance with Thévenin's theorem, this loading reduces the potential difference below the required open-circuit value. The higher the resistance of the voltmeter, the less the loading and a *figure of merit* is defined as the resistance of the voltmeter divided by its full-scale reading. For the voltmeter just mentioned the figure of merit would be 200 kΩ/10 V or 20 kΩ/V. Notice that this figure is fundamental to the sensing ammeter; without any external series resistance, 0.1 V would cause a f.s.d. of 50 μA, again corresponding to 2 kΩ/0.1 V or 20 kΩ/V.

Field-effect transistors and thermionic valves make it possible to accurately amplify small potential differences while maintaining low electrical noise and exceptionally high resistance of better than 10^{12} Ω between the sensing terminals. Such electronic systems front excellent voltmeters. Increasingly these instruments display the observed potential difference in digital rather than analogue form. Although the resolution or discrimination is thereby greatly enhanced, it must be appreciated that the accuracy does not necessarily match. A common way of converting an analogue potential difference to digital form is to compare it with a ramped potential difference that increases at a uniform rate. The moment when the ramp reaches the unknown potential difference is sensed by a comparator circuit (see section 10.6) and pulses generated at a fixed rate are counted during the ramp up to coincidence. This count gives a digital output proportional to the analogue input. A digital output is highly convenient for feeding into a computer for automatic processing.

A simple way of finding the resistance R between two points is to pass a current between them and use meters to measure the current I and corresponding potential difference V so that R is given by V/I. However, errors arise because of the noninfinite resistance of the voltmeter or nonzero resistance of the ammeter depending on whether the circuit arrangement of figure 3.18(a) or 3.18(b) is adopted. In the first arrangement, the potential difference across the resistance is observed but the ammeter indicates the sum of the currents through the resistance and voltmeter. In the second arrangement, the ammeter registers the current through the resistance only but the voltmeter registers the sum of the potential differences across the ammeter and resistance. Which arrangement is the better will depend on the resistances of the ammeter and voltmeter compared with the unknown R. Of course, suitable meter resistances can render the error negligible.

An important but somewhat cumbersome instrument that measures potential difference under truly open-circuit conditions is the *potentiometer*. Its essential features are displayed in figure 3.19. An e.m.f. $\mathscr{E}$ and variable resistor X deliver current to a resistor chain connected between points A and B. A galvanometer is connected through a switch G between A and a second switch K to allow comparison of either a standard e.m.f. $\mathscr{E}_s$ or unknown potential difference V with potential difference along the resistor chain. With switch K connected to the standard e.m.f., resistor X is adjusted until there is no deflection of the galvanometer on opening or closing G. Incidentally, it is sound practice to shunt the galvanometer to reduce sensitivity until approximate balance is attained. If I is the current through the resistor chain at balance of the standard e.m.f. and R_s is the resistance of the chain between A and the fixed connection S then, according to Kirchhoff's voltage law, the standard e.m.f. is given by

$$\mathscr{E}_s = R_s I \qquad (3.37)$$

After throwing switch K to connect the unknown potential difference V for comparison with the potential drop along the resistor chain between A and

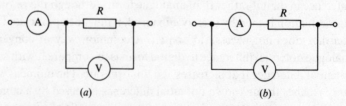

3.18 Resistance measurement using meters; A – ammeter, V – voltmeter.

3.8 Measurement of direct current, p.d. and resistance

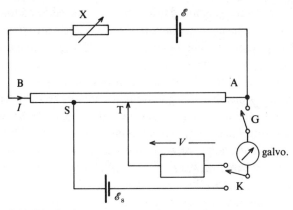

3.19 The basic arrangement of a potentiometer.

B, contact T is adjusted until balance is regained with no deflection of the galvanometer on opening or closing switch G. If at balance the current through the resistor chain is still I and the resistance between A and T is R_t then

$$V = R_t I \quad (3.38)$$

Elimination of I between equations (3.37) and (3.38) shows that V is given by

$$V = (R_t/R_s)\mathscr{E}_s \quad (3.39)$$

Of course, to obtain balance, the unknown must be connected into the potentiometer circuit with the appropriate polarity. Connection through a reversing switch facilitates this and is especially useful where it is necessary to reverse V to eliminate thermoelectric effects. As already asserted, the true open-circuit potential difference is measured by a potentiometer because the source of the unknown potential difference delivers no current through the galvanometer at balance. Adjustment of T can be achieved by means of a set of decade dials and suitable design enables these dials to display the unknown potential difference directly in volts at balance.

A potentiometer can be adapted to current measurement by using it to measure the potential difference V_s across a standard resistance R_s connected in series with the current so that the current is given by V_s/R_s. To determine an unknown resistance by means of a potentiometer involves its use to compare the potential difference across the unknown resistance with that across a standard resistor connected in series when a suitable current flows through the combination. If R and R_s are the unknown and standard resistances respectively and V and V_s the corresponding potential differences, R is obtained as $(V/V_s)R_s$. Potentiometric comparison of an unknown resistance with a standard resistance is enhanced by adoption of a *four-terminal* technique in which errors due to leads and contacts are avoided. In this approach, as illustrated in figure 3.20(*a*), current is delivered to a resistor through two large terminals situated outside two smaller terminals between which precisely the required resistance exists and the potential difference is measured. In good-quality standard resistors, the two pairs of terminals are mounted on substantial copper blocks and the resistor element connected between them as indicated in figure 3.20(*b*). The technique is particularly appropriate to measurement of the resistivity of semiconductors, for example, because it is difficult to make good ohmic contacts of low resistance to these materials. A specimen of known cross-section has end contacts that carry the current while probes at known positions along the length, as indicated in figure 3.20(*c*), enable the potential drop down a precise length to be measured. Notice that the possibly appreciable resistances of the contacts made by the fine potential probes only affect the sensitivity of measurement since no current flows through them when the potentiometer is balanced.

A feature of potentiometer operation common to all *null* methods is that the device sensing balance need not be linear. However, the ability of a potentiometer to resolve or discriminate depends on the sensitivity of the sensing device as well as the fineness with which the balance point may be adjusted. The latter is governed by the smallest resistance steps that are available compared with the total resistance between A and B. Accuracy depends on the accuracy of all the resistance ranges incorporated between A and B. Instruments are available with accuracies and discriminations as good as 0.001% and 0.0001% respectively and with a minimum potential step along AB as low as 0.1 μV. Despite their virtues, use of potentiometers

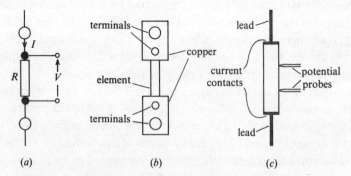

3.20 (*a*) Four-terminal technique for resistance measurement, (*b*) construction of a four-terminal standard resistor and (*c*) implementation of the four-terminal technique with respect to electrical measurements on semiconductors.

has greatly declined in recent years with the advent of cheap sensitive digital voltmeters featuring enormous input resistance.

3.9 The Wheatstone bridge

The bridge invented by Wheatstone provides a method of determining the resistance of a resistive element that is rapid and popular and perfectly satisfactory when the resistances of the end contacts of the element are either negligible or irrelevant so that the four-terminal technique discussed in the previous section is superfluous. Excellent discrimination is again feasible because operation of the bridge to measure resistance involves a null or balance technique as with the potentiometer.

Figure 3.21(a) depicts the usual circuit arrangement of a Wheatstone bridge. The four branches of the bridge AC, CB, BD and DA are energised by connecting the e.m.f. $\mathscr{E}$ through the switch K and resistance $R_\mathscr{E}$ to the nodes C and D. Operation of switch W allows a range of known values to be selected for the ratio n between the resistance R_2 included in branch AD and the resistance R_1 included in branch AC. Having selected the ratio n, the unknown resistance R_3 is measured by adjusting a known resistance R_4 until there is no potential difference between A and B as judged by null reading of a suitable detector included in branch AB. At balance there is no current in branch AB and so

$$\frac{V_1}{V_2} = \frac{R_1}{R_2} = \frac{V_3}{V_4} = \frac{R_3}{R_4} \tag{3.40}$$

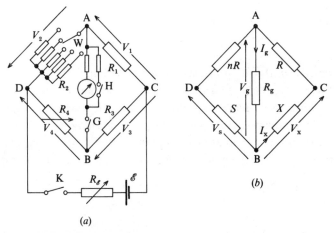

3.21 The Wheatstone bridge; (a) practical arrangement and (b) version analysed in text to deduce sensitivity.

Evidently, an unknown resistance R_3 can be found in terms of n and R_4 at balance as

$$R_3 = R_4/n \qquad (3.41)$$

Traditionally the detector in branch AB is a galvanometer or sensitive ammeter protected by a shunt H during initial approximate balancing. Off balance is sensed by gingerly closing and opening switch G and watching for a change of reading. Cautious operation is advisable to avoid damage to the detector as a result of inadvertent gross imbalance. One of many possible alternative means of detecting zero potential difference between A and B is a cathode-ray oscilloscope incorporating a sensitive direct amplifier.

The accuracy of measurement of an unknown resistance R_3 depends not only on the accuracy of the ratio arm resistances R_1 and R_2 and the comparison resistance R_4 but also on the fineness of adjustment of R_4 together with the sensitivity of detection of the null condition. It is therefore important to examine how the design of the bridge affects this sensitivity. Considering the usual case where the null detector is a galvanometer, it is required to achieve maximum current through it in branch AB for a given departure of R_4/n from the value that achieves balance. Obviously, the galvanometer current can be increased without limit by increasing the energising e.m.f. However, depending on the measuring circumstances, certain restrictions apply. Quite generally, increasing the e.m.f. $\mathscr{E}$ increases the Joule heating in all the resistors of the bridge which causes errors. A quite common overriding limitation is that the current in the unknown must not exceed a certain limiting value otherwise the error of measurement will be too great. While the known resistors R_1, R_2 and R_4 may be robust and exhibit low temperature coefficients of resistance, the unknown element R_3 might be a tiny specimen that exhibits a large temperature coefficient, as with semiconductors for example. The sensitivity of the bridge subject to a limiting current in the unknown will now be evaluated.

With reference to the simplified bridge circuit diagram of figure 3.21(b) in which the ratio arms have resistances R and nR, the balancing arm resistance S, the galvanometer resistance R_g and the unknown resistance X, the sensitivity subject to a limiting current in the unknown depends on the current ratio I_g/I_x. Applying Kirchhoff's current law to nodes A and B

$$\frac{V_g}{R_g} + \frac{V_s}{S} + \frac{V_x}{X} = 0 \qquad (3.42)$$

and

$$\frac{V_g - V_s}{nR} + \frac{V_g}{R_g} + \frac{V_g - V_x}{R} = 0$$

3.9 The Wheatstone bridge

or

$$\frac{V_g}{Y} - \frac{V_s}{nR} - \frac{V_x}{R} = 0 \qquad (3.43)$$

where

$$\frac{1}{Y} = \frac{1}{nR} + \frac{1}{R} + \frac{1}{R_g} \qquad (3.44)$$

From equations (3.42) and (3.43)

$$\frac{V_g}{\begin{vmatrix} 1/S & 1/X \\ -1/nR & -1/R \end{vmatrix}} = \frac{V_x}{\begin{vmatrix} 1/R_g & 1/S \\ 1/Y & -1/nR \end{vmatrix}}$$

so that

$$\frac{V_g}{V_x} = \frac{-(1/SR)+(1/nRX)}{-(1/nRR_g)-(1/SY)} = \frac{R_g}{X}\left[\frac{nX-S}{S+(nRR_g/Y)}\right]$$

and on substituting for $1/Y$ from equation (3.44)

$$\frac{I_g}{I_x} = -\frac{XV_g}{R_g V_x} = \frac{S-nX}{S+nR+(n+1)R_g} \qquad (3.45)$$

This equation shows that the bridge is balanced when, as expected from equation (3.41),

$$S/n = X \qquad (3.46)$$

However, it also shows that for a given deviation δ of S/n from X

$$\frac{I_g}{I_x} = \frac{\pm \delta}{X \pm \delta + R + (1+1/n)R_g} \qquad (3.47)$$

which is larger in magnitude the bigger n and the smaller R. Very interestingly, for small deviations δ

$$\left(\frac{I_g}{I_x}\right)_{n \to \infty, R \to 0} \to \frac{\pm \delta}{X + R_g} \qquad (3.48)$$

while

$$\left(\frac{I_g}{I_x}\right)_{n=1, R=X} = \frac{\pm \delta}{2(X + R_g)} \qquad (3.49)$$

so that the sensitivity of the bridge with all arms equal in resistance is only half the optimum sensitivity which in any case corresponds to an impractical arrangement involving negligible and infinite resistances. A rule of thumb of equal arm resistances for good sensitivity is therefore usually adhered to when designing Wheatstone bridges, although $n=10$, $R=0.1X$ is easy to arrange and provides a useful improvement to almost twice the sensitivity of that for $n=1$, $R=X$.

Having optimised the current I_g through the detecting arm AB, it is necessary to select the galvanometer that will give the best performance.

Design considerations of galvanometers are such that, other things like the permanent magnet being equal, the deflection of the moving coil for a given current is proportional to $R_g^{\frac{1}{2}}$. In terms of the galvanometer current I_g, the deflection θ is therefore proportional to $I_g R_g^{\frac{1}{2}}$, which is not unexpected since $I_g^2 R_g$ is the power dissipated in the galvanometer, and using equation (3.47) establishes that for a small deviation δ of S/n from X and the maximum permissible current I_x through the unknown

$$\theta \propto R_g^{\frac{1}{2}}/[X+R+(1+1/n)R_g]$$

Hence

$$d\theta/dR_g = \{\tfrac{1}{2}[X+R+(1+1/n)R_g]R_g^{-\frac{1}{2}} - R_g^{\frac{1}{2}}(1+1/n)\}/[X+R+(1+1/n)R_g]^2$$

and setting this equal to zero to find the resistance of galvanometer of given design that will register maximum deflection θ

$$R_g = \frac{X+R}{1+1/n} \qquad (3.50)$$

Equation (3.50) shows that, when the bridge arms are all equal to X at balance, the optimum galvanometer also has resistance X. When $R \ll X$ and $n \gg 1$, $R_g = X$ is again optimum while, provided $n \geq 1$ and $R \leq X$, which corresponds to good sensitivity, $X/2 \leq R_g \leq 2X$.

Alternating-current bridges often take the Wheatstone form, as will be discussed in section 7.3, and the Wheatstone lay-out is common in electronic instrumentation. One interesting electronic circuit of the Wheatstone form is the full-wave bridge rectifier.

3.10 Load-line analysis

So far this chapter has been concerned with the analysis of linear direct-current networks. When a nonlinear component is included in a direct-current network, analysis may proceed by graphical means.

Consider a linear network of direct sources and resistors connected, as shown in figure 3.22(a), to a component having two terminals between which the behaviour is nonlinear. Whatever the nature of the nonlinear component, its terminal behaviour $I = f(V)$ can be represented in the form of a static characteristic as discussed in section 2.3. Let the static characteristic be that shown in figure 3.22(b) for illustrative purposes. Whatever the linear direct network, it is equivalent to a Thévenin e.m.f. $\mathscr{E}_T$ in series with a Thévenin resistance R_T so that

$$I = (\mathscr{E}_T - V)/R_T \qquad (3.51)$$

This relationship can be represented on the same graphical plot as the static characteristic. As shown in figure 3.22(b), it is a straight line of slope $-1/R_T$

3.10 Load-line analysis

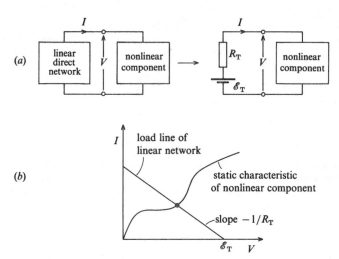

3.22 (a) Linear direct network connected to a nonlinear component and (b) load-line analysis of (a).

and intercept $\mathscr{E}_T$ on the potential difference V axis. Such a line is known as a *load line*; it is valid whatever the nature of the nonlinear component. Since the terminal current I and potential difference V must simultaneously satisfy the load line and the static characteristic, the values of I and V are given by the intersection of these plots which is known as the *operating point* of the nonlinear component.

Determination of the operating point is very important in electronics. Instead of a direct e.m.f. $\mathscr{E}_T$ there may be a time-varying e.m.f. or a time-dependent e.m.f. superimposed on a direct *bias* e.m.f. Another possibility is that R_T could vary with time. In all of these cases graphical solution of figure 3.22(b) gives the time dependences of I and V provided that the frequency of operation does not become high enough for reactive effects (see chapter 5) to render the static characteristic inappropriate. In the case of three-terminal nonlinear electronic devices such as transistors, the static characteristic for a pair of terminals is a family of curves, the relevant curve depending on the condition of the third terminal. Of great interest is the variation of the operating point as the third terminal condition is altered. This is illustrated in figure 3.23(a).

Sometimes it is useful to approximate the static characteristic by several linear regions. When the load line intersects an approximately linear region of the characteristic that passes through the origin, the device can be represented in this range by a constant resistance R. Since $I = V/R$ it follows from the geometry of figure 3.23(b) or equation (3.51) that $I = \mathscr{E}_T/(R_T + R)$.

The other approach to the analysis of circuits involving nonlinear

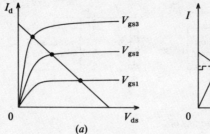 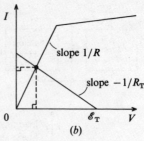

3.23 (a) Form of output characteristic of a junction field-effect transistor for three, input, gate potentials with load line superimposed. (b) A load line intersecting a region of a characteristic that is linear through the origin.

components is to represent the dependence of the terminal current I on the terminal potential difference V by a mathematical function. Often description of I as a power series in V is revealing. Simultaneous solution of $I = f(V)$ with equation (3.51) gives I and V and is especially simple when $R_T = 0$ so that $V = \mathscr{E}_T$.

4

Capacitance, inductance and electrical transients

4.1 Capacitance and capacitors

The act of placing electrical charge on a body requires the expenditure of work in overcoming the increasing electrical repulsive force as the charge builds up. In terms of the concept of electrical potential, a charge Q placed on a body raises its potential to some value V with respect to infinity. For reasons that will become apparent in a moment, the ratio of charge Q to potential V in this situation is defined as the *capacitance* C of the body so that

$$V = Q/C \qquad (4.1)$$

Clearly, to store charge on a body, it must be adequately insulated from its surroundings and a system capable of storing charge is described as a *capacitor*. Capacitive terminology originates in the fact that the larger C, the greater the charge that can be placed on a body before the rise in its potential leads to electrical breakdown of the surrounding insulation and charge leaking away unduly. Evidently, C partly describes the ability of a body to store charge but storage also depends on the quality of the insulation.

While both the potential and capacitance of a single body may be defined with respect to infinity, practical capacitors normally take the form of two conductors insulated from each other. In this case, the charging process involves the transfer of charge from one conductor or *plate* to the other so that a potential difference is established *between* them. Although equation (4.1) still applies, C and V are now respectively the capacitance and potential difference between the plates and corresponding to charge $+Q$ on one plate there is charge $-Q$ on the other. In the vast majority of cases V is proportional to Q so that C is constant and the capacitor can be described as a linear component. Nonlinearities, where they arise, stem from the location of the charge varying and the permittivity of the insulator changing during charging. In the particularly simple case of parallel plates of area A situated a small distance t apart compared with their lateral dimensions, the capacitance is given by

$$C = \varepsilon A/t \tag{4.2}$$

where ε is the permittivity of the insulating medium, that is, dielectric, between the plates. For a proper derivation of this equation the reader is referred to a standard text on electrostatics such as *Electricity and Magnetism* by W. J. Duffin, McGraw Hill, London (1973). However, it can be seen from equation (1.10) that for a given charge on the plates the potential difference between them is inversely proportional to ε so that C is proportional to ε. The potential difference is also proportional to t and inversely proportional to A.

Inspection of equation (4.1) reveals that the fundamental SI unit of capacitance is the coulomb/volt, which is precisely the practical *farad* unit usually written F for brevity. It follows from equation (4.2) that the fundamental SI unit of permittivity is the *farad/metre* or $F\,m^{-1}$ as asserted in section 1.3. To appreciate the difficulty of achieving significant capacitance in a component of modest volume, consider the capacitance that exists between parallel plates $1\,cm^2$ in area a distance of 1 mm apart in air. From equation (4.2), the capacitance of such an arrangement is just $(10^{-9}/36\pi)10^{-4}/10^{-3}\,F \approx 10^{-12}\,F = 1\,pF$! Over the years a variety of ways of achieving substantial capacitance has emerged. The most common form of construction has an exceedingly thin ($\sim 10\,\mu m$) film of plastic, such as polyethyleneterephthalate, polystyrene or polycarbonate, as the dielectric between either its own metallised surfaces or two sheets of aluminium foil as plates. For compactness this layer structure is rolled up into cylindrical form but it remains essentially parallel plate in character and gives capacitances of up to a few μF in dimensions of the order of 1 cm at working potential differences of up to about 500 V.

The thinner the plastic film is made in order to increase the capacitance, the lower is its resistance and direct breakdown voltage. Actually, the product of the resistance R and capacitance C between the plates of a capacitor is independent of the geometry and hence of the magnitude of the capacitance. It only depends on the nature of the dielectric and consideration of parallel plate capacitors demonstrates that the product is

$$CR = (\varepsilon A/t)(t/\sigma A) = \varepsilon/\sigma \tag{4.3}$$

which is just the permittivity to conductivity ratio of the dielectric. Interestingly, it emerges from the analysis of section 4.3 that $CR = \varepsilon/\sigma$ has the particular significance of being a measure of the time for charge placed

4.1 Capacitance and capacitors

on an isolated capacitor to leak away between its plates. Capacitors constructed with polystyrene dielectric exhibit an exceptionally high value of ε/σ in the region of 10^6 s and are also very stable. A point worth making about the maximum working voltages of capacitors is that when alternating potential differences are applied (see chapter 5), the voltage limit falls with increasing frequency due to rising thermal dissipation in the dielectric.

An alternative approach to achieving high capacitance in a small component is to incorporate a dielectric of high permittivity. Popular for this purpose is the ceramic, barium titanate, the anisotropic dielectric constant of which amounts to around 4000 along the single-crystal 'a' axis. Stack construction from very thin successive layers of metal and ceramic dielectric, alternate metal layers being electrically commoned, gives small 'ceramic chip' capacitors of up to about 1 μF capacitance.

Electrolytic capacitors provide very much larger capacitances still of up to 100 000 μF in volumes of up to ~ 200 cm^3. In this case an oxide film as little as 0.01 μm thick, formed by electrolytic action at the surface of a metallic foil, acts as the dielectric. One common construction comprises a rolled-up sandwich of aluminium plates interleaved with paper impregnated with ammonium borate that serves as the electrolyte. A polarising voltage forms the highly insulating oxide film at the positive foil which is often etched to increase its effective area. Even smaller electrolytic capacitors employ tantalum foil coated with a tantalum oxide film having a dielectric constant of 11. All electrolytic capacitors suffer from the limitation that the polarising voltage must be maintained in operation. Application of the opposite polarity will reduce the oxide with corresponding substantial conduction between the plates. Because these capacitors are sealed to prevent the electrolytic impregnation drying out, they can even explode as reduction proceeds due to excessive trapped gas pressure. Fortunately, many electronic applications do not violate the fixed polarity requirement. For example, electrolytic capacitors are suitable for many coupling applications (see section 4.4), in decoupling applications where a parallel capacitor reduces the alternating component of a potential difference but preserves the direct component and as reactive components in filters for direct power supplies. Even with the appropriate sign of polarising voltage there is nontrivial leakage and smaller size, larger value, electrolytic capacitors suffer from a very low breakdown voltage owing to the exceedingly thin dielectric. Types are made with breakdowns as low as 4 V.

Capacitors in modern, silicon chip, integrated circuits either make use of the nonlinear capacitance of a semiconducting P–N junction formed within the chip or employ the protective, silicon dioxide coating of the chip as dielectric with a deposited aluminium film as the upper plate and a heavily doped region of the silicon as the lower plate.

Variable capacitors generally function by altering the plate separation or overlap. For precision tuning, as in radio receivers for example, the overlap between two sets of metallic vanes in air is varied by rotating one set with respect to the other. Getting on for $1\,\mu\text{F}$ change of capacitance is possible with this system. Much smaller trimmers, other than miniature versions of the precision type, work by compression of metal/dielectric stacks or by sliding one metal tube inside another that is spaced by a suitable dielectric. For automatic tuning and many other purposes including parametric amplification, the variable capacitance available through varying the bias potential difference applied to a semiconducting P–N junction is invaluable.

To prevent subsequent deterioration, particularly due to penetration by water or water vapour, discrete fixed capacitors are coated with a protective lacquer or embedded in an insulating resin. They are manufactured in preferred ranges, which are often just a subset of the resistor preferred range, and their values together with other relevant information are displayed directly as printed figures or indirectly through a colour code. Reading from the end remote from the leads, the first three coloured bands of code indicate the capacitance in the same way as for resistors (see section 2.3). Further bands when present denote such things as tolerance and working voltage. Care must be exercised while reading, since successive same colour digits may not be separated on the body and there may be a background body colour.

The standard circuit symbols used to represent capacitance and capacitors are shown in figure 4.1. Figure 4.1(a) is the general British symbol for capacitance while 4.1(b) is its American equivalent. It is possible to indicate the positive plate of any polarised capacitor as shown in figure

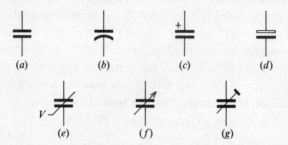

4.1 Circuit symbols representing capacitance and particular types of capacitor.

4.1 Capacitance and capacitors

4.1(c), but the symbol specifically for a polarised electrolytic capacitor is that shown in figure 4.1(d), the open plate being that requiring positive polarisation. A P–N junction capacitor should be denoted as in figure 4.1(e) in order to draw attention to the marked voltage nonlinearity. Variable and adjustable preset capacitors are represented as in figures 4.1(f) and 4.1(g) respectively. The preset symbol can of course be used to indicate the same function in other components.

It is important to appreciate that in all electrical circuits, besides capacitance due to any deliberately included capacitors, there is always unintentional *stray* capacitance between all the various metallic parts of the circuit, for example between lead wires, which may or may not be a nuisance. Much detailed electronic design is directed at reducing strays. Strays may be controlled by electrostatic *screening*. This involves placing the entire circuit and/or parts of it in earthed metal enclosures. Particularly useful are screened leads. These comprise an inner lead wire separated from a coaxial, braided, copper, outer, return lead by a plastic insulating dielectric. There is inevitable capacitance per metre length between the inner and outer leads. However, with the outer earthed, the capacitance is to earth which is hopefully harmless, there is no electromagnetic radiation when alternating current is carried and, not least, external electrical signals cannot create interfering electrical effects on the inner lead wire because the potential of the outer is fixed. Note that for capacitors in which an outer plate completely encloses an inner, if the outer is earthed the capacitance is just that between the inner and outer. If the inner is earthed, the capacitance is that between the outer and inner in parallel with that between the outer and the surroundings at earth potential.

By connecting capacitors in parallel or series it is possible to secure respectively a larger capacitance or a composite capacitor of smaller capacitance that will withstand a higher potential difference. For capacitors connected in parallel, the potential difference across them is common while the charge on the combination divides between the individual capacitors. The situation for three capacitors in parallel is depicted in figure 4.2(a) and the equivalent capacitance is

$$C = Q/V = (Q_1 + Q_2 + Q_3)/V = C_1 + C_2 + C_3$$

Clearly, this result extends such that the capacitance of a set of parallel capacitances C_i is

$$C = \sum_i C_i \tag{4.4}$$

Figure 4.2(b) shows the situation when three capacitors are connected in series. A charge Q transferred from one end to the other causes charges of

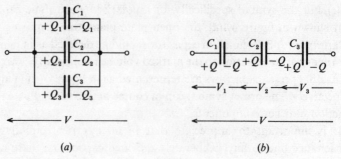

4.2 Capacitors (*a*) in parallel and (*b*) in series.

$+Q$ and $-Q$ to appear on opposite plates of the individual capacitors. The potential difference V across the series combination divides between the individual capacitors and the capacitance C of the combination is given by

$$1/C = V/Q = (V_1 + V_2 + V_3)/Q = 1/C_1 + 1/C_2 + 1/C_3$$

Extending the argument, the equivalent capacitance C of a set of series capacitances C_i is given by

$$1/C = \sum_i 1/C_i \tag{4.5}$$

Many combinations of capacitances can be organised into series and parallel combinations and equations (4.4) and (4.5) applied to find the equivalent capacitance. In the case of more complicated networks of capacitors, the effective capacitance between two points can be found by transferring charge Q between the points and examining how the charge distributes among the individual capacitors in accordance with potential difference constraints. Figure 4.3 illustrates this approach for a particular network of capacitances, potential difference constraints in meshes ABC and CDB respectively, giving

$$Q_1/C_1 + (Q_1 + Q_3 - Q)/C_2 - (Q - Q_1)/C_4 = 0$$
$$Q_3/C_3 - (Q - Q_3)/C_5 + (Q_1 + Q_3 - Q)/C_2 = 0$$

Solving these equations for Q_1 and Q_3 in terms of Q then gives the capacitance between A and D upon insertion in

$$C_{AD} = Q/V = Q/[Q_1/C_1 + (Q - Q_3)/C_5]$$

In general, the solution is tedious but the reader may care to verify that if all the capacitances are equal to C then $C_{AD} = C$ also.

Returning to the kind of discussion that introduced this section, whenever a capacitor is charged, there is an increase in potential energy associated with the charge stored on it; energy is said to be *stored* in the capacitor. If, at some moment during charging, the charge on a capacitor of

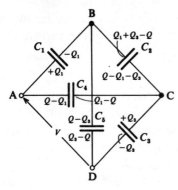

4.3 Example of a capacitive network for solution.

capacitance C is q, the corresponding potential difference across the capacitor is q/C. To transfer further charge dq between the plates requires the expenditure of work $q\,dq/C$. Thus when a capacitor is given a total charge Q, its potential energy increases by

$$W = \int_0^Q q\,dq/C$$

and if C is constant

$$W = Q^2/2C = QV/2 = CV^2/2 \tag{4.6}$$

where V is the potential difference corresponding to charge Q. Notice that no energy is dissipated on account of pure capacitance. In real charged capacitors, however (see also section 5.8), resistive losses do give rise to some dissipation.

In connection with energy it is interesting to consider a pure capacitance C charged from an e.m.f. $\mathscr{E}$. If Q is the final charge corresponding to potential difference $\mathscr{E}$ then the e.m.f. has expended energy $Q\mathscr{E}$ in the charging process. On the other hand, the energy stored in the capacitance is only $Q\mathscr{E}/2$. The balance of energy $Q\mathscr{E}/2$ has been dissipated in the connecting circuit as will be shown in section 4.3. It is misleading to suppose that the connecting circuit has zero resistance. Besides being impossible in practice, the charging current would be infinite so that despite zero resistance the dissipation would be finite.

4.2 Inductance and inductors

Before the circuit property of electrical inductance can be understood properly, appreciation of certain basic electromagnetic phenomena is required. When an electric current flows it exerts a force on any other current. The fundamental law of force for this *magnetic* type of interaction was established in a series of experiments conducted by Ampère around 1820. It transpires that the force $d\mathbf{F}_1$ on a current element $I_1\,d\mathbf{l}_1$ due to a current element $I_2\,d\mathbf{l}_2$ shows the dependence represented by

$$d\mathbf{F}_1 = C\left[\frac{I_1\,d\mathbf{l}_1 \times (I_2\,d\mathbf{l}_2 \times \mathbf{r}_{21})}{r^2}\right] \qquad (4.7)$$

where r is the separation of the current elements, $\mathbf{r}_{21}$ is unit vector pointing in the direction from $d\mathbf{l}_2$ to $d\mathbf{l}_1$ and C is a proportionality constant that depends on the unit system and the medium between the elements. The current elements $I_1\,d\mathbf{l}_1$ and $I_2\,d\mathbf{l}_2$ imply currents I_1 and I_2 over length elements $d\mathbf{l}_1$ and $d\mathbf{l}_2$ respectively. They may exist simply as elements of electrical wires carrying current but the law applies to any form of current element including, for example, that due to electronic motion in an atom. To set up a rationalised system of units, C is written $\mu/4\pi$ where μ is known as the *permeability* of the medium. In the special case of a vacuum between the elements, the permeability is written μ_0 and described as the *permeability of free space*. Clearly, the magnetic force between given currents flowing in a certain geometrical configuration in vacuum is a definite magnitude that can be measured experimentally and compared with the integrated prediction of equation (4.7). It turns out that to obtain the correct force in newtons from equation (4.7) with the ampere as the fundamental unit of current, μ_0 must be made exactly $4\pi \times 10^{-7}\,\text{N A}^{-2}$. In fact, it can now be appreciated that electrical units in the rationalised SI system are established by choosing μ_0 to be exactly $4\pi \times 10^{-7}\,\text{N A}^{-2}$ which makes the fundamental unit of current the ampere and that of charge consequently the coulomb, leaving ε_0 to be found experimentally to be close to $10^{-9}/36\pi\,\text{F m}^{-1}$ in accordance with the discussions of sections 1.3 and 2.1. Incidentally, the unit of permeability is normally referred to as the *henry/metre* which is entirely equivalent to the N A^{-2} as will be seen shortly.

By analogy with the introduction of the electric field vector $\mathbf{E}$ in electrostatics, it is convenient to consider that a current element produces something called *magnetic induction* in the space around it which then exerts a force on any other current element. Magnetic induction is defined properly by stating that a current element $I_1\,d\mathbf{l}_1$, situated at a position where an element of magnetic induction $d\mathbf{B}_2$ exists, experiences a force

$$d\mathbf{F}_1 = I_1\,d\mathbf{l}_1 \times d\mathbf{B}_2 \qquad (4.8)$$

If the induction $d\mathbf{B}_2$ is due to a current element $I_2\,d\mathbf{l}_2$, equation (4.7) shows that

$$d\mathbf{B}_2 = \frac{\mu(I_2\,d\mathbf{l}_2 \times \mathbf{r}_{21})}{4\pi r^2} \qquad (4.9)$$

Evidently the magnetic induction due to a current element is normal to the

4.2 Inductance and inductors

plane containing the current element and the line joining the element to the point in question. According to equation (4.8) the fundamental unit of magnetic induction is the N/A m which is equivalent to the N m/A m^2 or V s/m^2. Since the volt second, or V s, unit is called the *weber*, the fundamental unit of magnetic induction is also the weber/square metre or Wb/m^2, although it is more usually referred to as the *tesla* which abbreviates to T.

Equation (4.9) may be applied to determine the distribution of magnetic induction around given current arrangements, but the reader is referred to standard texts on electromagnetic theory such as *Electricity and Magnetism* by W. J. Duffin, McGraw Hill, London (1973), for such calculations. In the context of electrical networks, interest in magnetic induction pertains to development of the concept of *magnetic flux*. The integral of the magnetic induction over a surface is described as the magnetic flux through that surface and writing the flux as Φ

$$\Phi = \int_S \mathbf{B} \cdot d\mathbf{S} \tag{4.10}$$

from which the fundamental unit of magnetic flux is the weber, or Wb for brevity. It can be seen that the magnitude of the magnetic flux depends on the current, the geometry and the permeability of the medium. In the last connection, let it be noted that when a medium other than a vacuum surrounds a real current distribution, forces are exerted by the real current on the otherwise balanced circulating currents due to electrons in atoms of the medium to produce a net circulating current in it. This extra magnetisation current modifies the magnetic induction compared with that produced in a vacuum, which fact is taken into account by empirically assigning a different permeability. Most elements of matter only exhibit very weak magnetisation and their permeability is extremely close to μ_0. However, a few elements, notably those described as ferromagnetic, exhibit very large cooperative magnetic effects and their permeability can be more than two orders larger than μ_0. Addition of traces of certain other metals to ferromagnetic elements can even raise the initial permeability by a further two orders of magnitude!

Magnetic flux is relevant to electrical network analysis on account of the phenomenon of *electromagnetic induction*. Experiments begun by Faraday as long ago as 1831 showed that an electromotive force is induced in a circuit whenever the flux of magnetic induction through it is changing. Further experiments soon established that the magnitude of the inductive e.m.f. is proportional to the rate of change of flux and it matters not whether the flux changes through movement of the circuit or through time dependence of the magnetic induction. The sense or sign of the e.m.f. is given

Capacitance, inductance and electrical transients

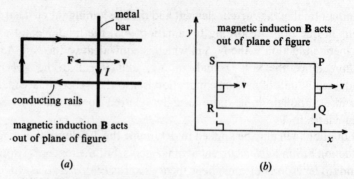

4.4 (a) Illustration of Lenz's law and (b) a closed circuit moving in a region of nonuniform magnetic induction.

by the law enunciated by Lenz which states that, in accordance with the conservation of energy, the e.m.f. acts so as to oppose the flux change causing it. Figure 4.4(a) illustrates this last point. It depicts a metal bar given an initial velocity v along a pair of conducting rails electrically connected at one end. Suppose a magnetic induction **B** acts out of the plane of the figure. The bar, rails and end connection form a circuit through which the magnetic flux increases as the bar moves. According to Lenz's law, the induced e.m.f. causes a current flow in the sense indicated so as to create a magnetic flux increment into the plane of the figure (see equation (4.9)) which opposes the flux change causing the e.m.f. That this sense of e.m.f. accords with the conservation of energy is clear on consideration of the force **F** acting on the moving bar. From equation (4.8) the force exerted by the induction **B** on the bar via the induced current is to the left so that the bar will come to rest. The opposite sign of induced e.m.f. is untenable since the bar would then continually accelerate and reach infinite velocity!

Consideration of a particular case establishes that the constant of proportionality between the induced e.m.f. $\mathscr{E}$ and the rate of change of flux $d\Phi/dt$ is precisely unity. Thus the laws of electromagnetic induction can be expressed with extreme simplicity by

$$\mathscr{E} = -\frac{d\Phi}{dt} = -\frac{d}{dt}\int_S \mathbf{B}\cdot d\mathbf{S} \qquad (4.11)$$

The negative sign here indicates that the e.m.f. is a back e.m.f. as stated in Lenz's law. To show that there is equality between $\mathscr{E}$ and $-d\Phi/dt$, consider a plane rectangular circuit PQRS moving with constant velocity v in the direction of RQ through a region of nonuniform magnetic induction **B** as illustrated in figure 4.4(b). Take orthogonal x and y axes parallel to RQ and RS respectively and let $PQ = RS = b$. Suppose further that the magnetic induction acts out of the plane of the figure such that B increases with x but

4.2 Inductance and inductors

is independent of y. If the inductions at RS and PQ are respectively B_{RS} and B_{PQ}, the flux gain in time dt is $(B_{PQ} - B_{RS})bv\,dt$ and

$$d\Phi/dt = (B_{PQ} - B_{RS})bv$$

The random motion of the mobile electrons in the circuit does not lead to any net force acting on them round the circuit through interaction with the magnetic induction **B**. However, the uniform velocity v imparted to the entire circuit leads to forces on the electrons in PS and RQ that are perpendicular to these sides and to forces of evB_{PQ} along QP and evB_{RS} along RS on electrons in sides PQ and RS respectively. Hence the net work done per unit charge taken round the loop amounts to $(B_{PQ} - B_{RS})bv$ in magnitude and it is as if such a magnitude of e.m.f. is acting in the circuit. This is just the same as the magnitude of $d\Phi/dt$ and the unity proportionality constant between $\mathscr{E}$ and $d\Phi/dt$ is consequently verified.

Having established the necessary background magnetic theory, the basic circuit concept of inductance can now be properly introduced. Whenever an electric current flows in a circuit it causes an associated magnetic induction given by equation (4.9) in the surrounding space and a certain magnetic flux Φ_L that is linked with the circuit itself and can be calculated from equation (4.10). The flux linkage depends on the current, the permeability and the geometry. If the permeability is constant, that is, independent of the current I, the flux linkage is proportional to the current and it is convention to write

$$\Phi_L = LI \tag{4.12}$$

where L is known as the *self inductance* or more simply the *inductance* of the circuit. This definition of inductance as just the flux linked with a circuit when unit current is flowing in it means that the fundamental unit of inductance is the weber/ampere, or Wb A^{-1}, which is termed the *henry* and written H for brevity. It should now be clear from equations (4.9), (4.10) and (4.12) why the unit of permeability is the henry/metre, or H m^{-1}, as asserted earlier.

The self inductance of a circuit becomes important when the current in it is changing because, in accordance with equation (4.11), a back e.m.f. $d\Phi_L/dt$ then occurs in it. If the inductance L defined by equation (4.12) is independent of the current I, the back e.m.f. is given by

$$\mathscr{E} = -\frac{d\Phi_L}{dI}\frac{dI}{dt} = -L\frac{dI}{dt} \tag{4.13}$$

and it follows that to establish a current I in a circuit against an inductive back e.m.f. involves work

$$W = \int_0^I L(dI/dt)I\, dt = \tfrac{1}{2}LI^2 \tag{4.14}$$

This expression is analogous to $\tfrac{1}{2}CV^2$ for the work to establish a potential V across capacitance C and potential energy $\tfrac{1}{2}LI^2$ is stored in the inductance when current I flows through it.

Alternative definitions of inductance to that based on equation (4.12) are generated by applying equations (4.13) and (4.14) whether or not $\Phi_L \propto I$. Thus inductance is also defined as the back e.m.f. per unit rate of change of current, which is certainly the most convenient definition from the circuit point of view, or as twice the work involved in establishing unit current. If $\Phi_L \propto I$, all three definitions are of course equivalent. Although this is the case for most media, for certain media such as ferromagnetic iron, Φ_L is far from proportional to I and the various definitions for L differ significantly. Figure 4.5 shows the form of dependence of Φ_L on I for a ferromagnet undergoing initial magnetisation. Fur current I amounting to OB in the figure, the first definition of inductance gives

$$L = AB/OB$$

while from the second definition

$$L = \frac{d\Phi_L}{dt} \bigg/ \frac{dI}{dt} = \frac{d\Phi_L}{dI} = \frac{AB}{DB}$$

In the case of the third definition

$$W = \int_{\Phi_L=0}^{\Phi_L=OC} (d\Phi_L/dt)I\, dt = \text{area OAC}$$

and

$$L = 2(\text{area OAC})/(OB)^2$$

The order of magnitude of inductance that occurs in practical situations can be assessed by estimating the inductance of a single circular loop of wire in air. Application of equation (4.9) establishes that the magnetic induction at the centre of such a loop of radius a is $\mu_0 I/2a$. Its inductance is therefore

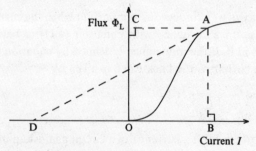

4.5 Illustration of the various definitions of self inductance.

4.2 Inductance and inductors

of order $(\pi a^2)(\mu_0 I/2a)(1/I) = \pi a \mu_0/2$ which amounts to 20 nH when the radius is 1 cm. The actual inductance, of course, differs somewhat from $\pi a \mu_0/2$ because the magnetic induction varies over the plane of the loop, but only the order of inductance is of interest here. Much greater inductance can be obtained by increasing the size of the loop, winding coils with many turns or inserting a core of high permeability. The last two procedures are particularly effective. With regard to increasing the turns N, the flux Φ is proportional to N but this flux links with each turn so that the flux Φ_L linked with the circuit is proportional to N^2. Iron-cored coils give high inductance at the mains frequency (50 Hz in the United Kingdom) while ferrite cores provide permeabilities of up to a few hundred at frequencies of up to around 100 MHz. Ferrites have the structure Fe_2O_3MO, where M represents a divalent metal, and huge initial permeabilities are exhibited by mixed ferrites in which nickel and zinc act as M.

Unfortunately, no matter how the inductance is increased there is an accompanying increase in resistive loss and the fact that a much closer approach to pure capacitance is possible than to pure inductance is worthy of special comment. Increase of size or turns introduces straightforward resistance with energy loss due to Joule heating. Interestingly, for a given winding cross-section, halving the cross-section of the wire and doubling the number of turns multiplies both the inductance and resistance by four times so that the ratio of inductance to resistance L/R is unaltered. The significance of L/R will become apparent in the following section; it determines the rate at which current grows or decays. When a core is present, additional resistive losses arise due to magnetic hysteresis and eddy currents. Hysteresis losses are associated with irrecoverable energy loss on magnetisation or demagnetisation. Eddy current losses occur on account of electromagnetic induction in conducting paths in the core causing currents in it. These currents can be reduced by increasing the resistance of the core either by laminating it, making it from tiny particles or adding suitable impurities, for example, silicon to iron. Losses due to eddy currents are much more serious at higher frequencies because $d\Phi_L/dt$ is higher. Ferrites are fortunately highly insulating and also show negligible hysteresis loss.

A circuit component designed to exhibit a certain inductance is described as an *inductor*. Some colour-coded preferred ranges of inductors are manufactured but small air-cored inductors are often home-made. Of course, air-cored inductors are linear components, whereas inductors with ferromagnetic cores are nonlinear. Variable and preset inductors are available through adjustment of a moving contact or, as is more often the case, of the extent of insertion of a core.

Figure 4.6 features the standard symbols by which inductance or

78   *Capacitance, inductance and electrical transients*

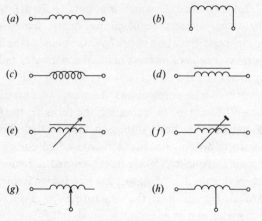

4.6 Circuit symbols for inductance and inductors; (*a*) and (*b*) general, preferred, (*c*) general, not now preferred, (*d*) with core, (*e*) continuously variable with core, (*f*) preset with core, (*g*) with adjustable contact and (*h*) with fixed tapping connection.

inductors are depicted in circuit diagrams. The first three are general symbols, the first two being preferred to the older third version. Presence of a core is indicated by an extra line as in the fourth symbol. Representation of variables and presets is just as for resistors while a fixed tapping is denoted as shown in the final symbol.

Wherever possible capacitors are used to implement electronic functions rather than inductors because inductors are generally bigger, more expensive and less ideal in operation. Inductors are particularly avoided in integrated circuits. Like stray capacitance, stray inductance can be a problem. Especially sensitive circuits or parts of them are often screened from the effects of stray inductive coupling by inserting them inside an enclosure of high magnetic permeability such as a mumetal can.

The effective inductance of a completely series or parallel combination of inductances is easily found. With reference to figure 4.7(*a*) showing three inductances connected in series, the current is common so that

$$V = V_1 + V_2 + V_3 = L_1 \, dI/dt + L_2 \, dI/dt + L_3 \, dI/dt$$
$$= (L_1 + L_2 + L_3) \, dI/dt$$

which means that the effective inductance is $(L_1 + L_2 + L_3)$. Extending this argument to any number of inductances L_i connected in series, the effective inductance is

$$L = \sum_i L_i \qquad (4.15)$$

The potential difference across inductances connected in parallel is

4.2 Inductance and inductors

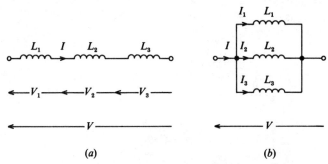

4.7 Inductances connected (*a*) in series and (*b*) in parallel.

common and with reference to the three shown in figure 4.7(*b*)

$$V = L_1\, dI_1/dt = L_2\, dI_2/dt = L_3\, dI_3/dt$$

But, according to Kirchhoff's current law, $I = I_1 + I_2 + I_3$ and so

$$\frac{dI}{dt} = \frac{dI_1}{dt} + \frac{dI_2}{dt} + \frac{dI_3}{dt} = \left(\frac{1}{L_1} + \frac{1}{L_2} + \frac{1}{L_3}\right)V$$

The effective inductance L of the parallel combination defined by $V = L(dI/dt)$ is therefore given by $(1/L) = (1/L_1) + (1/L_2) + (1/L_3)$ and again extending the argument to any number of inductances L_i connected in parallel, the effective inductance L is given by

$$1/L = \sum_i 1/L_i \qquad (4.16)$$

Many combinations of inductances can be broken down into series and parallel arrangements so that equations (4.15) and (4.16) can be applied to find the equivalent inductance. In other cases Kirchhoff's laws must be applied directly to find the equivalent inductance. Suppose, for example, that it is required to find the effective inductance L_{AD} between nodes A and D of the network shown in figure 4.8, which is a replica of that of figure 4.3 with all the capacitances replaced by inductances. Kirchhoff's current law enables the branch currents to be labelled as indicated and applying the voltage law to meshes ABC and BCD respectively gives

$$L_1 \frac{dI_1}{dt} + L_2 \frac{dI_2}{dt} - L_4 \frac{d(I - I_1)}{dt} = 0$$

$$L_2 \frac{dI_2}{dt} + L_3 \frac{d(I - I_1 + I_2)}{dt} - L_5 \frac{d(I_1 - I_2)}{dt} = 0$$

Solving these equations for I_1 and I_2 in terms of I gives L_{AD} upon insertion in

$$L_{AD} = V \bigg/ \frac{dI}{dt} = \left[L_1 \frac{dI_1}{dt} + L_5 \frac{d(I_1 - I_2)}{dt} \right] \bigg/ \frac{dI}{dt}$$

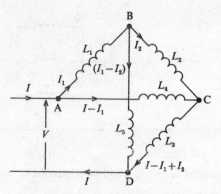

4.8 Example of an inductive network for solution.

and tedious algebraic manipulation yields

$$L_{AD} = \frac{L_2(L_1+L_5)(L_3+L_4) + L_1L_4(L_3+L_5) + L_3L_5(L_1+L_4)}{(L_3+L_5)(L_1+L_2+L_4) + L_2(L_1+L_4)}$$

The reader may care to verify that when $L_2 = 0$, this expression correctly reduces to the inductance of $L_1 \| L_4$ in series with $L_3 \| L_5$. Also when $L_4 = L_5 = 0$, it correctly reduces to the inductance of $L_1 \| L_2 \| L_3$.

4.3 Transient responses of C–R and L–R circuits to a step e.m.f.

When an e.m.f. is suddenly applied to a purely resistive circuit, the current rises almost instantaneously to its steady-state value. However, subsequent to the sudden application of an e.m.f. to circuits containing resistance and capacitance or resistance and inductance, there is a slower approach to the steady state. A correspondingly slow approach to equilibrium also occurs upon the sudden removal of an e.m.f. from C–R and L–R circuits and the behaviour of a circuit during its approach to equilibrium is appropriately termed its *transient response*. In this section the transient responses of C–R and L–R circuits to step e.m.f.s will be analysed.

To determine the transient responses, Kirchhoff's laws must be adapted to take account of potential differences that occur through the presence of capacitance and inductance. The laws must also be taken to apply to instantaneous currents and potential differences. Even with these generalisations, conventional circuit analysis as developed in chapter 3 is restricted to situations where the transient response is not too rapid. Electromagnetic theory and practical experience both show that when time-dependent currents exist, electromagnetic waves are emitted into the surrounding space (again see, for example, *Electricity and Magnetism* by W. J. Duffin, McGraw Hill, London (1973)). This constitutes an energy loss not taken into account in conventional circuit analysis. According to

4.3 C–R and L–R response to step e.m.f.

Fourier analysis (see section 11.2), if the transient behaviour extends over time t, this is equivalent to sinusoidal variations at frequencies up to $\sim 1/t$ and electromagnetic theory shows that there will then be waves with wavelengths down to $\sim ct$ where c is the velocity of light. Fortunately, electromagnetic radiation is only emitted in significant amounts when the wavelength is smaller than a few times the circuit dimensions. For a typical circuit of about 0.1 m size, the critical wavelength is ~ 1 m which corresponds to a frequency of 300 MHz and a transient response time of only 3 ns. Apparently conventional circuit analysis is applicable in all but extremely rapid response situations and its use, although restricted, is not unduly limited. A particularly clear reason why conventional circuit theory does not apply when the wavelength becomes small compared with the size of the circuit is that the current round a series circuit then not only changes in magnitude but even reverses in sign; it is certainly not constant round the circuit. In the remainder of this chapter it will be assumed that the transient response is never so rapid as to cause significant electromagnetic radiation so that conventional circuit analysis, using suitably modified forms of Kirchhoff's laws as discussed above, can be applied.

Consider the simple C–R and L–R series circuits shown in figure 4.9 and suppose that initially the capacitive element in the former is uncharged and the inductive element in the latter carries no current. These conditions can be ensured sufficiently closely in practice by shorting the capacitive element in the capacitive circuit and opening switch K in the inductive circuit for long enough. At some origin of time $t=0$ let the switch in the capacitive circuit be connected to A and the switch in the inductive circuit be closed. Taking account of potential differences across the capacitive and inductive elements, Kirchhoff's voltage law gives

$$\left. \begin{array}{l} \mathscr{E}=Q/C+RI \\ \mathscr{E}=L\,\mathrm{d}I/\mathrm{d}t+RI \end{array} \right\} \qquad (4.17)$$

for the two circuits, where Q is the charge on the capacitive element and I is

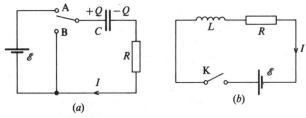

4.9 Introduction of a step e.m.f. to (a) a C–R and (b) an L–R series circuit.

the current in each case. But $I = dQ/dt$ in the C–R circuit and therefore

$$RC\, dQ/dt + Q = C\mathscr{E}$$

$$\frac{L}{R}\frac{dI}{dt} + I = \frac{\mathscr{E}}{R}$$

It is particularly noticeable that these last two equations have the same form. Indeed, writing Q in the first and I in the second as x, $C\mathscr{E}$ in the first and $\mathscr{E}/R$ in the second as x_0, RC in the first and L/R in the second as τ and d/dt as $\mathscr{D}$

$$(\tau\mathscr{D} + 1)x = x_0 \tag{4.18}$$

represents both equations. Substitution in this differential equation reveals that the solution is

$$x = A \exp(-t/\tau) + x_0$$

where A is an arbitrary constant. However, because of the imposed initial conditions, in both cases $x = 0$ when $t = 0$ so that $A = -x_0$. Hence

$$x = x_0[1 - \exp(-t/\tau)] \tag{4.19}$$

The behaviour of x represented by equation (4.19) is displayed in figure 4.10(a). There is a gradual increase of x as time progresses culminating in an asymptotic approach to a steady-state value x_0. Although x never quite reaches x_0, its difference from x_0 becomes negligible for most practical purposes when $t \gg \tau$. The time τ characterises the growth rate and is called the *time constant* of the circuit. For a C–R circuit

$$\tau = RC \tag{4.20}$$

whereas for an L–R circuit

$$\tau = L/R \tag{4.21}$$

Capacitor–resistor combinations can readily be formed with time constants ranging from less than 1 ps to as long as 10^6 s, but it is rare for the time constant of an inductive circuit to exceed a few seconds.

Physically, on applying the e.m.f. $\mathscr{E}$ in the C–R circuit, because the

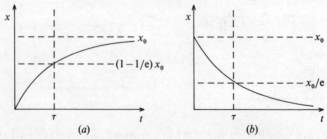

4.10 (a) Growth and (b) decay in C–R and L–R series circuits.

4.3 C–R and L–R response to step e.m.f.

capacitive element is uncharged initially, all the e.m.f. initially appears across the resistive element so that an initial current $\mathscr{E}/R$ flows in agreement with equation (4.17). However, as the capacitive element charges, the potential difference across it grows while that across the resistive element falls so that the current also falls. Eventually the current reaches zero when all the e.m.f. is dropped across the capacitive element and a charge $C\mathscr{E}$ is carried by it in accordance with equation (4.17) or (4.19).

On applying the e.m.f. $\mathscr{E}$ to the L–R circuit, because the current is initially zero, no potential difference is initially dropped across the resistive element and all the e.m.f. appears across the inductive element. The initial rate of change of current is therefore $\mathscr{E}/L$, but as the current grows an increasing potential difference V develops across the resistive element and the rate of growth of current, $(\mathscr{E} - V)/L$, falls. Eventually the growth rate approaches zero, heralding the steady state in which the e.m.f. $\mathscr{E}$ is entirely across the resistive element and the current is $\mathscr{E}/R$ as given in equation (4.17) or (4.19).

Suppose now that, after sufficient time has elapsed for the steady state to have been established in the two circuits, the switch is thrown to B in the C–R circuit of figure 4.9(a) and the e.m.f. is somehow shorted in the L–R circuit of figure 4.9(b). Application of Kirchhoff's voltage law shows that, subsequent to such switching, the behaviour of the two circuits is governed by the relations

$$\left. \begin{array}{l} 0 = Q/C + RI \\ 0 = L\, dI/dt + RI \end{array} \right\} \quad (4.22)$$

Adopting the same notation as before allows both of these to be expressed as

$$(\tau \mathscr{D} + 1)x = 0 \quad (4.23)$$

and, this time, substitution shows that the solution is

$$x = B \exp(-t/\tau)$$

where B is an arbitrary constant. Taking a new origin of time at the onset of this process, $x = x_0$ when $t = 0$ so that $B = x_0$ and

$$x = x_0 \exp(-t/\tau) \quad (4.24)$$

Figure 4.10(b) displays the behaviour of x with time represented by equation (4.24). There is a gradual fall with asymptotic approach to zero. Although x never quite reaches zero, it effectively does so for most purposes after a time $t \gg \tau$. The same time τ characterises the decay on removal of an e.m.f. as characterises the growth on application of an e.m.f. Physically, decay occurs in the C–R circuit because the potential difference due to charge on the capacitive element causes current to flow. As the capacitive element discharges, the potential difference and current fall so that the rate of discharge decreases and there is an asymptotic approach to the uncharged condition. When the e.m.f. is removed from the inductive circuit, the current falls at a rate such that the inductive back e.m.f. matches the resistive potential drop due to the current flowing. As the current falls, so does the resistive potential drop and hence the rate of fall of current, again giving an asymptotic approach to the equilibrium state of zero current.

Notice that if at the new origin of time, rather than the e.m.f. $\mathscr{E}$ being shorted, switch K is opened in the L–R circuit of figure 4.9(b), a large rate of fall of current is imposed. In this case a large inductive voltage surge appears across the inductive element. Such surges can damage inductors by breaking down the insulation between the windings. They can also cause problems in electronic circuits where other components with inadequate voltage ratings are included. Of course, the voltage surge tends to cause arcing at the opening switch, which limits the rate of change of current.

Application of the analysis of this section confirms, as asserted in section 4.1, that when an e.m.f. $\mathscr{E}$ is connected to capacitance C through *any* resistance R (including $R \approx 0$), half of the energy $C\mathscr{E}^2$ delivered by the e.m.f. to the circuit is dissipated in the resistive connections while the other half is stored in the capacitive element as potential energy. From equation (4.19), the charge on the capacitive element of a C–R circuit during charging is

$$Q = C\mathscr{E}[1 - \exp(-t/RC)]$$

Hence the charging current is

$$I = dQ/dt = (\mathscr{E}/R)\exp(-t/RC) \qquad (4.25)$$

from which the energy dissipated in the connecting circuit is

$$W = \int_0^\infty RI^2\, dt = (\mathscr{E}^2/R)\int_0^\infty [\exp(-2t/RC)]\, dt$$
$$= (-C\mathscr{E}^2/)[\exp(-2t/RC)]_0^\infty$$

that is

$$W = C\mathscr{E}^2/2 \qquad (4.26)$$

independent of R. Upon subsequent discharge the current is again from equation (4.24) given by equation (4.25) and the same calculation shows that the other half of the energy delivered by the e.m.f., that was stored in the capacitive element during charging, is now dissipated in the connecting circuit.

4.4 Basic four-terminal C–R networks

As already mentioned, because of their cumbersome size, high cost and far from ideal nature, inductors are avoided in favour of capacitors as far as possible in circuit applications. Two very simple C–R networks of

4.4 Basic four-terminal C–R networks

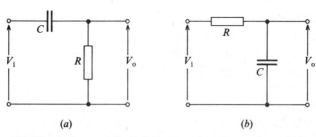

4.11 Simple four-terminal C–R networks; (a) coupling, differentiating or high-pass filter network and (b) integrating or low-pass filter network.

exceptional practical importance are shown in figure 4.11. Each features a series C–R combination connected to a pair of input terminals and a pair of output terminals. Notice that these four-terminal arrangements are achieved by making one terminal common to the input and output in each case. The operations performed by these two circuits are expressed through the relations that exist between the input and output potential differences V_i and V_o respectively. These relations will now be investigated.

Application of Kirchhoff's voltage law to the circuit of figure 4.11(a) gives

$$V_i = Q/C + V_o$$

where Q is the charge associated with capacitance C. Assuming that the output terminals remain open circuit as indicated, or that any load connected between them is such that the current drawn is negligible compared with that through resistance R

$$dQ/dt = V_o/R$$

and

$$dV_i/dt = V_o/RC + dV_o/dt \qquad (4.27)$$

If additionally the time constant RC is arranged to be small enough, equation (4.27) reduces to

$$V_o = RC\, dV_i/dt \qquad (4.28)$$

to a sufficiently good approximation and the circuit acts as a *differentiator* between its input and output terminals. This behaviour is much used in electronics; for example, to derive sharp triggering pulses from square-wave signals.

The question naturally arises regarding how small RC has to be to achieve reasonably accurate differentiation. For a sinusoidal input of angular frequency ω, the output is also sinusoidal at angular frequency ω (see chapter 5) and the amplitude of dV_o/dt amounts to ω times the amplitude of V_o. In this case, comparison of the terms on the right-hand side of equation (4.27) reveals that the condition for good differentiation is

$$\omega RC \ll 1 \tag{4.29}$$

As will be shown in chapter 11, any complicated input wave or pulse can be Fourier analysed into a spectrum of sinewave inputs. A periodic wave analyses into sinewaves of the fundamental frequency and higher harmonics while a pulse analyses into a continuous spectrum of sinewaves with appreciable amplitudes at frequencies up to a few times the reciprocal pulse width. Clearly for good differentiation in any particular case, the condition represented by the inequality (4.29) must be satisfied at the highest frequency present and a rule of thumb requires its satisfaction at ten times the fundamental frequency for sharp-cornered waves, or ten times the frequency corresponding to the reciprocal duration for pulses. Inspection of equation (4.28) shows that making RC small to satisfy inequality (4.29) makes the output potential difference V_o small. How this drawback can be overcome by using active versions of this circuit is discussed in section 10.5.

Under the constraint of the opposite imposed condition

$$\omega RC \gg 1 \tag{4.30}$$

the first term on the right-hand side of equation (4.27) becomes negligible compared with the second so that to a sufficient approximation

$$dV_i/dt = dV_o/dt$$

or

$$V_o = V_i + \text{constant} \tag{4.31}$$

Here, the constant of integration simply represents the fact that the capacitor can accommodate a steady potential difference. Overall, relations (4.30) and (4.31) show that the circuit of figure 4.11(a) passes signals of angular frequency greater than $1/RC$ between its input and output while accommodating a direct potential difference. This feature is put to widespread use in electronic circuitry where potential differences usually comprise a signal plus a direct or bias component and there is a need to transfer signals between points where the bias levels are necessarily different. When the circuit of figure 4.11(a) is used in this way it is said to *couple* a signal between two parts of a circuit. C–R coupling is common but not universal at the inputs and outputs of electronic equipment. It will be understood later that the circuit of figure 4.11(a) is actually an example of a *high-pass filter*; it passes signals at frequencies above $\sim 1/RC$.

According to the outcome of the analysis just presented, the response of the circuit of figure 4.11(a) to a square-wave input would be as shown in figure 4.12(a) at low and high-enough frequencies. However, the treatment has involved neglecting one or other term on the right-hand side of equation (4.27). In fact, precise analysis of the response to square-wave inputs is possible by the transient approach of the preceding section and

4.4 Basic four-terminal C–R networks

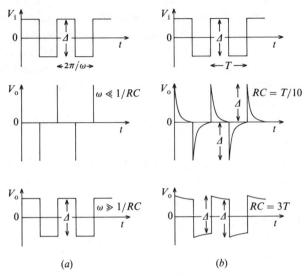

4.12 Response of the circuit of figure 4.11(a) to a square-wave input; (a) according to approximate theory showing differential action when $\omega \ll 1/RC$ and passing behaviour when $\omega \gg 1/RC$ and (b) according to precise transient analysis for $RC = T/10 = 2\pi/10\omega$ and $RC = 3T = 6\pi/\omega$.

this is very worthwhile since certain important differences emerge compared with the approximate approach although there are of course broad similarities. The square-wave input is treated as alternately switched direct e.m.f.s. Let the difference between these e.m.f.s be Δ. When V_i instantaneously steps by Δ, V_o follows instantaneously because the capacitor takes time to charge. The way in which the potential difference builds up across the capacitor as it charges was considered in the previous section and is shown in figure 4.10(a). There is gradual growth characterised by the time constant $\tau = RC$. Following the simultaneous steps Δ in V_i and V_o, the potential difference V_o across the resistor decays towards zero with time constant RC in complementary fashion to the growth across the capacitor, for the potential difference V_i remains constant across the series combination until the next step occurs. If the time constant RC is small compared with the period T, the decay is almost total before the next step in the input arrives and, on account of the alternate steps in V_i being of opposite sign, the output is of the form shown in figure 4.12(b) for $RC = T/10$. The reality of alternate positive and negative pulses of height Δ and duration $\sim RC$ should be compared with the behaviour expected according to the differential action predicted by the approximate analysis. Differential response suggests infinitely narrow alternate positive and negative pulses of infinite amplitude coincident in occurrence with the steps in the input square wave. If the time constant RC is long compared with the period T, there is little decay of V_o before the next step arrives and the output is much as shown in figure 4.12(b) for $RC = 3T$. Note that here again the output steps by Δ. However, this time, the output is not quite square compared with an expected square-wave output according to the coupling action predicted by the approximate analysis.

Application of Kirchhoff's voltage law to the circuit of figure 4.11(b) gives

$$V_i = V_o + RI$$

where I is the current through resistance R. Again assuming zero or negligible loading of the output so that any current between the output terminals can be neglected compared with I

$$I = dQ/dt = C \, dV_o/dt$$

and

$$V_i = V_o + RC \, dV_o/dt \tag{4.32}$$

By the same argument as before, provided

$$\omega RC \gg 1 \tag{4.33}$$

for the lowest frequency present, the first term on the right-hand side of equation (4.32) can be neglected compared with the second and to a good approximation

$$V_o = \frac{1}{RC} \int V_i \, dt \tag{4.34}$$

In these circumstances the circuit is acting as an *integrator* between its input and output terminals but note that, as for the differentiator, V_o is small compared with V_i owing to the requirement to make RC large enough to satisfy inequality (4.33). Where this is a problem, it can be avoided by using the active version of this circuit mentioned in section 10.5. Integrating circuits are greatly used in electronics. Examples of applications include the derivation of ramp waveforms from square-wave signals, the measurement of time and charge, analogue computation and analogue-to-digital conversion.

Under the opposite imposed condition

$$\omega RC \ll 1 \tag{4.35}$$

for the highest frequency present, the second term on the right-hand side of equation (4.32) can be neglected compared with the first and to a good approximation

$$V_o = V_i \tag{4.36}$$

It will be understood later that the circuit of figure 4.11(b) is actually an example of a *low-pass filter*; it passes signals at frequencies below $\sim 1/RC$.

4.5 L–C–R response to step e.m.f.

Its electronic applications are referred to when its filtering action is discussed in section 8.2.

According to the outcome of the analysis of the previous two paragraphs, the response of the circuit of figure 4.11(b) to a square-wave input would be as shown in figure 4.13(a) at low and high-enough frequencies. However, the treatment has involved neglecting one or other term on the right-hand side of equation (4.32). It is again interesting to compare the precise response deduced by considering the transients. Here the capacitor charges towards the input potential difference V_i through the series resistor as considered in section 4.3. If the time constant RC is short compared with the period T, the charging process is almost complete following a step in V_i before it steps again and, bearing in mind the opposite sign of alternate steps in V_i, the output is of the form shown in figure 4.13(b) for $RC = T/10$. If the time constant RC is long compared with the period, there is little charging and consequently an almost linear rise of V_o before the next step arrives so that the output is much as shown in figure 4.13(b) for $RC = 3T$.

4.5 Transient response of an L–C–R circuit to a step e.m.f.

Consider next the sudden application of an e.m.f. $\mathscr{E}$ to a series circuit comprising capacitance C, inductance L and resistance R, as could be accomplished by closing the switch K in the circuit of figure 4.14(a).

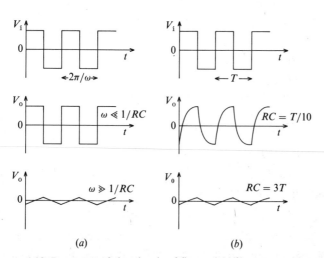

4.13 Response of the circuit of figure 4.11(b) to a square-wave input (a) according to approximate theory showing passing action when $\omega \ll 1/RC$ and integrating behaviour when $\omega \gg 1/RC$ and (b) according to precise transient analysis for $RC = T/10 = 2\pi/10\omega$ and $RC = 3T = 6\pi/\omega$.

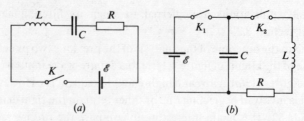

4.14 Circuits for (a) the sudden application of an e.m.f. to a series L–C–R combination and (b) discharging a charged capacitor through an inductor and resistor in series.

According to Kirchhoff's voltage law, at any time t after the application of the e.m.f.

$$L\,dI/dt + RI + Q/C = \mathscr{E} \tag{4.37}$$

where I is the current in the circuit and Q the charge associated with the capacitance so that $I = dQ/dt$. Hence

$$L\,d^2Q/dt^2 + R\,dQ/dt + Q/C = \mathscr{E} \tag{4.38}$$

or in terms of the charge difference $\Delta Q = Q - Q_0$ where Q_0 is the time-independent quantity $C\mathscr{E}$

$$LC\,d^2(\Delta Q)/dt^2 + RC\,d(\Delta Q)/dt + \Delta Q = 0 \tag{4.39}$$

Insertion of the function $\exp(\gamma t)$ for ΔQ in equation (4.39), where γ is independent of t, shows that it is a solution provided that

$$LC\gamma^2 + RC\gamma + 1 = 0$$

or

$$\gamma = -\frac{R}{2L} \pm \left(\frac{R^2}{4L^2} - \frac{1}{LC}\right)^{\frac{1}{2}} \tag{4.40}$$

Evidently in terms of the convenient parameters

$$\alpha = R/2L \tag{4.41}$$

and

$$\beta = \left(\frac{R^2}{4L^2} - \frac{1}{LC}\right)^{\frac{1}{2}} \tag{4.42}$$

the general solution for ΔQ is

$$\Delta Q = A\exp(-\alpha + \beta)t + B\exp(-\alpha - \beta)t \tag{4.43}$$

where A and B are arbitrary constants determined by the boundary conditions. As was discussed in section 4.3, it takes time for charge to accumulate on a capacitor or for current to grow in an inductive circuit. Thus, assuming there is no charge on the capacitor or current flowing just before the e.m.f. is applied at time $t=0$, the boundary conditions are that

4.5 L–C–R response to step e.m.f.

both Q and I are zero when $t=0$. According to equation (4.43) this means that

$$A+B=-Q_0$$
$$(-\alpha+\beta)A-(\alpha+\beta)B=0$$

from which

$$A=-\frac{(\alpha+\beta)}{2\beta}Q_0$$

and

$$B=\frac{(\alpha-\beta)}{2\beta}Q_0$$

Substitution of these expressions for A and B into equation (4.43) gives

$$Q=Q_0\{1-\tfrac{1}{2}[(\alpha/\beta+1)\exp(-\alpha+\beta)t-(\alpha/\beta-1)\exp(-\alpha-\beta)t]\} \quad (4.44)$$

It can now be appreciated that the essential behaviour of Q depends on whether the parameter β is real or imaginary. Reference to equation (4.42) shows that β is imaginary if $R^2/4L^2 < 1/LC$. Writing $\beta=j\omega_0$ in such circumstances, where $j=\sqrt{-1}$, equation (4.44) becomes

$$Q=Q_0\{1-(\tfrac{1}{2}\exp-\alpha t)[(\alpha/\beta+1)\exp j\omega_0 t-(\alpha/\beta-1)\exp-j\omega_0 t]\} \quad (4.45)$$

in which ω_0 is real and is given by

$$\omega_0=\left(\frac{1}{LC}-\frac{R^2}{4L^2}\right)^{\frac{1}{2}} \quad (4.46)$$

Clarification of the nature of relation (4.45) for Q follows from use of the identities

$$\exp\pm j\omega_0 t=\cos\omega_0 t\pm j\sin\omega_0 t$$

(see appendix 2) which leads to

$$Q=Q_0\{1-(\exp-\alpha t)[\cos\omega_0 t+(\alpha/\beta)j\sin\omega_0 t]\}$$

or

$$Q=Q_0\left\{1-(\exp-\alpha t)\left[\frac{\sin(\omega_0 t+\phi)}{\sin\phi}\right]\right\} \quad (4.47)$$

where

$$\tan\phi=\omega_0/\alpha=(4L/R^2C-1)^{\frac{1}{2}} \quad (4.48)$$

Apparently, when $R^2/4L^2 < 1/LC$, following the application of a step e.m.f. $\mathscr{E}$, the charge on the capacitor exhibits decaying simple harmonic oscillations at angular frequency $(1/LC-R^2/4L^2)^{\frac{1}{2}}$ during the approach to the steady state of constant charge $Q_0=C\mathscr{E}$. The time constant of the decay is $2L/R$ and, since $I=dQ/dt$, the current in the circuit also oscillates with

92   *Capacitance, inductance and electrical transients*

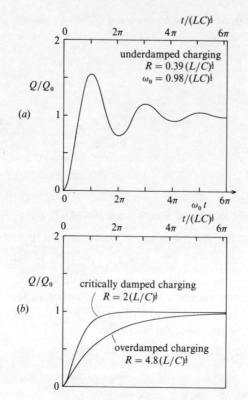

4.15 Charge associated with capacitance C in a series L–C–R circuit following the application of a step e.m.f. (*a*) When underdamped with $R=0.39(L/C)^{\frac{1}{2}}$, (*b*) when overdamped with $R=4.8(L/C)^{\frac{1}{2}}$ and critically damped with $R=2(L/C)^{\frac{1}{2}}$.

decaying amplitude. It is common to refer to the behaviour that occurs subject to the condition $R^2/4L^2 < 1/LC$ as *underdamped*. The first peak in Q can be up to twice Q_0. Figure 4.15(*a*) shows the time dependence of Q that occurs when $\alpha = 0.2\omega_0$, which corresponds to $R=0.39(L/C)^{\frac{1}{2}}$. When the resistance is small enough to satisfy $R \ll 2(L/C)^{\frac{1}{2}}$, the angular frequency of oscillation is close to $(1/LC)^{\frac{1}{2}}$ and the decay per period is tiny. The fact that circuits containing both inductance and capacitance can exhibit natural oscillation is of great importance to electronics, particularly in connection with the transmission and reception of high-frequency signals in communication links. At other times the occurrence of oscillations upon application of pulses to circuits is a nuisance and in these circumstances the oscillatory behaviour is described as *ringing*.

In the alternative situation to that discussed so far, namely that corresponding to $R^2/4L^2 \geqslant 1/LC$, reference to equations (4.41) and (4.42)

4.5 L–C–R response to step e.m.f.

shows that β is real and less than α. In this case both exponents in equation (4.44) are real and negative so that, as t increases, the two exponential terms fall steadily towards zero and Q asymptotically approaches Q_0 at large enough t. Not surprisingly the time constants that determine the rate of growth of Q are combinations of the L/R and RC time constants encountered in L–R and C–R circuits. Figure 4.15(b) shows the particular time dependence of Q that takes place when $\alpha = 1.1\beta$ which corresponds to $R = 4.8(L/C)^{\frac{1}{2}}$.

The behaviour when $R^2/4L^2 > 1/LC$, that is, $R > 2(L/C)^{\frac{1}{2}}$, is described as *overdamped*. When $R = 2(L/C)^{\frac{1}{2}}$, β reaches zero and oscillation is only just avoided. In this condition the circuit is said to be *critically damped*. Especial care must be exercised in deducing the behaviour of Q from equation (4.44) when $\beta = 0$ because β is present in numerator and denominator terms. For sufficiently small β, the exponential terms may be expanded to first order in βt and so

$$Q = Q_0\{1 - (\tfrac{1}{2}\exp -\alpha t)[(\alpha/\beta + 1)(1 + \beta t) - (\alpha/\beta - 1)(1 - \beta t)]\}$$

which shows that in the limit, as β goes to zero,

$$Q = Q_0[1 - (1 + \alpha t)\exp -\alpha t] \tag{4.49}$$

If so desired, the validity of equation (4.49) as a solution can be checked by substitution back into equation (4.38). The critically damped behaviour of Q represented by equation (4.49) is presented in figure 4.15(b) for comparison with the overdamped and underdamped responses.

It is relatively easy to extend the theory developed in this section to find the time dependence of the discharge of a capacitor through a series combination of an inductor and resistor. Suppose that the capacitor is fully charged initially by an e.m.f. $\mathscr{E}$ as could be achieved in practice by, for example, closing switch K_1 in the circuit of figure 4.14(b) for long enough with K_2 open. The discharge through the L–R combination would then be initiated by opening K_1 and immediately afterwards closing K_2. Application of Kirchhoff's voltage law shows that at any time t during the discharge

$$L\,dI/dt + RI + Q/C = 0$$

or

$$LC\,d^2Q/dt^2 + RC\,dQ/dt + Q = 0 \tag{4.50}$$

with the same notation as before. Conveniently, equation (4.50) in Q is identical to equation (4.39) in ΔQ and so by comparison with equation (4.43)

$$Q = A\exp(-\alpha + \beta)t + B\exp(-\alpha - \beta)t$$

where α and β are given by equations (4.41) and (4.42) as before. The initial conditions of the discharge are that $Q = C\mathscr{E} = Q_0$ when $t = 0$ and $I = 0$ when $t = 0$, the latter again applying because it takes time for current to change in an inductive circuit. Thus

$$A + B = Q_0$$
$$(-\alpha + \beta)A - (\alpha + \beta)B = 0$$

from which A and B may be found and substituted in the equation for Q to yield

$$Q = \tfrac{1}{2}Q_0\{(1 + \alpha/\beta)\exp(-\alpha + \beta)t + (1 - \alpha/\beta)\exp(-\alpha - \beta)t\} \quad (4.51)$$

If $R < 2(L/C)^{\frac{1}{2}}$, β is imaginary, in which case, writing $\beta = j\omega_0$ as before, equation (4.51) can be rearranged as

$$Q = Q_0(\exp -\alpha t)[\cos \omega_0 t + (\alpha/\omega_0)\sin \omega_0 t]$$

or

$$Q = Q_0(\exp -\alpha t)\left[\frac{\sin(\omega_0 t + \phi)}{\sin \phi}\right] \quad (4.52)$$

where $\tan \phi$ is given by equation (4.48) again. Clearly, in these circumstances the discharge is oscillatory with decaying amplitude as the uncharged state $Q = 0$ is approached. Such behaviour is once more described as underdamped and figure 4.16(a) shows the time dependence of Q for the particular case $\alpha = 0.2\omega_0$ or $R = 0.39(L/C)^{\frac{1}{2}}$. When $R \ll 2(L/C)^{\frac{1}{2}}$, the frequency of oscillation is close to $(1/LC)^{\frac{1}{2}}$ and the decay in amplitude per period is tiny. Actually, the term *decrement* is applied to the ratio of successive maxima and equation (4.52) shows that the decrement δ is given by

$$\delta = \exp(2\pi\alpha/\omega_0) \quad (4.53)$$

The *logarithmic decrement* is accordingly

$$\ln \delta = 2\pi\alpha/\omega_0 = 2\pi/(4L/R^2C - 1)^{\frac{1}{2}} \quad (4.54)$$

If, on the other hand, $R \geqslant 2(L/C)^{\frac{1}{2}}$, β is real and less than α so that both exponents in equation (4.51) are real and negative and the discharge proceeds steadily. The term overdamped is applied again when $R > 2(L/C)^{\frac{1}{2}}$ and figure 4.16(b) shows an example of overdamped response corresponding to $R = 4.8(L/C)^{\frac{1}{2}}$. Careful consideration of equation (4.51) reveals that in the critically damped condition corresponding to $R = 2(L/C)^{\frac{1}{2}}$

$$Q = Q_0(1 + \alpha t)\exp -\alpha t \quad (4.55)$$

and figure 4.16(b) also shows this critical response.

There are many physical systems which obey the same form of differential equations as those considered in this section and consequently exhibit

4.5 L–C–R response to step e.m.f.

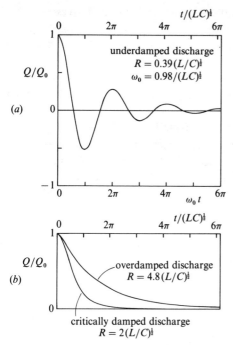

4.16 Charge associated with capacitance C in a series L–C–R circuit during discharge. (a) When underdamped with $R=0.39(L/C)^{\frac{1}{2}}$ and (b) when overdamped with $R=4.8(L/C)^{\frac{1}{2}}$ and critically damped with $R=2(L/C)^{\frac{1}{2}}$.

similar forms of response. The motion of a mass m subject to a restoring force proportional to the distance x from some origin and a damping force proportional to velocity obeys the differential equation

$$m\frac{d^2x}{dt^2} + b\frac{dx}{dt} + kx = 0 \qquad (4.56)$$

for example, where b and k are damping and restoring force proportionality constants. Analogy between this mechanical system and the electrical circuit analysed is expressed by correspondence between x and Q, dx/dt and I, m and L, b and R and k and $1/C$. The existence of finite b or R is responsible for energy dissipation in the two situations. With b or R zero, oscillatory energy is conserved and there are undamped harmonic oscillations. Damped harmonic oscillations occur if $0 < b < 2(mk)^{\frac{1}{2}}$ or $0 < R < 2(L/C)^{\frac{1}{2}}$.

Finally, although this section has dealt with the transient response of series L–C–R circuits to step e.m.f.s, the analysis with slight modification is applicable to parallel L–C–R combinations. In the absence of any e.m.f., application of Kirchhoff's current law to a parallel circuit gives

$$C\frac{dV}{dt}+\frac{V}{R}+\frac{1}{L}\int V\,dt=0$$

or

$$LC\frac{d^2V}{dt^2}+\frac{L}{R}\frac{dV}{dt}+V=0 \tag{4.57}$$

Comparison of equation (4.57) with equation (4.50) reveals that the behaviour of V in the parallel circuit is precisely that of Q in a series circuit comprising capacitance L, inductance C and resistance $1/R$. The parallel L–C–R circuit is said to be the *dual* of the series L–C–R circuit.

5

Introduction to the steady-state responses of networks to sinusoidal sources

5.1 Sinusoidal sources and definitions

A sinusoidal source is one which delivers an e.m.f. or load-independent current that varies with time t according to the mathematical expression

$$x = x_0 \sin(\omega t + \phi) \tag{5.1}$$

This particularly well-known function of time (see appendix 1) is plotted in the graph of figure 5.1, the main feature being that it repeats at regular intervals. Its peak value x_0 is called the *amplitude*. By definition, the function is given by the projection of a line of length x_0, rotating at an angular rate of ω radians per second, onto an appropriate fixed direction as shown in figure 5.1. The parameter ω is known as the *pulsatance* or *angular frequency* while the time interval T between repetitions of the function is termed the *period*. Clearly, the function repeats whenever ωt increases by 2π and therefore T is related to ω by $\omega T = 2\pi$ or

$$T = 2\pi/\omega \tag{5.2}$$

The *frequency* of repetition is evidently

$$f = 1/T = \omega/2\pi \tag{5.3}$$

Although frequency has the dimension of inverse time so that its fundamental unit could be s^{-1}, the basic unit of frequency is normally given the special description *hertz* or Hz, after the physicist Hertz, to indicate the repetitive character involved.

The quantity ϕ is called the *phase*. In general, two sinusoidal functions of the same frequency will have different phases and will peak at differing times. For instance, $x_1 = x_{10} \sin(\omega t + \phi_1)$ and $x_2 = x_{20} \sin(\omega t + \phi_2)$ peak when $\omega t = [(2n+1)\pi/2] - \phi_1$ and $\omega t = [(2n+1)\pi/2] - \phi_2$ respectively. The angular interval between neighbouring corresponding values of two sinusoidal functions, such as successive positive peaks, is known as the

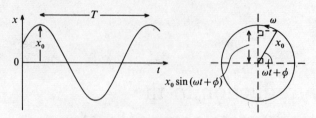

5.1 The function $x = x_0 \sin(\omega t + \phi)$.

phase difference. In the case of x_1 and x_2 just defined, the phase difference is of course $(\phi_1 - \phi_2)$ compared with a time difference of $(\phi_1 - \phi_2)/\omega$. That periodic function which peaks first is said to *lead* the other in phase while that which peaks afterwards is said to *lag* in phase.

Since the sinusoidal function x varies symmetrically about zero, its average value over an integral number of periods is zero. However, it turns out that a convenient measure of its magnitude for practical purposes is the *root-mean-square* (r.m.s.) value defined by

$$x_{\text{rms}} = \left[\frac{1}{T} \int_0^T x^2 \, dt \right]^{\frac{1}{2}} \tag{5.4}$$

There is a simple relationship between the r.m.s. value x_{rms} and the amplitude. Evaluating the integral gives

$$x_{\text{rms}}^2 = \frac{x_0^2}{\omega T} \int_0^{2\pi} \sin^2(\omega t + \phi) \, d(\omega t)$$

$$= \frac{x_0^2}{2\omega T} \int_0^{2\pi} [1 - \cos 2(\omega t + \phi)] \, d(\omega t)$$

$$= \frac{x_0^2}{4\pi} \left[\omega t - \frac{\sin 2(\omega t + \phi)}{2} \right]_0^{2\pi}$$

the second term of which has the same value at both limits so that

$$x_{\text{rms}} = x_0/\sqrt{2} \tag{5.5}$$

The reason why the r.m.s. value is useful with regard to electrical networks is that the heat developed in time dt in a resistance R by an applied e.m.f. $\mathscr{E}$ is $\mathscr{E}^2 \, dt/R$. Thus when the applied e.m.f. is sinusoidal, the heat generated per period is

$$W = (1/R) \int_0^T \mathscr{E}_0^2 \sin^2(\omega t + \phi) \, dt$$

which reduces, on using definition (5.4), to

$$W = \mathscr{E}_{\text{rms}}^2 T/R \tag{5.6}$$

5.1 Sinusoidal sources and definitions

The r.m.s. e.m.f. or potential difference across a resistor is evidently the magnitude of direct e.m.f. or potential difference that would cause the same heat dissipation in it. The heat generated in a resistance R by a sinusoidal current over a period is similarly given in terms of the r.m.s. current I_{rms} as

$$W = RI_{rms}^2 T \tag{5.7}$$

Certain other mathematical aspects of sinusoidal functions are considered in appendix 1.

In practice, sinusoidal e.m.f.s are generated by electronic oscillators and by electrical machines, the latter giving rise to induced sinusoidal e.m.f.s by virtue of electrical coils rotating in magnetic fields. To appreciate the latter method of generation, consider a plane coil of area **A** rotating at an angular frequency ω about an axis in its plane and suppose, for simplicity, that a uniform magnetic induction **B** is applied perpendicular to this axis. The laws of electromagnetic induction expressed by equation (4.11) show that an e.m.f. will be induced in the coil given by

$$\mathscr{E} = -(d/dt)[BA\cos(\omega t + \phi)] = BA\omega \sin(\omega t + \phi)$$

where ϕ is the angle that the vector **A** makes with the magnetic induction **B** at time $t = 0$ in the sense of rotation. A prime example of this type of e.m.f. is the alternating-current mains, the r.m.s. e.m.f. being around 240 V and the frequency 50 Hz in the United Kingdom. Explaining how electronic oscillators produce sinusoidal e.m.f.s over a vast range of frequency is a feature of most electronics textbooks. Usually, sinusoidal oscillation is achieved by applying frequency-selective positive feedback between the output and input of an electronic amplifier (see sections 10.4 and 10.6). The positive feedback causes any fluctuation to grow and the frequency-selective action filters out all fluctuations except the sinusoidal variation at the desired frequency. Filters that provide the necessary frequency-selective action are considered in chapter 8.

Deducing the responses of networks to sinusoidal sources is of paramount importance because, quite apart from the common occurrence of sources featuring outputs that closely approximate to sinusoidal in time, whatever the time dependence of a signal, it can be analysed into a frequency spectrum of purely sinusoidal signals, as will be proved in chapter 11. An *alternating* or periodic signal analyses into a *harmonic* frequency spectrum while a *pulse* or nonrepetitive signal analyses into a *continuous* frequency spectrum of sinusoidal signals. The feasibility of such spectral signal analysis means that, in principle, the response of any electrical network to any signal can be deduced from its response to sinusoidal signals.

5.2 Responses of purely resistive, purely capacitive and purely inductive circuits to sinusoidal e.m.f.s

Throughout this chapter it will be assumed that the frequency of the source is not so high that conventional circuit theory, as developed in the earlier chapters, breaks down. As discussed at the beginning of section 4.3, this imposition merely restricts the frequency to the range where the wavelength of corresponding electromagnetic radiation is much greater than the dimensions of the circuit under discussion. Typically the restriction is only to frequencies below a few hundred MHz.

Consider first a sinusoidal e.m.f. $\mathscr{E}_0 \sin \omega t$ connected in series with just an ideal resistor of constant resistance R as shown in figure 5.2(a). The time origin has been chosen here to make the phase angle ϕ of the source conveniently equal to zero. According to Kirchhoff's voltage law, the current will be given everywhere in the circuit by

$$I = (\mathscr{E}_0/R) \sin \omega t \tag{5.8}$$

The implications of this equation are that the current is in phase with the potential difference across the resistor as illustrated in figure 5.2(b) and that the ratio of the amplitude of the potential difference across the resistor to the amplitude of the current is simply the resistance R.

Turning to the case of a sinusoidal e.m.f. connected in series with just an ideal capacitor of constant capacitance C as illustrated in figure 5.2(c), the

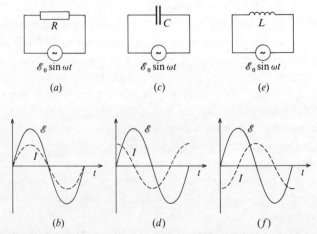

5.2 A sinusoidal e.m.f. connected in series with (a) an ideal resistor only, (c) an ideal capacitor only and (e) an ideal inductor only. Note the circuit diagram symbol for an alternating e.m.f. Graphs (b), (d) and (f) show the steady-state responses for circuits (a), (c) and (e) respectively.

5.2 Steady-state sinusoidal R, C and L response

charge on the capacitor is from equation (4.1)

$$Q = C\mathscr{E}_0 \sin \omega t$$

so that the current is

$$I = dQ/dt = \omega C \mathscr{E}_0 \cos \omega t \tag{5.9}$$

Clearly, the potential difference across the capacitor *lags* 90° in phase behind the current as illustrated in figure 5.2(d). Where there is a phase difference of 90° between the potential difference across a circuit element and the current, the ratio of the amplitude of the potential difference across the element to the amplitude of the current is called the *reactance*. From equation (5.9) the reactance corresponding to capacitance C is

$$X_C = 1/\omega C \tag{5.10}$$

Of course, in accordance with equation (5.5), reactance can also be defined as the ratio of r.m.s. potential difference to r.m.s. current when a phase difference of 90° occurs between the potential difference and current.

When a sinusoidal e.m.f. is connected in series with just an ideal inductor of constant inductance L as illustrated in figure 5.2(e), the current I is obtained by equating the applied e.m.f. to the magnitude of the back e.m.f. given by equation (4.13), which yields

$$\mathscr{E}_0 \sin \omega t = L\, dI/dt$$

The solution of this differential equation that is of interest here is

$$I = -(\mathscr{E}_0/\omega L) \cos \omega t \tag{5.11}$$

Although formation of the complete solution demands the addition of a constant to the right-hand side of equation (5.11), the point is that any practical inductive circuit inevitably includes a finite though perhaps very small resistance. As might be suspected from section 4.3, and as will be verified in the next section, the presence of finite resistance leads to the constant term being replaced by a decaying transient. The characteristic time constant of the decay is L/R, as before, so that when R is negligibly small the transient decays infinitely slowly and in the theoretical limit of $R = 0$ becomes a constant. Such discussion demonstrates that equation (5.11) is in fact the steady-state solution for any practical inductive circuit of negligible resistance. Clearly, in the steady state, the potential difference across the ideal inductor in the circuit of figure 5.2(e) *leads* the current by a phase angle of 90° as illustrated in figure 5.2(f). Further, the reactance corresponding to inductance L is

$$X_L = \omega L \tag{5.12}$$

Although the effect of applying a sinusoidal e.m.f. to a purely resistive, capacitive or inductive circuit has been deduced, the same phase and amplitude relations occur when a sinusoidal constant-current source is applied. The potential difference developed across a resistance R by a constant-current source $I_0 \sin \omega t$ is $RI_0 \sin \omega t$ so that again the potential difference and current are in phase and the ratio of the amplitudes of the potential difference and current is the resistance R. When current $I_0 \sin \omega t$ is fed into a capacitor of capacitance C

$$I_0 \sin \omega t = C \, dV/dt$$

and therefore

$$V = -(I_0/\omega C) \cos \omega t \qquad (5.13)$$

As for an applied sinusoidal e.m.f., the potential difference across the capacitor *lags* in phase 90° behind the current and the ratio of potential difference amplitude to current amplitude is $1/\omega C$. It is left as an exercise for the reader to show that phase and amplitude relationships in a purely inductive circuit are the same whether a sinusoidal e.m.f. or constant-current source is applied.

It follows immediately from the definition of reactance that its fundamental SI unit is the same as that of resistance, namely the ohm denoted by Ω. The SI unit of phase is, of course, the radian normally abbreviated to rad, but in practice phase angles are usually expressed in degrees. Some appreciation of the order of magnitude of reactance is valuable at this stage. Capacitance normally encountered spans the vast range from a fraction of a pF to $10\,000\,\mu$F! The capacitive reactance corresponding to a modest capacitance of 1 nF is as high as approximately 1.59 MΩ at a frequency of 100 Hz but plunges to only 159 Ω at 1 MHz. This illustrates why the shunting effect of even tiny stray capacitance can be extremely significant at the high frequencies encountered in electronics. The input of a cathode-ray oscilloscope used in the observation and measurement of electrical signals is typically equivalent to a resistance of 1 MΩ in parallel with a capacitance of 30 pF and, although capable of operation at very high frequencies (up to several hundred MHz), must be used with care as a voltmeter in such circumstances because of the parallel reactance. Inductance commonly experienced ranges from a fraction of a μH to a few H. The inductive reactance corresponding to a modest inductance of 1 mH, although soaring to 6.3 kΩ at 1 MHz, amounts to only approximately 0.63 Ω at 100 Hz. This shows why, for example, inductive chokes for operation at mains and audio frequencies (50 Hz and 20 Hz–20 kHz respectively) are bulky.

5.3 Sinusoidal response through differential equation solution

Three different techniques exist by which Kirchhoff's laws can be applied to find the steady-state response of an electrical network to a

5.3 Sinusoidal response from differential equation

sinusoidal source. In the most direct method of analysis, Kirchhoff's laws are expressed in differential-equation form and the resulting equation or equations solved. In a second approach, Kirchhoff's laws are expressed in the form of a phasor diagram and a geometrical solution of the diagram obtained. A third very neat procedure involves using the imaginary operator $j = \sqrt{-1}$ in applying Kirchhoff's laws and leads to a solution through complex algebraic manipulation. While the techniques are of general applicability, each method of analysis will be explained by applying it in turn to a particular series circuit comprising, as illustrated in figure 5.3(a), inductance L, resistance R and e.m.f. $\mathscr{E}_0 \sin \omega t$.

Considering the most direct method, application of Kirchhoff's voltage law to the circuit of figure 5.3(a) yields the differential equation

$$L(dI/dt) + RI = \mathscr{E}_0 \sin \omega t \tag{5.14}$$

in the current I at time t. The solution is (consult appendix 4 if necessary)

$$I = A \exp(-Rt/L) + \frac{1}{L\mathscr{D} + R} \mathscr{E}_0 \sin \omega t \tag{5.15}$$

where A is an arbitrary constant and $\mathscr{D}$ the operator d/dt. The first term represents a transient which gradually disappears with the expected time constant L/R to leave the steady-state response represented by the second term. Omitting the first term for the time being, while interest centres on the steady state, and carrying out some algebraic manipulation,

$$I = (R - L\mathscr{D}) \frac{\mathscr{E}_0 \sin \omega t}{R^2 + \omega^2 L^2}$$

since $\mathscr{D}^2 \sin \omega t = -\omega^2 \sin \omega t$. Hence

$$I = \frac{\mathscr{E}_0}{R^2 + \omega^2 L^2} (R \sin \omega t - \omega L \cos \omega t)$$

But, with reference to figure 5.3(b), $R/(R^2 + \omega^2 L^2)^{\frac{1}{2}}$ and $\omega L/(R^2 + \omega^2 L^2)^{\frac{1}{2}}$ can respectively be expressed as $\cos \phi$ and $\sin \phi$ so that

$$I = \frac{\mathscr{E}_0}{(R^2 + \omega^2 L^2)^{\frac{1}{2}}} \sin(\omega t - \phi) \tag{5.16}$$

5.3 (a) Sinusoidal e.m.f. applied to a series L–R circuit and (b) the phase angle ϕ between the current and e.m.f. in this circuit.

where

$$\tan \phi = \omega L/R \tag{5.17}$$

Equation (5.16) reveals that the current lags behind the e.m.f. by a phase angle ϕ given by equation (5.17). Quite generally, the amplitude of the potential difference divided by that of the current between two terminals of a network comprising resistive and reactive components is described as the *impedance* between the terminals. Apparently the impedance corresponding to an inductance L in series with a resistance R is

$$|Z| = (R^2 + \omega^2 L^2)^{\frac{1}{2}} \tag{5.18}$$

From the definition of impedance, its SI unit is the ohm, like that of reactance and resistance.

Now consider the current I that flows during the approach to the steady state following the sudden application of an e.m.f. $\mathscr{E}_0 \sin \omega t$ to a series circuit containing inductance L and resistance R. It is given by the complete solution (5.15) of differential equation (5.14) subject to the boundary condition that $I = 0$ when $t = 0$. Making use of the steady-state solution represented by equations (5.16) and (5.17), it follows from equation (5.15) that the complete solution is

$$I = A \exp(-Rt/L) + \frac{\mathscr{E}_0}{(R^2 + \omega^2 L^2)^{\frac{1}{2}}} \sin(\omega t - \phi)$$

Inserting the condition $I = 0$ when $t = 0$ leads to

$$0 = A - \frac{\mathscr{E}_0}{(R^2 + \omega^2 L^2)^{\frac{1}{2}}} \sin \phi$$

Consequently

$$I = \frac{\mathscr{E}_0}{(R^2 + \omega^2 L^2)^{\frac{1}{2}}} [\sin \phi \exp(-Rt/L) + \sin(\omega t - \phi)] \tag{5.19}$$

The behaviour of this function when the time constant L/R equals π/ω so that $\tan \phi = \pi$ or $\phi = 72.34°$ and

$$I = \frac{\mathscr{E}_0}{3.3R} [0.953 \exp(-\omega t/\pi) + \sin(\omega t - 72.34°)]$$

is presented in figure 5.4 to illustrate what happens during the approach to the steady state. It shows that the current gradually settles down to sinusoidal variation. The time taken to settle down is a few time constants. In the particular case under consideration, the period is just twice the time constant so that it only takes about two periods to almost reach the steady state. The larger L compared with R, the longer the current will take to settle down to sinusoidal variation.

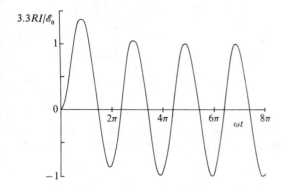

5.4 Behaviour of the current I in a series L–R circuit immediately following the sudden application of a sinusoidal e.m.f. $\mathscr{E}_0 \sin \omega t$, given that the time constant L/R is $\pi/\omega = T/2$ in terms of the pulsatance ω or period T.

5.4 Steady-state sinusoidal response from phasor diagram

When sinusoidal sources of just a single frequency are present in a linear network, in the steady state all the currents and potential differences also vary sinusoidally at the same frequency so that the response is entirely defined by the amplitudes and phases of the currents and potential differences. In the previous section, theoretical expressions were obtained for the amplitude and phase of the steady-state current in a simple series L–R circuit connected to a single sinusoidal e.m.f. The potential differences across the resistance R and inductance L are given respectively by just RI and $-L\,dI/dt$ so that their steady-state amplitudes are closely related to the amplitude of the current and they are respectively in and 90° out of phase with the current.

In general, application of Kirchhoff's laws to find the steady-state response of a linear network to sources of a single frequency demands the summation of sinusoidal terms differing in amplitude and phase but having the same frequency. Now a sinusoidal term such as $x_0 \sin(\omega t + \phi)$ can be represented by the projection of a line of length x_0, rotating at angular frequency ω, onto a suitable reference direction in the plane of rotation, as shown in figure 5.1 and again in figure 5.5(a). The line being projected, through its length and orientation at some moment such as $t=0$, clearly represents the amplitude and phase of the sinusoidal term in question. Lines, the length and orientation of which respectively represent the amplitude and phase of quantities that vary sinusoidally with time *at a common frequency*, are termed *phasors*. This terminology distinguishes them from vectors which they are not. Further discussion of phasors appears in appendix 3. The important point here is that, rather than

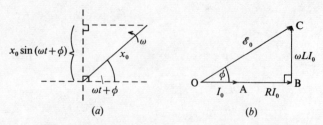

5.5 (a) Representation of a sinusoidal quantity by a phasor and (b) phasor diagram for a series L–R circuit connected to a sinusoidal e.m.f.

summing individual line projections of the form $x_0 \sin(\omega t + \phi)$ in applying Kirchhoff's laws, when the frequency is common the resultant of the set of lines, which rotates at the common frequency of course, can first be found and then be projected. Moreover, since it is known that the projected resultant varies sinusoidally at the common source frequency, the final projection is unnecessary and all that is really required is to find the amplitude and phase of the resultant of an appropriate set of phasors. This can be achieved extremely conveniently simply by drawing a fixed *phasor diagram* corresponding to time $t = 0$.

To illustrate the phasor diagram method of steady-state sinusoidal analysis, consider again the steady-state response of a series circuit containing inductance L and resistance R connected to a sinusoidal e.m.f. $\mathscr{E}_0 \sin \omega t$. In this case there is only one mesh. Let the current in it be represented by a phasor OA of arbitrary length and direction, for this current phasor will merely serve as a reference for other phasors of the circuit. Although the current phasor could be drawn in any direction in the diagram, it is customary to draw it in the horizontal direction as shown in figure 5.5(b). Now the net potential difference dropped across the purely resistive parts of the circuit has amplitude RI_0 in terms of the current amplitude I_0 and the same phase as the current. It can therefore be represented in the phasor diagram by a phasor OB which is R times longer than and parallel to the phasor representing the current, as shown in figure 5.5(b). The net potential difference dropped across the purely inductive parts of the circuit has amplitude $\omega L I_0$ and leads the current by 90° in phase. Accordingly it can be represented in the phasor diagram by a phasor BC which is ωL times longer than and rotated anticlockwise by 90° with respect to the phasor representing the current, also as shown in figure 5.5(b). It is convention to show phase lead by anticlockwise and phase lag by clockwise rotation in a phasor diagram; note that this complies with the complete solution being given by a projection of the phasor diagram rotating anticlockwise at angular frequency ω as shown in figure 5.5(a).

Completion of the phasor diagram for the circuit under consideration follows from Kirchhoff's voltage law requiring that the e.m.f. equals the sum of the potential differences across the inductive and resistive parts. Thus the e.m.f. can be represented in the phasor diagram by a phasor which is the resultant (vector sum) of the phasors representing the resistive and inductive potential differences, that is, by the phasor OC in figure 5.5(b). Making use of Pythagorus' theorem in the phasor diagram, it is seen that the amplitude $\mathscr{E}_0$ of the e.m.f. is related to the current amplitude by

$$\mathscr{E}_0 = (R^2 + \omega^2 L^2)^{\frac{1}{2}} I_0 \tag{5.20}$$

The phasor diagram also reveals that the current lags behind the e.m.f. by a phase angle ϕ where

$$\tan \phi = \omega L / R \tag{5.21}$$

Of course, equations (5.20) and (5.21) represent precisely the same steady-state response as equations (5.16) and (5.17) obtained in the previous section by solution of the appropriate differential equation.

5.5 Steady-state sinusoidal response through complex representation

Multiplication of a vector by -1 is equivalent in its effect to rotating the vector through an angle of 180°. Let j represent an operation which if repeated gives rise to multiplication by -1 so that

$$j(jx) = -x$$

and $j^2 = -1$ or

$$j = \sqrt{-1}$$

Since the operation of j repeated is equivalent to rotation through 180° then the operation of j alone must be equivalent to rotation through 90°. It immediately follows that j can be used to denote a phase lead of 90° and $-j$ a phase lag of 90°. This greatly facilitates deduction of the sinusoidal responses of networks. In general terms, introduction of the operator j forms the basis of what is known as *complex algebra* and, if necessary, the treatment of this topic presented in appendices 2 and 3 or some alternative treatment should be consulted at this point.

As already hinted at, making use of the operator j, the information contained in a phasor diagram can be expressed in algebraic form. Denoting phasors by heavy type, the particular phasor diagram of figure 5.5(b) representing the application of Kirchhoff's voltage law is expressible as

$$\mathscr{E} = R\mathbf{I} + j\omega L \mathbf{I} = (R + j\omega L)\mathbf{I} \tag{5.22}$$

In this equation $\mathscr{E}$ and $\mathbf{I}$ are of course the relevant e.m.f. and current phasors respectively. Quite generally, for any two-terminal linear network, the ratio of the phasor **V** representing the potential difference between the terminals to the phasor **I** representing the current between them is a *complex* quantity known as the *complex impedance*. It has a *real* and *imaginary* part, the latter being that part which is multiplied by the 'imaginary' operator j. In the particular case of a series circuit containing resistance R and inductance L, it is seen from equation (5.22) that the complex impedance is

$$Z = \mathbf{V}/\mathbf{I} = \mathscr{E}/\mathbf{I} = R + j\omega L \tag{5.23}$$

the real part being the resistance R and the imaginary part the reactance ωL. All the information needed to find the amplitude ratio $|\mathbf{V}|/|\mathbf{I}|$, already termed the impedance and written $|Z|$, and the phase relationship between **V** and **I** is contained in the complex impedance. The amplitude ratio is the square root of the sum of the squares of the real and imaginary parts of the complex impedance while the ratio of the imaginary to real part is the tangent function of the phase angle ϕ by which the terminal potential difference leads the current. For the particular case considered, these rules give

$$|Z| = (R^2 + \omega^2 L^2)^{\frac{1}{2}}; \quad \tan \phi = \omega L/R$$

in agreement with the results of the previous two sections. As a specific example of their application consider a series circuit comprising 1 kΩ resistance, 100 mH inductance and a sinusoidal e.m.f. of 10 V amplitude and 1 kHz frequency. The impedance is evidently

$$(10^6 + 4\pi^2 10^6 \times 10^{-2})^{\frac{1}{2}} \Omega \approx (1 + 0.395)^{\frac{1}{2}} 10^3 \, \Omega \approx 1.18 \, \text{k}\Omega$$

the e.m.f. leads the current by

$$\tan^{-1}(2\pi \times 10^3 \times 10^{-1}/10^3) \approx \tan^{-1} 0.628 \approx 32.1°$$

and the amplitude of the current is

$$10 \text{ V}/1.18 \text{ k}\Omega \approx 8.5 \text{ mA}$$

By now it should be clear that the complex impedance corresponding to a resistance R is

$$Z = \mathbf{V}/\mathbf{I} = R \tag{5.24}$$

to an inductance L is

$$Z = \mathbf{V}/\mathbf{I} = j\omega L \tag{5.25}$$

to a capacitance C is

$$Z = \mathbf{V}/\mathbf{I} = -j/\omega C = 1/j\omega C \tag{5.26}$$

and to a resistance R in series with a reactance X is

$$Z = \mathbf{V}/\mathbf{I} = R + jX \tag{5.27}$$

In the last case (indicating amplitude and r.m.s. magnitudes by subscripts 0 and rms as usual)

$$|Z| = V_0/I_0 = V_{\text{rms}}/I_{\text{rms}} = (R^2 + X^2)^{\frac{1}{2}} \tag{5.28}$$

and **V** leads **I** by a phase angle ϕ where
$$\tan \phi = X/R \tag{5.29}$$
Applying these results to a circuit comprising e.m.f. $\mathscr{E}_0 \sin \omega t$ in series with resistance R and capacitance C, Kirchhoff's voltage law gives
$$\mathscr{E} = Z\mathbf{I} = (R - j/\omega C)\mathbf{I} \tag{5.30}$$
Accordingly
$$|Z| = (R^2 + 1/\omega^2 C^2)^{\frac{1}{2}} \tag{5.31}$$
and $\mathscr{E}$ lags **I** by a phase angle ϕ where
$$\tan \phi = 1/\omega CR \tag{5.32}$$
Relationships (5.30)–(5.32) are illustrated in the phasor diagram of figure 5.6.

Where a set of complex impedances Z_i are connected entirely in series, the total potential difference phasor **V** is related to the current phasor **I** by
$$\mathbf{V} = Z_1 \mathbf{I} + Z_2 \mathbf{I} + Z_3 \mathbf{I} + \cdots = (Z_1 + Z_2 + Z_3 + \cdots)\mathbf{I} = Z\mathbf{I}$$
so that their effective complex impedance is simply
$$Z = \sum_i Z_i \tag{5.33}$$
When analysing the response of essentially parallel circuits to sinusoidal sources, the reciprocal of the complex impedance called the complex *admittance* is often more convenient. In terms of the complex impedance $Z = R + jX$, the complex admittance Y is
$$Y = \frac{1}{Z} = \frac{1}{R + jX} = \frac{R - jX}{R^2 + X^2} = G - jB \tag{5.34}$$
where
$$G = R/(R^2 + X^2) \tag{5.35}$$
is known as the *conductance* and
$$B = X/(R^2 + X^2) \tag{5.36}$$
is known as the *susceptance*. In a purely resistive circuit, $X = 0$, so that $B = 0$ and $G = 1/R$ in conformity with the definition of conductance introduced

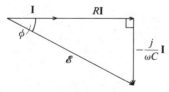

5.6 Phasor diagram for a series C–R circuit connected to a sinusoidal e.m.f.

earlier in section 2.2. For a purely reactive network, $R=0$ so that $G=0$ and $B=1/X$, that is, the susceptance is the reciprocal of the reactance. Usefully in the case of a set of parallel impedances, Z_i, where Y_i are the corresponding admittances, the total complex admittance Y is by the method of section 3.2

$$Y = \frac{1}{Z} = \sum_i \frac{1}{Z_i} = \sum_i Y_i \tag{5.37}$$

which constitutes the sum of the individual complex admittances.

The steady-state sinusoidal responses of networks involving multiple meshes or nodes can be found with the help of complex representation through the methods of mesh and node-pair analysis developed in chapter 3 in connection with direct-current networks. Also applicable to linear circuits with sinusoidal sources and impedances are the superposition and reciprocity theorems of chapter 3. The Thévenin and Norton theorems introduced in chapter 3 carry over to circuits with sinusoidal sources and impedances on just substituting the word impedance for resistance everywhere and interpreting e.m.f.s and constant-current sources as being sinusoidal as a function of time.

5.6 Series resonant circuit

The sinusoidal responses of circuits containing series and parallel combinations of resistors, capacitors and inductors is of profound theoretical interest and great practical importance. Consider first the series circuit shown in figure 5.7. The complex algebraic representation of Kirchhoff's voltage law for this circuit is

$$\mathscr{E} = (R + j\omega L - j/\omega C)\mathbf{I} \tag{5.38}$$

where $\mathscr{E}$ and $\mathbf{I}$ denote the e.m.f. and current phasors. Thus the complex impedance is

$$Z = R + j(\omega L - 1/\omega C) \tag{5.39}$$

showing that

$$|Z| = [R^2 + (\omega L - 1/\omega C)^2]^{\frac{1}{2}} \tag{5.40}$$

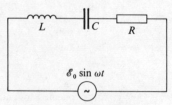

5.7 Series resonant circuit.

5.6 Series resonant circuit

and that the e.m.f. leads the current by a phase angle ϕ where
$$\tan \phi = (\omega L - 1/\omega C)/R \tag{5.41}$$
When $\omega L > 1/\omega C$, the e.m.f. leads the current in phase, but when $\omega L < 1/\omega C$, the e.m.f. lags behind the current in phase. Under the special condition $\omega L = 1/\omega C$, not only is $\phi = 0$ from equation (5.41) so that the e.m.f. and current are in phase, but $|Z| = R$ from equation (5.40) which is the minimum value of the impedance and purely resistive. This means that the amplitude of the current reaches a maximum value when $\omega L = 1/\omega C$ and *resonance* is said to occur. The frequency at which resonance occurs is naturally called the *resonant frequency* and is given by
$$f_r = \omega_r/2\pi = 1/2\pi(LC)^{\frac{1}{2}} \tag{5.42}$$
In terms of the resonant pulsatance ω_r, the impedance at any pulsatance ω is from equation (5.40)
$$|Z| = \left\{ R^2 + \left[\left(\frac{L}{C}\right)^{\frac{1}{2}} \left(\frac{\omega}{\omega_r} - \frac{\omega_r}{\omega}\right) \right]^2 \right\}^{\frac{1}{2}}$$

or
$$|Z| = R\left\{ 1 + \left[Q\left(\frac{\omega}{\omega_r} - \frac{\omega_r}{\omega}\right) \right]^2 \right\}^{\frac{1}{2}} \tag{5.43}$$

where the parameter
$$Q = \frac{1}{R}\left(\frac{L}{C}\right)^{\frac{1}{2}} \tag{5.44}$$

is known as the *quality factor*. This nomenclature is most appropriate because for a given shift of ω from ω_r, the larger Q the larger the change in $|Z|$ from its resonant value R and the *sharper* the resonance. This is illustrated in figure 5.8(a) where the ratio of the amplitude of the current to

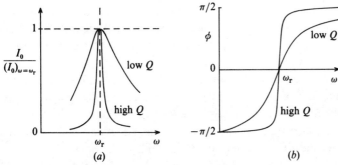

5.8 (a) Frequency response of the amplitude of the current in a series resonant circuit and (b) the corresponding behaviour of the phase lag of the current with respect to the e.m.f.

that at resonance

$$I_0/(I_0)_{\omega=\omega_r}=(\mathcal{E}_0/|Z|)/(\mathcal{E}_0/|Z|_{\omega=\omega_r})=R/|Z|$$

is plotted as a function of the pulsatance ω. With an electronic oscillator as the source of e.m.f., the pulsatance ω can be tuned through ω_r and the resonant response observed. Of course, any internal resistance of the source contributes to the resistance R and lowers the quality or Q-*factor* as it is often known.

Practical series L–C–R combinations can be constructed with Q-factors as high as a few hundred and, together with complementarily responding parallel resonant circuits to be discussed in the following section, are of great importance in electronics. The resonant response serves, for instance, to select the frequency of oscillation of many electronic oscillators or the frequency of detection or amplification. Usefully, the frequency selected can be altered by tuning the inductor or capacitor, the latter being much more common since variable capacitance is more easily realised. Frequency-selective detection and amplification through L–C–R resonant circuits are encountered in radio and television reception. Clearly crucial to satisfactory reception is the discrimination provided by the resonant response against stations with transmission frequencies differing from that of the station it is desired to receive.

It is interesting to observe that the Q-factor of a series resonant circuit can be alternatively expressed from equations (5.42) and (5.44) as

$$Q=\omega_r L/R = 1/\omega_r RC \tag{5.45}$$

Significantly, at resonance, the ratio of the amplitude of the potential difference across the inductance to that across the resistance is $\omega_r L(I_0)_{\omega=\omega_r}/R(I_0)_{\omega=\omega_r}=Q$ and the ratio of the amplitude of the potential difference across the capacitance to that across the resistance is $(1/\omega_r C)(I_0)_{\omega=\omega_r}/R(I_0)_{\omega=\omega_r}$ which is also equal to Q. But at resonance the net reactance $(\omega_r L - 1/\omega_r C)$ is zero and from equation (5.38), $\mathcal{E}_0 = R(I_0)_{\omega=\omega_r}$, that is, the amplitude of the potential difference across the resistance at resonance is equal to the amplitude of the applied e.m.f. Thus the Q-factor can further be described in terms of the factor by which the amplitude of the applied e.m.f. is magnified at resonance as potential difference across the inductance or capacitance. This aspect is vividly illustrated in the phasor diagram of a resonating series resonant circuit which is as depicted in figure 5.9(*a*). The phases of the potential differences across the inductance and capacitance are opposite, the former leading and the latter lagging the current by 90°. At resonance, the amplitudes $\omega_r L I_0$ and $(1/\omega_r C)I_0$ of these potential differences are equal so that they balance, making the phasor potential difference $\mathcal{E}$ across the L–C–R combination just that across the resistance

5.6 Series resonant circuit

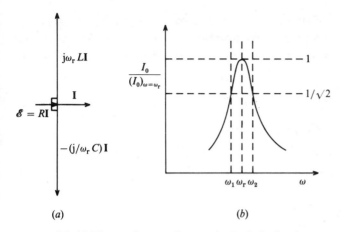

(a)   (b)

5.9 (a) Phasor diagram for a series L–C–R circuit at resonance and (b) resonant response of a series L–C–R circuit showing the half-power points at $\omega = \omega_1, \omega_2$.

which is RI. Clearly, the amplitudes of the inductive and capacitive potential differences can be much greater than the amplitude of the applied e.m.f. and the ratio is the Q-factor.

Equation (5.41) for the phase difference between the e.m.f. and current at any pulsatance ω can be written in terms of the Q-factor and resonant pulsatance ω_r as

$$\tan \phi = Q\left(\frac{\omega}{\omega_r} - \frac{\omega_r}{\omega}\right) \qquad (5.46)$$

Figure 5.8(b), showing the phase shift ϕ as a function of the pulsatance ω, is easily understood in terms of this equation. Note carefully that the figure shows the phase angle ϕ by which the *current lags the applied e.m.f.*

Yet another way of defining the Q-factor, and one which particularly emphasises its indication of the sharpness of resonance, is in terms of what are known as the *half-power* points. It will be proved in section 5.8 that the average power dissipated in a circuit of impedance Z by a source e.m.f. $\mathscr{E}_0 \sin \omega t$ is $\mathscr{E}_0^2 \cos \phi / 2|Z|$ where ϕ is the phase difference between the e.m.f. and current. Accepting this result, the average power dissipated at resonance is $\mathscr{E}_0^2/2R$ in agreement with equation (5.6). On the other hand, when

$$Q\left(\frac{\omega}{\omega_r} - \frac{\omega_r}{\omega}\right) = \pm 1 \qquad (5.47)$$

then $(\omega L - 1/\omega C) = \pm R$ and $|Z| = \sqrt{2}R$ from equation (5.40) or (5.43) while $\phi = \pm 45°$ from equation (5.41) or (5.46) so that the average power is $\mathscr{E}_0^2 \cos(\pm 45°)/2\sqrt{2}R = \mathscr{E}_0^2/4R$. This means that at the pulsatances ω_1 and

ω_2 that satisfy relation (5.47), the average power is half of that dissipated at resonance. Appropriately, pulsatances ω_1 and ω_2 are known as the *half-power points*. They are given from equation (5.47) by (taking $\omega_2 > \omega_1$)

$$\frac{\omega_2}{\omega_r} - \frac{\omega_r}{\omega_2} = \frac{1}{Q} = \frac{\omega_r}{\omega_1} - \frac{\omega_1}{\omega_r}$$

that is

$$(\omega_2^2 - \omega_r^2)\omega_1 = (\omega_r^2 - \omega_1^2)\omega_2$$

which reduces to

$$\omega_1 \omega_2 = \omega_r^2 \qquad (5.48)$$

Thus

$$Q\left(\frac{\omega_2}{\omega_r} - \frac{\omega_1 \omega_2}{\omega_r \omega_2}\right) = 1$$

or

$$Q = \frac{\omega_r}{\omega_2 - \omega_1} = \frac{f_r}{f_2 - f_1} \qquad (5.49)$$

This last equation reveals that Q can be defined as the ratio of the resonant frequency to the frequency interval between the half-power points. The locations of the half-power points on the current amplitude response of a series resonant circuit are shown in figure 5.9(b).

As a specific example of a series resonant circuit, consider one comprising inductance of 10 mH, capacitance of 10 nF and resistance of 10 Ω. The resonant pulsatance is $(LC)^{-\frac{1}{2}} = (10^{-2} \times 10^{-8})^{-\frac{1}{2}} = 10^5$ rad s^{-1} corresponding to a resonant frequency of approximately 15.9 kHz. The Q-factor is $(L/C)^{\frac{1}{2}}/R = (10^{-2}/10^{-8})^{\frac{1}{2}}/10 = 100$ and the separation of the half-power frequencies is $f_r/Q = 159$ Hz. Fed from a source of e.m.f. of r.m.s. magnitude 1 V and negligible internal impedance compared with 10 Ω, the r.m.s. current at resonance is 100 mA.

5.7 Parallel resonant circuits

The entirely parallel resonant circuit shown in figure 5.10(a) behaves in a complementary fashion to the entirely series resonant circuit treated in the previous section. Application of Kirchhoff's current law to the parallel circuit gives

$$\mathbf{I} = \left(\frac{1}{R} + \frac{1}{j\omega L} - \frac{\omega C}{j}\right)\mathbf{V} \qquad (5.50)$$

where **I** and **V** respectively denote phasor representations of the load-independent current source and potential difference across the parallel components. Thus the complex admittance is

$$Y = 1/R + j(\omega C - 1/\omega L) \qquad (5.51)$$

5.7 Parallel resonant circuits

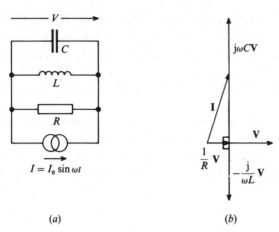

5.10 (a) Completely parallel resonant circuit and (b) its phasor diagram.

which is similar to the expression for the complex impedance of the series resonant circuit, showing the same kind of frequency dependence. Indeed, equation (5.38) for the series resonant case transforms into equation (5.50) for the parallel resonant case upon substituting $1/R$ for R and interchanging the roles of L and C and of potential difference across the L–C–R combination and current. It follows that resonance occurs in the entirely parallel circuit at the same frequency

$$f_r = \omega_r/2\pi = 1/2\pi(LC)^{\frac{1}{2}} \tag{5.52}$$

as in the series resonant circuit. However, it is now the admittance

$$|Y| = [(1/R)^2 + (\omega C - 1/\omega L)^2]^{\frac{1}{2}} \tag{5.53}$$

that exhibits a resonant minimum and the potential difference across the parallel combination a corresponding maximum.

From the transformation relationship between the parallel and series resonant circuits, the Q-factor of the parallel form is

$$Q = \omega_r C/(1/R) = R/\omega_r L = R(C/L)^{\frac{1}{2}} \tag{5.54}$$

Evidently, to maintain the Q-factor of a parallel L–C–R combination high, the circuit must not be connected in parallel with a load or source that reduces the overall parallel resistance significantly. Figure 5.10(b) presents the phasor diagram that prevails for a completely parallel resonant circuit at a general pulsatance ω. This diagram shows, in conjunction with equation (5.54), that, at resonance, the amplitudes of the currents flowing in the inductive and capacitive branches are both Q times the amplitude of the current flowing in the resistive branch and that the amplitude of the current in the resistive branch is equal to the amplitude of the total current delivered to the combination. Apparently, at resonance, the current delivered to the combination is magnified Q times in the reactive branches. The phase difference of π between the currents in the capacitive and inductive branches actually means that a magnified version of the source current is oscillating backwards and forwards round the reactive mesh at resonance.

Figures 5.11(a) and (b) respectively present the frequency responses of the amplitude and phase of the potential difference across a completely parallel resonant circuit fed from a load-independent, sinusoidal, current source. From equation (5.50) or (5.51) the current delivered by the source leads the potential difference across the combination by a phase angle ϕ where

$$\tan \phi = (\omega C - 1/\omega L)R \tag{5.55}$$

In the vicinity of resonance there is a progressive changeover of the phase from a lag to a lead of $90°$ as shown.

Practical, parallel, resonant circuits are normally constructed by connecting a capacitor of capacitance C and negligible loss in parallel with an inductor of inductance L' and series resistance R' as indicated in figure 5.12(a). Their response to a sinusoidal current source can be deduced by first finding a completely parallel L–C–R circuit that behaves identically. To be equivalent, the complex impedance of the completely parallel circuit (shown again in figure 5.12(b) for easy comparison) must be identical to that of the practical circuit. This requires that

$$R' + j\omega L' = (1/R + 1/j\omega L)^{-1}$$

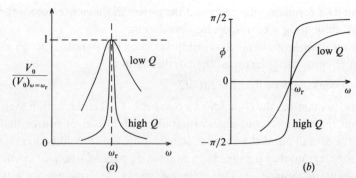

5.11 (a) Frequency response of the amplitude of the potential difference across a parallel resonant circuit fed from a sinusoidal load-independent current source and (b) the corresponding behaviour of the phase lag of the potential difference with respect to the current source.

5.7 Parallel resonant circuits

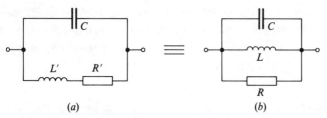

5.12 (a) Practical form of parallel resonant circuit and (b) its completely parallel equivalent, where L and R are given in terms of L' and R' by equations (5.56) and (5.57) of the text.

or
$$\frac{R'-j\omega L'}{(R')^2+(\omega L')^2}=\frac{\omega L-jR}{\omega LR}$$

from which
$$R=[(R')^2+(\omega L')^2]/R' \qquad (5.56)$$
$$\omega L=[(R')^2+(\omega L')^2]/\omega L' \qquad (5.57)$$

Notice particularly that R and L are frequency dependent; the behaviours of the circuits of figures 5.12(a) and (b) are of course quite different if the resistances, capacitances and inductances in each are frequency independent.

Since resonance occurs in the completely parallel case when $\omega L=1/\omega C$, it follows from equation (5.57) that a corresponding resonance occurs in the practical L–C–R circuit of figure 5.12(a) when
$$\omega C=\omega L'/[(R')^2+(\omega L')^2]$$
or
$$(\omega L')^2=L'/C-(R')^2$$

Thus the practical circuit resonates at pulsatance
$$\omega'_r=\frac{1}{(L'C)^{\frac{1}{2}}}\left[1-\frac{(R')^2C}{L'}\right]^{\frac{1}{2}} \qquad (5.58)$$

The Q-factor of the resonance will be given by $R/\omega_r L$ where ω_r is the resonant pulsatance of the completely parallel equivalent. Making use of equations (5.56) and (5.57), the Q-factor of the practical circuit turns out to be
$$Q=R/\omega_r L=\omega_r L'/R' \qquad (5.59)$$

While a large Q-factor for a completely parallel circuit requires that the parallel resistance be maintained large, to obtain a practical circuit of the form shown in figure 5.12(a) with a large Q-factor demands that the series resistance R' be kept small. This resembles the requirement for a series resonant circuit to exhibit a large Q-factor. In fact, $R' \ll \omega_r L$ must be satisfied to achieve a high Q-factor for the practical circuit of figure 5.12(a).

At the resonant pulsatance ω_r of the completely parallel circuit, equation (5.57) can be expressed in terms of the Q-factor as

$$\omega_r L = (1 + 1/Q^2)\omega_r L'$$

This shows that when the Q-factor is high enough for $Q^2 \gg 1$ to hold then $\omega_r L$ is very close to $\omega_r L'$ in value and

$$\omega_r = 1/(LC)^{\frac{1}{2}} \approx 1/(L'C)^{\frac{1}{2}} \tag{5.60}$$

so that, from equation (5.59),

$$Q \approx \frac{1}{R'}\left(\frac{L'}{C}\right)^{\frac{1}{2}} \tag{5.61}$$

Equation (5.61) is identical in form to equation (5.44) for the Q-factor of a series resonant circuit, a convenient result for recall purposes since it is practical circuits having high Q-factors that are of greatest interest and widest application. One consequence of equation (5.61) is that, when the Q-factor is high, equation (5.58) for the resonant pulsatance of the practical form of parallel resonant circuit can be expressed alternatively as

$$\omega'_r \approx \frac{1}{(L'C)^{\frac{1}{2}}}\left(1 - \frac{1}{Q^2}\right)^{\frac{1}{2}} \approx \frac{1}{(L'C)^{\frac{1}{2}}} \tag{5.62}$$

Actually, the resonance condition considered so far in the practical form of parallel resonant circuit is that of *phase resonance*. The impedance at phase resonance is just

$$|Z| = R = R' + (\omega_r L')^2/R' = (1 + Q^2)R' \tag{5.63}$$

and is much greater than R' when $Q^2 \gg 1$. While the condition for *amplitude resonance* (maximum amplitude) is identical to that for phase resonance in completely series or completely parallel L–C–R circuits, this is not the case for the practical circuit of figure 5.12(a). Actually, however, when the Q-factor is high, the amplitude and phase resonances almost coincide even in the practical form of circuit. To find the frequency or pulsatance at which the amplitude resonates, the condition $d|Z|/d\omega = 0$ for the turning points of the impedance must be imposed. For the practical circuit of figure 5.12(a), the complex impedance is

$$Z = \frac{(R' + j\omega L')(-j/\omega C)}{(R' + j\omega L' - j/\omega C)} = \frac{L'/C - jR'/\omega C}{R' + j(\omega L' - 1/\omega C)}$$

and so

$$|Z| = \left[\frac{(\omega L')^2 + (R')^2}{(\omega R'C)^2 + (\omega^2 L'C - 1)^2}\right]^{\frac{1}{2}} \tag{5.64}$$

Writing $\omega^2 L'C$ as a new variable x and $(L')^{\frac{1}{2}}/R'C^{\frac{1}{2}}$ as Q' since this latter

5.8 Power dissipation with sinusoidal current

quantity has been shown to be the Q-factor when it is large

$$|Z| = \left[\frac{(Q')^2 x + 1}{(1/Q')^2 x + (x-1)^2}\right]^{\frac{1}{2}} R' \tag{5.65}$$

The maximum occurs in the impedance $|Z|$ when $(|Z|/R')^2$ is a maximum and differentiation with respect to x shows that this happens when

$$[(1/Q')^2 x + (x-1)^2](Q')^2 - [(Q')^2 x + 1][(1/Q')^2 + 2(x-1)] = 0$$

which reduces to

$$x^2 + \frac{2}{(Q')^2} x + \left[\frac{1}{(Q')^4} - \frac{2}{(Q')^2} - 1\right] = 0$$

The solutions of this quadratic equation are

$$x = -\frac{1}{(Q')^2} \pm \left[1 + \frac{2}{(Q')^2}\right]^{\frac{1}{2}}$$

and since only positive frequencies are physically meaningful, maximum impedance occurs when

$$\omega^2 = \frac{1}{L'C}\left[\left(1 + \frac{2}{(Q')^2}\right)^{\frac{1}{2}} - \frac{1}{(Q')^2}\right] \tag{5.66}$$

Comparison of this equation with equation (5.62) for phase resonance confirms that the phase and amplitude resonances occur at almost the same frequency in the practical circuit of figure 5.12(a) when the Q-factor is high, the two frequencies being very close to that represented by the much simpler expression $1/2\pi(L'C)^{\frac{1}{2}}$.

Notice that from equations (5.57) and (5.59)

$$1 + Q^2 = [(R')^2 + (\omega_r L)^2]/(R')^2 = \omega_r^2 L L'/(R')^2 = L'/(R')^2 C$$

and so

$$(1 + Q^2) = (Q')^2 \tag{5.67}$$

Also from equation (5.58), at phase resonance

$$x = 1 - (1/Q')^2$$

and therefore from equation (5.65) the impedance at phase resonance is

$$|Z| = (Q')^2 R' = (1 + Q^2) R'$$

in agreement with equation (5.63).

5.8 Power dissipation associated with sinusoidal current

The concept of electrical power was introduced in section 2.3 in connection with the dissipation of energy in resistors carrying direct current. It is a straightforward matter to extend this concept to circuits in which the current varies with time t. If at some moment a portion of a circuit is carrying current I and dropping potential difference V, the instantaneous power in it is simply

$$P = VI$$

Now it has been established in this chapter that when sinusoidal current of the form $I_0 \sin \omega t$ flows through any linear electrical network connected between two points, the steady-state potential difference between the points is also sinusoidal but of the form $V_0 \sin(\omega t + \phi)$. It follows that the instantaneous power in a linear network carrying sinusoidal current can be expressed in terms of such notation as

$$P = V_0 I_0 \sin(\omega t + \phi) \sin \omega t \tag{5.68}$$

Integrating over a period and dividing by the periodic time, the corresponding *average power* is

$$P_{av} = \frac{V_0 I_0}{T} \int_0^T \sin(\omega t + \phi) \sin \omega t \, dt$$

$$= \frac{V_0 I_0}{\omega T} \int_0^{2\pi} (\cos \phi \sin^2 \omega t + \sin \phi \sin \omega t \cos \omega t) \, d(\omega t)$$

$$= \frac{V_0 I_0}{2\pi} \int_0^{2\pi} \left[\cos \phi \left(\frac{1 - \cos 2\omega t}{2} \right) + \sin \phi \left(\frac{\sin 2\omega t}{2} \right) \right] d(\omega t)$$

$$= \frac{V_0 I_0}{4\pi} \left[\cos \phi \left(\omega t - \frac{\sin 2\omega t}{2} \right) - \sin \phi \left(\frac{\cos 2\omega t}{2} \right) \right]_0^{2\pi}$$

Since $\sin 2\omega t$ and $\cos 2\omega t$ each have the same value at the limits of integration, this reduces to the simple expression

$$P_{av} = \tfrac{1}{2} V_0 I_0 \cos \phi \tag{5.69}$$

which is equivalent in terms of r.m.s. magnitudes, through use of equation (5.5), to

$$P_{av} = V_{rms} I_{rms} \cos \phi \tag{5.70}$$

The quantities $V_{rms} I_{rms}$ and $\cos \phi$ in equation (5.70) are respectively known as the *apparent power* and *power factor*.

Only when a network is purely resistive does $\phi = 0$, $\cos \phi = 1$ and the true average power equal the apparent power, that is,

$$(P_{av})_{\phi=0} = V_{rms} I_{rms} = R I_{rms}^2 \tag{5.71}$$

in accordance with equation (5.7). Figure 5.13(a) shows the current I, potential difference V and power P plotted as a function of time for a purely resistive circuit subject to sinusoidal stimulus. The power fluctuates at twice the frequency of the current and for this reason it is not permissible to calculate it using phasor or complex representation. Note in passing that

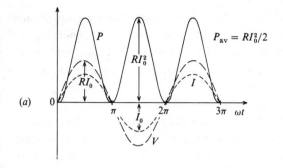

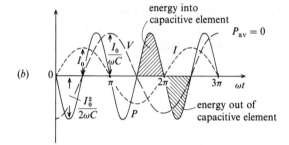

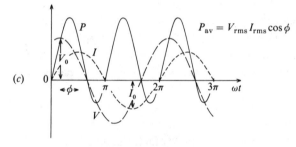

5.13 Plots of the dependences on time of current I, potential difference V and power dissipation P for a linear circuit subject to sinusoidal stimulus; (a) a purely resistive circuit, (b) a purely capacitive circuit and (c) a circuit in which the potential difference leads the current by a phase angle ϕ.

the power dissipation is at all times positive leading to the average represented by equation (5.71).

In the case of a purely inductive or purely capacitive network, $\phi = \pm 90°$ and P_{av} is exactly zero. The significance of this can be properly appreciated on studying figure 5.13(b) showing the current I, potential difference V and power P plotted as a function of time for a purely capacitive circuit subject to sinusoidal stimulus. Again the power fluctuates at twice the frequency of the current but this time it is symmetrically positive and negative during alternate half-periods of the power. These periods of positive and negative power dissipation correspond to alternate periods of charging and discharging, that is, increasing and decreasing of the magnitude of the potential difference, during which energy is alternately stored and released by the capacitive element as discussed in section 4.1. Similar alternate storage and release of energy occurs on account of inductance.

When the potential difference leads the current by a phase angle ϕ between $0°$ and $90°$, the time dependences of the current I, potential difference V and power P are as shown in figure 5.13(c). Once more the power dissipation goes alternately positive and negative but on balance the power is positive with the average value given by equation (5.69) or (5.70). If the circuit under consideration is considered to comprise resistance R in series with reactance X then in accordance with equation (5.29)

$$\tan \phi = X/R$$

so that

$$\cos \phi = R/(R^2 + X^2)^{\frac{1}{2}} \tag{5.72}$$

and

$$P_{av} = V_{rms} I_{rms} R/(R^2 + X^2)^{\frac{1}{2}} = R I_{rms}^2 \tag{5.73}$$

This alternative expression for the average power emphasises the fact that power is only dissipated on average in resistance and not in reactance. Indeed, equation (5.70) for the average power can be interpreted as stating that only that component of r.m.s. current that is in phase with the r.m.s. potential difference gives rise to power dissipation over a period in the steady state.

Practical inductors dissipate electrical energy by a variety of mechanisms as discussed in section 4.2. Representing the overall loss by some resistance R in series with the inductance L, the power factor of an inductor is, from equation (5.72), $R/(R^2 + \omega^2 L^2)^{\frac{1}{2}}$. Often R is small compared with ωL, in which case the power factor approximates to $R/\omega L$. Variable inductors having low power factors can be used to control alternating current very efficiently with substantial power saving compared with resistive control by rheostats. An example of this type of application is dimming of lights powered from the 50 Hz mains supply. Often the highly appropriate term *choke* is used to refer to an inductor that restricts alternating current.

The obvious way to represent a practical capacitor when direct potential differences are applied is as its capacitance C_p in parallel with some large resistance R_p to account for the small direct conductance of the dielectric. However, the series resistance of the leads is finite and under alternating-current conditions there is considerable energy loss associated with the polarisation of the dielectric not being able to follow the electric field

5.8 Power dissipation with sinusoidal current

instantaneously so that a delay is introduced. In these circumstances the capacitor is often represented by capacitance C_s in series with some resistance R_s, making the power factor $R_s/(R_s^2 + 1/\omega^2 C_s^2)^{\frac{1}{2}}$ according to equation (5.72). Normally energy loss in capacitors is small enough to allow the approximation $R_s \ll 1/\omega C_s$ which means that $C_s \approx C_p$ and that the power factor is close to $\omega C_s R_s$ or $\omega C_p R_s$. It is noteworthy that extremely small power factors can be achieved and that in certain cases the power factor is constant over a considerable range of frequency. Typical power factors for mica and polystyrene capacitors are around 0.0002.

The fundamental unit of power has already been shown to be the watt, usually abbreviated to W. In the case of circuits carrying alternating current it is important, from both the point of view of the supply and equipment operated, to know the current rating besides that of real power. If the power factor is small, there is a danger of vastly underestimating the r.m.s. current from the real power rating. For this reason the apparent power $V_{rms} I_{rms}$ is often stated, but in V A units to distinguish it from the real power rating in W. Sometimes the amplitude $V_{rms} I_{rms} \sin \phi$ of the reactive power flowing in and out of the reactive component of a circuit is given, this time in reactive V A units or *vars*.

It was proved in section 2.5 that the matched condition in which maximum power is transferred from a direct source to a load resistance occurs when the load resistance equals the source resistance. Consider now the matching of a sinusoidal source of e.m.f. $\mathscr{E}_0 \sin \omega t$ and internal complex impedance $Z_S = R_S + jX_S$ to a complex load impedance $Z_L = R_L + jX_L$. Using equation (5.73), the power in the load is

$$P_{av} = R_L I_{rms}^2 = \frac{R_L \mathscr{E}_{rms}^2}{(R_S + R_L)^2 + (X_S + X_L)^2} \tag{5.74}$$

In the *unusual* situation where the complex load impedance is fixed but the source resistance and reactance are independently adjustable, consideration of equation (5.74) immediately shows that matching is achieved if $X_S = -X_L$ and $R_S = 0$. The reactance condition here corresponds of course to series resonance. For the *more usual* situation of the source fixed and the resistance and reactance of the load independently adjustable, matching clearly requires $X_L = -X_S$ again under which condition the power becomes $R_L \mathscr{E}_{rms}^2/(R_S + R_L)^2$ and differentiation with respect to R_L reveals that matching necessitates $R_L = R_S$ also. Adjustment of some reactance to reduce the net reactance to zero is described as tuning out the reactance. Sometimes *only the reactance* of the load is adjustable for maximum power. When *only the resistance* of the load can be varied, maximum power corresponds to setting dP_{av}/dR_L equal to zero, that is

$$(R_S+R_L)^2+(X_S+X_L)^2-2R_L(R_S+R_L)=0$$

or

$$R_L=[R_S^2+(X_S+X_L)^2]^{\frac{1}{2}} \tag{5.75}$$

Where transformers are used to match a load to a source, as discussed in section 6.3, it is the load impedance $|Z_L|=(R_L^2+X_L^2)^{\frac{1}{2}}$ that is adjustable through choice of the turns ratio. Rewriting equation (5.74) in terms of the load impedance $|Z_L|$ and phase angle ϕ by putting $R_L=|Z_L|\cos\phi$ and $X_L=|Z_L|\sin\phi$ so that $\cos\phi$ is the power factor of the load, it becomes

$$P_{av}=\frac{|Z_L|\cos\phi\,\mathscr{E}_{rms}^2}{(R_S+|Z_L|\cos\phi)^2+(X_S+|Z_L|\sin\phi)^2} \tag{5.76}$$

The maximum power transfer that can be achieved by varying $|Z_L|$ occurs when $dP_{av}/d|Z_L|=0$, that is, when

$$(R_S+|Z_L|\cos\phi)^2+(X_S+|Z_L|\sin\phi)^2$$
$$=2|Z_L|[(R_S+|Z_L|\cos\phi)\cos\phi+(X_S+|Z_L|\sin\phi)\sin\phi]$$

or

$$|Z_L|=(R_S^2+X_S^2)^{\frac{1}{2}}=|Z_S| \tag{5.77}$$

5.9 Sinusoidal sources in nonlinear circuits

The first eight sections of this chapter have been devoted to considering the steady-state responses of linear networks to sinusoidal sources. In this last section a brief introduction will be given to the host of interesting phenomena that arise when sinusoidal sources act in nonlinear circuits. For simplicity, attention will be restricted to effects that occur in purely resistive circuits as a result of the particular nonlinearity represented by

$$I=bV+cV^2 \tag{5.78}$$

where I is the current, V the potential difference and b and c are constants that are independent of V. This form of dependence, besides being approximately applicable in certain situations whatever the magnitude of current and potential difference, is highly relevant to small fluctuations in the current and potential difference about static bias levels, a common occurrence in the field of electronics. The fluctuations, often sinusoidal, are then the signals of interest and the relevance comes about because, no matter what form the static characteristic $I=f(V)$ takes, for small-enough changes denoted according to convention by lower-case letters i and v, to a first order of approximation

$$i=bv+cv^2 \tag{5.79}$$

The linear case would correspond of course to the constant c being zero.

5.9 Sinusoidal sources in nonlinear circuits

Consider, to begin with, the effect of applying a single small sinusoidal e.m.f. $\mathscr{E}_0 \sin \omega t$ between the terminals of a biased circuit that exhibits the nonlinearity inherent in equation (5.79). Similar behaviour occurs, of course, for an unbiased circuit obeying equation (5.78) irrespective of the amplitude of the e.m.f. The current is

$$b\mathscr{E}_0 \sin \omega t + c(\mathscr{E}_0 \sin \omega t)^2 = b\mathscr{E}_0 \sin \omega t + \tfrac{1}{2}c\mathscr{E}_0^2(1 - \cos 2\omega t)$$

and contains a component of twice the frequency of the source. Clearly, as in the case of electrical power, the phasor and complex algebraic methods of linear circuit analysis are inappropriate.

Of greater interest is the response of a circuit obeying equation (5.79) when two sinusoidal e.m.f.s of differing frequency are applied. If these e.m.f.s are represented by $\mathscr{E}_{10} \sin \omega_1 t$ and $\mathscr{E}_{20} \sin \omega_2 t$, the current is

$$b(\mathscr{E}_{10} \sin \omega_1 t + \mathscr{E}_{20} \sin \omega_2 t) + c(\mathscr{E}_{10} \sin \omega_1 t + \mathscr{E}_{20} \sin \omega_2 t)^2$$

which can be rearranged as

$$\tfrac{1}{2}c(\mathscr{E}_{10}^2 + \mathscr{E}_{20}^2) + b(\mathscr{E}_{10} \sin \omega_1 t + \mathscr{E}_{20} \sin \omega_2 t)$$
$$- \tfrac{1}{2}c(\mathscr{E}_{10}^2 \cos 2\omega_1 t + \mathscr{E}_{20}^2 \cos 2\omega_2 t)$$
$$+ c\mathscr{E}_{10}\mathscr{E}_{20}(\cos(\omega_1 - \omega_2)t - \cos(\omega_1 + \omega_2)t)$$

This formulation reveals that the nonlinearity introduces components into the current at several frequencies not present in the applied e.m.f.s. In particular, frequencies that are the sum and difference of the input e.m.f. frequencies are produced so that *mixing* is said to occur. Mixing is normally implemented in electronic circuitry by using a diode to introduce the required nonlinearity while an appropriately tuned resonant circuit selects the component at the sum or difference frequency.

One situation where mixing gives a marked advantage is in the reception of electromagnetic waves. This includes radio and television reception. To *discriminate directly* between two different stations transmitting at, say, 10 MHz and 10.1 MHz requires a tunable resonant circuit with a Q-factor of much greater than 10 MHz/0.1 MHz = 100 within the receiver. However, on mixing the incoming signals with another from a tunable oscillator in the receiver so that one station creates a difference-frequency component at a selected relatively low *intermediate frequency* of, say, 500 kHz, the other produces a difference-frequency component shifted by 100 kHz from the intermediate frequency. The tuned circuits in an intermediate frequency amplifier need now only exhibit a Q-factor that is, say, 10 MHz/500 kHz or 20 times smaller than before to obtain the same discrimination between stations. There is also the added advantage of amplification at a much lower fixed intermediate frequency rather than over a vast range of high transmission frequencies. To receive any particular transmitting station,

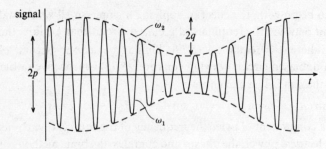

5.14 Amplitude modulation.

the frequency of the *local* oscillator is adjusted to change the frequency of the incoming signal of interest to the intermediate frequency.

With the aid of a nonlinear circuit, it is also possible to produce *amplitude modulation* by means of which information is transmitted in one mode of electromagnetic communication. Suppose two sinusoidal e.m.f.s are applied to a nonlinear circuit as before, but $\omega_2 \ll \omega_1$ and a tuned circuit selects an output corresponding to those components of the current that are close in pulsatance to ω_1. The components of current close in pulsatance to ω_1 are

$$b\mathscr{E}_{10} \sin \omega_1 t + c\mathscr{E}_{10}\mathscr{E}_{20}[\cos(\omega_1 - \omega_2)t - \cos(\omega_1 + \omega_2)t]$$
$$= (p + q \sin \omega_2 t) \sin \omega_1 t$$

where p and q are constants. Thus the output is as depicted in figure 5.14; a carrier wave of relatively high pulsatance ω_1 carries amplitude modulation $q \sin \omega_2 t$ at the relatively low pulsatance ω_2. In audio communication by amplitude modulation, speech or music, comprising a spectrum of audio-frequency pressure sinewaves, is first transformed into matching audio-frequency electrical signals in a microphone. These are then transmitted as audio-frequency amplitude modulation of a radio-frequency carrier wave.

Remember finally that the principle of superposition does not apply in nonlinear circuits; as was noted in section 3.6, for this principle to hold, the circuit must behave linearly.

6

Transformers in networks

6.1 Mutual inductance

Consider any two electrical meshes which may or may not be electrically connected together. For descriptive purposes it is convenient standard practice to refer to one mesh as the primary and the other as the secondary. A current in the primary produces magnetic flux, some fraction of which, depending on the intervening media and geometry, links with the secondary. For most intervening media, the flux Φ_s linked with secondary is proportional to the primary current I_p and it is customary to write

$$\Phi_s = M_{ps} I_p \tag{6.1}$$

where M_{ps} is called the *mutual inductance* between the primary and secondary, in analogy with the definition of self inductance. In a similar way, current I_s in the secondary causes flux Φ_p to be linked with the primary and this is expressed by writing

$$\Phi_p = M_{sp} I_s \tag{6.2}$$

where M_{sp} is the mutual inductance between the secondary and primary. In fact it turns out, as will be proved in a moment, that $M_{sp} = M_{ps}$ and so there is only one mutual inductance between circuits, normally written M, and defined by

$$\Phi_s = M I_p; \quad \Phi_p = M I_s \tag{6.3}$$

The property of mutual inductance becomes important from a circuit point of view when there are changing currents, for it then gives rise to induced e.m.f.s. Combining equations (4.11) and (6.3) reveals that mutual inductance M between a primary and secondary gives rise to induced e.m.f.s in the secondary and primary of

$$\mathscr{E}_s = -M \, dI_p/dt; \quad \mathscr{E}_p = -M \, dI_s/dt \tag{6.4}$$

respectively, assuming that the mutual inductance is independent of the current. The negative signs here once more indicate that the e.m.f.s are back e.m.f.s acting so as to oppose the cause producing them. As in the case of self inductance, alternative definitions of M arise through applying equations (6.3) and (6.4) irrespective of whether the flux is proportional to current. When the current steadily varies sinusoidally with time at pulsatance ω, equations (6.4) can be alternatively expressed in terms of phasors $\mathscr{E}_p$, $\mathscr{E}_s$, $\mathbf{I}_p$ and $\mathbf{I}_s$ as

$$\mathscr{E}_s = -j\omega M \mathbf{I}_p; \quad \mathscr{E}_p = -j\omega M \mathbf{I}_s \tag{6.5}$$

showing that mutual inductance gives rise to a coupling reactive impedance ωM.

To prove that $M_{sp} = M_{ps}$, think about the magnetic energy that is stored when currents are established in a pair of magnetically coupled meshes. Let the primary and secondary exhibit self inductances L_p and L_s respectively. Suppose that initially an e.m.f. connected in the primary establishes steady current I_{pm} in it while the secondary is maintained open circuit so that no energy is transferred to it. At the conclusion of this process, the only energy stored is associated with the self inductance of the primary and in accordance with equation (4.14) amounts to $L_p I_{pm}^2/2$. Next, suppose that some source subsequently connected in the secondary establishes steady current I_{sm} in it so that there is energy stored in association with the self inductance of the secondary amounting to $L_s I_{sm}^2/2$. To keep the primary current constant at I_{pm} during this second stage, extra e.m.f. $M_{sp} \, dI_s/dt$ must act in the primary to balance the back e.m.f. $-M_{sp} \, dI_s/dt$. The extra e.m.f. delivers current I_{pm} during growth of the secondary current so that it provides extra energy

$$\int_{I_s=0}^{I_s=I_{sm}} I_{pm} M_{sp} (dI_s/dt) \, dt = M_{sp} I_{pm} I_{sm}$$

Evidently, the stored energy associated with steady primary and secondary currents I_{pm} and I_{sm} is

$$W = \tfrac{1}{2} L_p I_{pm}^2 + \tfrac{1}{2} L_s I_{sm}^2 + M_{sp} I_{pm} I_{sm}$$

Now the same circuit condition can be reached by first causing I_s to grow to I_{sm} with the primary open circuit and then making I_p grow to I_{pm} keeping I_s constant at I_{sm} with extra secondary e.m.f. $M_{ps} \, dI_p/dt$. Following this procedure, the stored energy is seen to be given by

$$W = \tfrac{1}{2} L_s I_{sm}^2 + \tfrac{1}{2} L_p I_{pm}^2 + M_{ps} I_{sm} I_{pm}$$

and comparison with the previous expression establishes that $M_{sp} = M_{ps}$ as anticipated.

Clearly, from equations (6.3) or (6.4), the fundamental SI unit of mutual inductance is the same as that of self inductance, already shown to be the

6.1 Mutual inductance

henry or H in section 4.2. The mutual inductance between two meshes can be increased by forming part of each mesh into a multiple-turn coil and closely coupling the two coils so obtained. Close coupling can be realised simply by bringing the coils near each other or interwinding one with the other. Winding the two coils round a common core of material of high magnetic permeability not only achieves close coupling but also increases the total magnetic flux. Obviously, magnitudes of mutual inductance can be obtained similar to those of self inductance.

A selection of symbols used to represent mutual inductance in circuits is shown in figure 6.1. Numerical or algebraic indication of the magnitude of mutual inductance is given between those parts of the symbol that also represent the inevitable associated self inductances, values of which are indicated to the side. The first form of symbol is preferred to the older but quite similar second symbol. Presence of a core is indicated by one or more lines in the middle as in the third figure, while fixed and variable tappings are denoted as for inductors (see figure 4.6). Polarities of e.m.f.s induced in networks by mutual inductances may be reversed by reversing either the direction of the input current, the sense of one winding or the output terminal connections. In a given situation, the polarity of the mutually inductive effect is indicated by a pair of dots as illustrated in figure 6.1(a). The convention is that if current flows *into the dot end* of one coil and is *increasing positively*, the e.m.f. induced in the other coil is *positive at its dot end* with respect to the undotted end.

Consider now a primary and secondary circuit between which mutual inductance exists on account of a primary and secondary coil wound such that all the magnetic flux Φ that passes through the primary also passes through the secondary and vice versa. As already intimated, this can virtually be achieved in practice by closely interwinding the two coils or by linking them with a magnetic core of high-permeability material. If the total numbers of turns on the primary and secondary coils are N_p and N_s, respectively, the ratio of flux linked with the primary to that linked with the

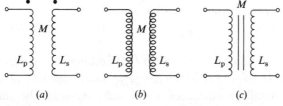

6.1 Mutual inductance symbols; (a) modern preferred, (b) older version and (c) indicating presence of a core of magnetic material.

secondary when current I_p flows in the primary is
$$\Phi_p/\Phi_s = L_p I_p/MI_p = N_p\Phi/N_s\Phi$$
so that
$$L_p/M = N_p/N_s \tag{6.6}$$
Similarly, from the ratio of flux linkages with the secondary and primary for current I_s in the secondary, it follows that
$$L_s/M = N_s/N_p \tag{6.7}$$
Combining equations (6.6) and (6.7) gives the valuable relation
$$M^2 = L_p L_s \tag{6.8}$$
When the flux linkage is less than perfect, equations (6.6) and (6.7) become
$$L_p/M > N_p/N_s; \quad L_s/M > N_s/N_p$$
from which it may be deduced that
$$M^2 < L_p L_s \tag{6.9}$$
It is customary to write that, in general,
$$M = k(L_p L_s)^{\frac{1}{2}} \tag{6.10}$$
where k, known as the *coefficient of coupling*, obeys
$$0 \leqslant k \leqslant 1 \tag{6.11}$$
Notice in passing that, because self inductance is proportional to the square of the number of turns, when there is perfect coupling,
$$L_s/L_p = (N_s/N_p)^2 \tag{6.12}$$
This useful relation is, of course, just that obtained by dividing equation (6.7) by equation (6.6).

To illustrate the effect that mutual inductance can have in a circuit, the impedance between the terminals of the circuit shown in figure 6.2 will be found, first assuming $M = 2$ H and then that M is zero. When $M = 2$ H, in terms of phasor notation Kirchhoff's voltage law gives
$$j\omega 4 \mathbf{I}_p + j\omega 2 \mathbf{I}_s = j\omega 3 \mathbf{I}_s + j\omega 2 \mathbf{I}_p$$
and so $\mathbf{I}_s = 2\mathbf{I}_p$. Since Kirchhoff's current law gives $\mathbf{I} = \mathbf{I}_p + \mathbf{I}_s$ this means that $\mathbf{I}_p = \mathbf{I}/3$. Hence, from Kirchhoff's voltage law,
$$\mathbf{V} = j\omega \tfrac{1}{3}\mathbf{I} + j\omega 4 \mathbf{I}_p + j\omega 2 \mathbf{I}_s = j\omega 3 \mathbf{I}$$
revealing that the impedance between the terminals corresponds to an inductance of 3 H. By comparison, when $M = 0$ there is an inductance of $\tfrac{1}{3}$ H in series with parallel inductances of 4 H and 3 H. Again the impedance between the terminals corresponds to an inductance but, this time, from equations (4.15) and (4.16) the inductance is $[\tfrac{1}{3} + (4 \times 3)/(4+3)]$ H $= 2.04$ H.

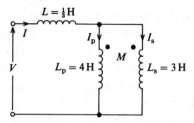

6.2 Circuit with both self and mutual inductance.

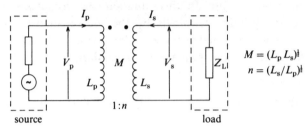

6.3 Perfectly coupled, lossless transformer connected to a source and load.

6.2 Transformers

When the magnetic flux Φ through two perfectly magnetically coupled coils is changing at a rate $d\Phi/dt$, the e.m.f.s $\mathscr{E}_p$ and $\mathscr{E}_s$ respectively induced in the primary and secondary are given by

$$\mathscr{E}_p = -N_p \, d\Phi/dt; \quad \mathscr{E}_s = -N_s \, d\Phi/dt$$

where N_p and N_s are the turns on the primary and secondary. It follows that these e.m.f.s are always in the simple ratio of the turns, that is,

$$\mathscr{E}_s/\mathscr{E}_p = N_s/N_p = (L_s/L_p)^{\frac{1}{2}} = n \tag{6.13}$$

where n is the turns ratio between the secondary and primary. Now suppose that such varying magnetic flux in a pair of coils is associated with some time-dependent source connected across the primary and some complex load impedance Z_L connected across the secondary. If it were possible to entirely represent the arrangement of coils by just self inductances L_p and L_s and a mutual inductance M as illustrated in figure 6.3, then the instantaneous potential differences V_s and V_p across the secondary and primary would *exactly* equal the corresponding back e.m.f.s and consequently be in the same ratio

$$V_s/V_p = \mathscr{E}_s/\mathscr{E}_p = n \tag{6.14}$$

In practice, pairs of coils can be constructed such that they are well represented by the circuit arrangement shown in figure 6.3, and equation (6.14) applies in such cases to a good approximation. From the practical point of view, the important conclusion to be drawn is that two closely coupled coils can be used to *transform* the magnitude of a time-dependent potential difference without affecting its time dependence. A pair of coils specifically designed for this purpose is described as a *transformer*. Particularly simple and interesting operation occurs when, compared with power dissipation in the load, there is within the transformer neither appreciable power dissipation on account of loss mechanisms, which will be discussed shortly, nor significant rate of change of energy storage through variation of magnetic flux. In this situation, the *instantaneous* power delivered to the primary terminals by the source almost equals the *instantaneous* power delivered by the secondary to the load. Thus if I_p and I_s represent the instantaneous primary and secondary currents

$$V_p I_p = -V_s I_s \tag{6.15}$$

to a high degree of approximation. Combining equations (6.14) and (6.15) yields the interesting result

$$V_s/V_p = n; \quad I_s/I_p = -1/n \tag{6.16}$$

In other words, when the potential difference is transformed *up* from primary to secondary, the current is correspondingly transformed *down* between primary and secondary. The negative sign here merely means of course that when the current flows *into* the top terminal of the primary, the secondary current flows *out* of the top terminal of the secondary for the winding sense shown. A transformer for which equations (6.16) would apply exactly can only be imagined and is therefore described as *ideal*. From the foregoing discussion, an ideal transformer would exhibit perfect magnetic coupling, zero loss and infinite inductance, the latter to prevent energy storage through finite magnetic flux. These points will be emphasised again during the analysis of section 6.3. For many applications, it is possible to construct transformers that closely approach the theoretical ideal just defined in its entirety. What is more, real transformers can be described by equivalent circuits based on the ideal form.

A great advantage of sinusoidal alternating sources of electrical power compared with direct sources is that they can be transformed up or down very efficiently by transformers with little power loss. In the transmission of electrical power, energy loss occurs mainly on account of ohmic losses in the transmitting cables. Series resistance R causes loss RI_{rms}^2 when the current is sinusoidal. By distributing power round the country through a grid of overhead cables at very high voltage and correspondingly low current, the ohmic loss in the transmission cables is minimised. Adequate insulation keeps losses in the insulation, which are proportional to V_{rms}^2, negligible. Power stations generate alternating electrical power at 11–33 kV

6.2 Transformers

in the United Kingdom which is then stepped up through transformers to 400 kV for feeding into the grid. For each megawatt distributed through the grid, the grid current is only 2.5 A r.m.s. so that, even where the cable resistance is of the order of ohms, the power wasted in the cable is only ~ 10 W! Because of safety, insulation and a whole host of other reasons, relatively low voltages are required by industrial and domestic equipment at the consuming end. The operating voltage is therefore stepped down through transformers in stages at substations in the neighbourhood of towns and generally distributed to consumers at around 240 V r.m.s. At each transformation in these installations, a power transformer only wastes power of the order of 1% of its full-load rating.

In the field of electronics, transformers are sometimes used to *couple* electrical signals between parts of circuits while blocking direct voltages. Here they fulfil the same function as the simple C–R network described in section 4.4. An attractive feature of transformer coupling is that, by choosing the turns ratio, the secondary load can be matched to the primary source so that optimum transfer of signal power between the source and load is achieved. This particularly important aspect is carefully analysed in section 6.3. Adjustment of the turns ratio is also useful to provide the appropriate load for some purpose other than optimum power transfer. Such a situation arises at the input of a sensitive electronic amplifier where a suitable coupling transformer can provide the optimum source resistance from the point of view of minimum electrical noise. Actually, the cost and size of transformers at other than radio frequencies, where they comprise just a few small turns in air, means that C–R coupling is adopted rather than transformer coupling except in tuned radio-frequency amplifiers or where a certain impedance transformation is critical. Sometimes just the isolating property of a transformer is valuable by itself.

Practical transformers can be divided into three categories: mains-frequency (50 Hz) transformers, audio-frequency (20 Hz–20 kHz) transformers and radio-frequency (~ 30 kHz–~ 300 MHz) transformers. The first of these is concerned with handling electrical power, the second with handling low-frequency electrical signals including those directly corresponding to audio information such as speech or music and the last with handling high-frequency electrical signals including those carrying information through modulation (see section 5.9). Close coupling in each case demands tight interwinding of the primary and secondary coils but its achievement is greatly assisted by winding the coils round a common core material of high magnetic permeability. Inclusion of such a core is also very helpful in achieving sufficiently high inductive reactance for almost ideal transformation, especially at lower frequencies. Remarks made regarding core materials in section 4.2 in connection with the achievement of high inductance in inductors apply equally well to transformers of course, and the reader is referred back to that section on this point.

Worthy of attention is the fact that the magnetic flux in ferromagnetic types of cores is not proportional to the magnetising current but shows a complicated dependence. This means that the mutual inductance between coils wound on such cores varies with current. Fortunately the mutually inductive behaviour is linear (mutual inductance constant) for all other types of core and even in the case of ferromagnetic cores is effectively linear with regard to small signals about any bias level. It will be taken for granted throughout the remainder of this book that wherever mutual inductance arises, it is constant.

Losses that arise in transformers are *copper losses* due to the inevitable resistances of the windings and *core losses* due to hysteresis and eddy current heating. In addition, there may be appreciable flux leakage and significant capacitance between the turns. For a very brief discussion of the reduction of core losses by such techniques as lamination to reduce eddy currents and choice of material with small hysteresis loop, the reader is referred back to section 4.2 on inductors. In an equivalent circuit, copper losses are represented by series resistances r_p and r_s of the primary and secondary windings, while core losses are represented by a resistance r_c in parallel with the primary. Flux leakage can be represented by segregating a fraction $(1-k)$ of the inductance of each winding, such that only current in the remaining fraction k creates magnetic flux that links the two windings. A complete equivalent circuit according to these ideas is shown in figure 6.4(a). Figures 6.4(b) and (c) show how to represent the coupled inductances in terms of an ideal transformer. The circuits of figures 6.4(b) and (c) are equivalent because, for the circuit of figure 6.4(b), applying Kirchhoff's voltage law to the primary and secondary gives

$$\mathbf{V}_p = j\omega k L_p \mathbf{I}_p + j\omega M \mathbf{I}_s = j\omega k [L_p \mathbf{I}_p + (L_p L_s)^{\frac{1}{2}} \mathbf{I}_s]$$

$$\mathbf{V}_s = j\omega k L_s \mathbf{I}_s + j\omega M \mathbf{I}_p = (L_s/L_p)^{\frac{1}{2}} \mathbf{V}_p$$

On the other hand, applying Kirchhoff's current law to the node on the primary side of the circuit of figure 6.4(c) gives, since it is an ideal transformer

$$\mathbf{V}_p = j\omega k L_p [\mathbf{I}_p + (L_s/L_p)^{\frac{1}{2}} \mathbf{I}_s] = j\omega k [L_p \mathbf{I}_p + (L_p L_s)^{\frac{1}{2}} \mathbf{I}_s]$$

while

$$\mathbf{V}_s = (L_s/L_p)^{\frac{1}{2}} \mathbf{V}_p$$

again.

6.3 Reflected impedance and matching

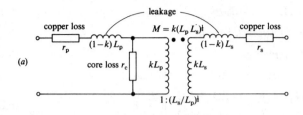

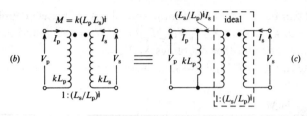

6.4. (a) Equivalent circuit of a transformer including representation of losses and leakage and, (b) and (c), how to represent the transforming part in terms of an ideal transformer.

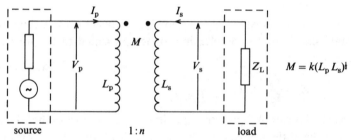

6.5 Lossless transformer connected to a source and load. Note that, although this figure is identical in form to that of figure 6.3, it represents the situation for *any* coefficient of coupling k in the range 0–1 rather than for just k = 1 and is useful here for ease of reference.

6.3 Reflected impedance and matching by transformers

Consider any lossless transformer, represented by mutual inductance M and primary and secondary self inductances L_p and L_s, connected to a sinusoidal source on the primary side and a complex load impedance Z_L on the secondary side as shown in figure 6.5. Application of Kirchhoff's voltage law to the primary and secondary meshes yields the mesh equations

$$\mathbf{V}_p = j\omega L_p \mathbf{I}_p + j\omega M \mathbf{I}_s \tag{6.17}$$
$$0 = j\omega M \mathbf{I}_p + j\omega L_s \mathbf{I}_s + Z_L \mathbf{I}_s \tag{6.18}$$

in the steady-state current and potential difference phasors. In addition

$$\mathbf{V}_s = -Z_L \mathbf{I}_s \tag{6.19}$$

and equation (6.10) applies, that is,

$$M = k(L_p L_s)^{\frac{1}{2}} \tag{6.20}$$

Combining equations (6.17)–(6.20) reveals that

$$\frac{\mathbf{V}_s}{\mathbf{V}_p} = \frac{M\mathbf{I}_p + L_s \mathbf{I}_s}{L_p \mathbf{I}_p + M\mathbf{I}_s} = \left(\frac{kL_p^{\frac{1}{2}}\mathbf{I}_p + L_s^{\frac{1}{2}}\mathbf{I}_s}{L_p^{\frac{1}{2}}\mathbf{I}_p + kL_s^{\frac{1}{2}}\mathbf{I}_s}\right)\left(\frac{L_s}{L_p}\right)^{\frac{1}{2}} \tag{6.21}$$

from which it can be seen that only in the special case of perfect coupling corresponding to $k=1$ does $\mathbf{V}_s/\mathbf{V}_p = (L_s/L_p)^{\frac{1}{2}} = N_s/N_p = n$ as found in the previous section. For all other couplings $\mathbf{V}_s/\mathbf{V}_p \neq n$.

Rearranging equation (6.18) shows that the ratio of secondary to primary current is

$$\frac{\mathbf{I}_s}{\mathbf{I}_p} = \frac{-j\omega M}{j\omega L_s + Z_L} = -\left(\frac{j\omega k L_s}{j\omega L_s + Z_L}\right)\left(\frac{L_p}{L_s}\right)^{\frac{1}{2}} \tag{6.22}$$

Clearly only when there is perfect coupling and additionally $|Z_L|$ can be neglected compared with the secondary inductive reactance ωL_s does $\mathbf{I}_s/\mathbf{I}_p = -(L_p/L_s)^{\frac{1}{2}} = -1/n$.

Substituting in equation (6.17) for $\mathbf{I}_s$ in terms of $\mathbf{I}_p$ from equation (6.22) establishes that the complex impedance presented to the source by the primary is

$$Z_p = \frac{\mathbf{V}_p}{\mathbf{I}_p} = j\omega L_p + \frac{\omega^2 M^2}{j\omega L_s + Z_L} \tag{6.23}$$

Clearly this complex primary input impedance comprises the self-inductive reactance of the primary winding in series with some complex impedance that depends on the mutual inductance M and the total complex self impedance $j\omega L_s + Z_L$ of the secondary mesh. Complex impedance appearing at the terminals of a transformer over and above the self-inductive reactance expected is said to be *reflected* through the transformer and is called the *reflected impedance*. Rewriting the total complex self impedance $j\omega L_s + Z_L$ of the secondary as $R_s + jX_s$, X_s may have either sign depending on the extent of the capacitive or inductive part of the load. In terms of R_s and X_s, the reflected *series* impedance in the primary is

$$Z_{pr} = \frac{\omega^2 M^2 R_s}{R_s^2 + X_s^2} - \frac{j\omega^2 M^2 X_s}{R_s^2 + X_s^2} \tag{6.24}$$

For a passive load, R_s is positive so that the resistive part of Z_{pr} is also positive. The reactive part of Z_{pr} has the opposite sign to that of X_s. It is said that an inductive load reflects a series capacitive reactance into the primary while sufficient capacitive load reflects a series inductive reactance. By this statement it is only meant to convey whether the sign of the reflected

6.3 Reflected impedance and matching

reactance is the same as that of a capacitive or inductive reactance as the case may be. The frequency dependence of the second term in equation (6.24) is quite different from that for an actual inductive or capacitive reactance. Also rather interesting is the fact that when $X_s = 0$, the reflected series resistance is inversely proportional to R_s, but when $X_s \gg R_s$ the reflected series resistance is directly proportional to R_s.

Regarding the behaviour of a lossless transformer in the special case when the coupling is perfect, sometimes appropriately referred to as the unity-coupled condition since it corresponds to $k = 1$, in such circumstances from equation (6.21)

$$\mathbf{V}_s/\mathbf{V}_p = (L_s/L_p)^{\frac{1}{2}} = n \tag{6.25}$$

as already pointed out. When $k = 1$, it also follows from equation (6.22) that

$$\frac{\mathbf{I}_s}{\mathbf{I}_p} = -\left(\frac{j\omega L_s}{j\omega L_s + Z_L}\right)\left(\frac{1}{n}\right) \tag{6.26}$$

and from equation (6.23) that

$$Z_p = j\omega L_p + \frac{\omega^2 L_p L_s}{j\omega L_s + Z_L}$$

which reduces to

$$Z_p = \frac{j\omega L_p Z_L}{j\omega L_s + Z_L}$$

This last result may be illuminatingly re-expressed as

$$Z_p = \frac{j\omega L_p Z_L/n^2}{j\omega L_p + Z_L/n^2} \tag{6.27}$$

which is the complex impedance corresponding to the reactance of the primary self inductance L_p in parallel with complex impedance Z_L/n^2. Evidently, when the coupling is perfect in a lossless transformer, the reflected impedance is just a *parallel* impedance

$$Z_{pr} = Z_L/n^2 \tag{6.28}$$

Choosing a time origin so that $\mathbf{V}_p$ is the phasor representation of potential difference $V_{p0} \sin \omega t$, $\mathbf{V}_s$ represents sinusoidal potential difference $nV_{p0} \sin \omega t$ in a unity-coupled lossless transformer and the instantaneous power developed in a load resistance R_L is simply $(nV_{p0} \sin \omega t)^2/R_L$ corresponding to average power dissipation of $(nV_{p0})^2/2R_L$. Since the primary looks like resistance R_L/n^2 in parallel with inductive reactance ωL_p, the average input power supplied amounts to $V_{p0}^2/(2R_L/n^2)$ which is the same as the power developed in the load. In addition, energy is stored and released during alternate half-cycles on account of the primary inductance L_p as the magnetic flux alternately grows and diminishes. *On average* there is no energy dissipation associated with the inductance and the lossless, unity-coupled transformer transfers energy between the primary and secondary with 100% efficiency.

When a transformer is not only lossless and unity-coupled but its inductances are large enough to satisfy the condition $\omega L_s \gg |Z_L|$,

$$\mathbf{V}_s/\mathbf{V}_p = (L_s/L_p)^{\frac{1}{2}} = n \tag{6.29}$$

again, while from equation (6.22)

$$\mathbf{I}_s/\mathbf{I}_p \approx -(L_p/L_s)^{\frac{1}{2}} = -1/n \tag{6.30}$$

and from equations (6.27) and (6.28), since $\omega L_s \gg |Z_L|$ corresponds to $\omega L_p \gg |Z_L|/n^2$,

$$Z_p \approx Z_{pr} = Z_L/n^2 \tag{6.31}$$

Equations (6.29) and (6.30) confirm the correctness of equations (6.16) previously derived for this situation in section 6.2. Large inductance makes the core magnetising current, which in this case is just the primary current with the secondary open-circuit, very small compared with the total current and hence the magnetic energy stored in the core negligible. Equations (6.30) and (6.31) become exact, of course, for an imaginary ideal transformer having unity coupling, zero loss and infinite inductances.

A very important consequence of equation (6.31) from a practical point of view is that, as stated in the previous section, inclusion of a near ideal transformer between a source and a load enables the load to be matched to the source, or the optimum impedance to be presented for some other purpose, through selection of the turns ratio n. When, as often occurs, the load and source are purely resistive, precise matching is possible through choice of the ratio n. Where there is a complex load impedance but the source is still purely resistive, the reactance of the load may sometimes be tuned out and the turns ratio selected to match the resistances. As proved in section 5.8, the best that can be done to transfer maximum power from a source of given complex impedance to a load of given complex impedance is to insert a transformer between them of turns ratio n such that n is the square root of the ratio of the modulus $|Z_L|$ of the complex load impedance to the modulus $|Z_S|$ of the complex source impedance.

Insight into the effect of making $\omega L_s \gg |Z_L|$ to approach the ideal behaviour is provided by thinking about what happens when a load $Z_L = R_L$ is progressively reduced from infinity in a lossless, unity-coupled transformer circuit. Infinite R_L corresponds to the secondary mesh being open circuit and the secondary current being zero. In this condition, infinite resistance is reflected into the primary so that the primary current $\mathbf{I}_p$ is $\mathbf{V}_p/j\omega L_p$ and it is obvious that the secondary current cannot be given by

$-\mathbf{I}_p/n$; it is of course given by equation (6.26). With open-circuit secondary and a sinusoidal source in the primary, magnetic flux is alternately established and removed in association with the primary inductance but there is no power dissipation, only alternate energy storage and release. When R_L becomes finite, extra primary and secondary currents of $\mathbf{V}_p/R_L/n^2$ and $-\mathbf{V}_s/R_L = -n\mathbf{V}_p/R_L$ flow in the reflected and actual loads which create cancelling magnetic fluxes since the effectiveness of secondary current in producing flux is $N_s/N_p = n$ times that of primary current. Although there is no additional fluctuation in stored energy on account of the extra currents, extra power is delivered from the primary source to the load. Eventually when R_L becomes small enough to satisfy $R_L \ll \omega L_s$, that is, $R_L/n^2 \ll \omega L_p$, the power delivered to the load swamps the fluctuation of stored energy in the transformer. In this case, therefore, the instantaneous power delivered by a source to the primary virtually equals the instantaneous power dissipated in the secondary load.

6.4 Critical coupling of resonant circuits

The behaviour of an isolated series L–C–R circuit was analysed in section 5.6 and its resonant frequency response noted. What happens when two such circuits are magnetically coupled by means of a transformer will now be studied. Interest will focus on the response of the coupled circuits as a function of the coefficient of coupling of the primary and secondary coils as well as the frequency. To simplify the analysis while preserving the features of the behaviour of interest, identical primary and secondary L–C–R circuits will be assumed. Figure 6.6 presents the circuit arrangement to be analysed, the aspect of interest being its response to sinusoidal input.

Putting $R + j(\omega L - 1/\omega C) = Z$, Kirchhoff's voltage law in the primary and secondary meshes may be expressed by

$$\mathbf{V}_i = Z\mathbf{I}_p + j\omega M \mathbf{I}_s \tag{6.32}$$

$$0 = Z\mathbf{I}_s + j\omega M \mathbf{I}_p \tag{6.33}$$

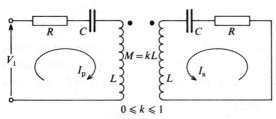

6.6 Magnetically coupled, identical, series resonant circuits.

Elimination of the primary current $\mathbf{I}_p$ between equations (6.32) and (6.33) therefore establishes that the secondary current is given by

$$\mathbf{I}_s = \left(\frac{-j\omega M}{Z^2 + \omega^2 M^2}\right)\mathbf{V}_i = \left\{\frac{-j\omega M}{[R + j(\omega L - 1/\omega C)]^2 + \omega^2 M^2}\right\}\mathbf{V}_i$$

As the coupling between the primary and secondary becomes very weak, the response of each is expected to approach that of an isolated series resonant circuit. It consequently makes sense to write the equation for the secondary current $\mathbf{I}_s$ in terms of the resonant pulsatance $\omega_r = 1/(LC)^{\frac{1}{2}}$ and Q-factor $Q = \omega_r L/R = 1/\omega_r CR$ of the uncoupled primary or secondary circuit. Taking this step, since $M = kL$,

$$\mathbf{I}_s = \left\{\frac{-j\omega kL}{[1 + j(Q\omega/\omega_r - Q\omega_r/\omega)]^2 R^2 + \omega^2 k^2 L^2}\right\}\mathbf{V}_i$$

and introducing the normalised pulsatance $x = \omega/\omega_r$

$$\mathbf{I}_s = -\frac{j}{R}\left\{\frac{kQx}{[1 + jQ(x - 1/x)]^2 + k^2 Q^2 x^2}\right\}\mathbf{V}_i \tag{6.34}$$

It follows from equation (6.34) that the amplitude I_{s0} of the secondary current is related to the amplitude V_{i0} of the input potential difference by

$$\frac{I_{s0}}{V_{i0}} = \frac{kQx}{\{[1 - Q^2(x - 1/x)^2 + k^2 Q^2 x^2]^2 + [2Q(x - 1/x)]^2\}^{\frac{1}{2}} R} \tag{6.35}$$

The dependence of the function I_{s0}/V_{i0} on normalised pulsatance x for various coefficients of coupling k is complicated. In the following discussion it should be borne in mind that in practical applications of this circuit arrangement the Q-factor would be high (say, $Q \geqslant 10$).

To begin with, consider the dependence of I_{s0}/V_{i0} on k when $x = 1$, that is, at the uncoupled resonant pulsatance $\omega = \omega_r$. When $x = 1$ equation (6.35) reduces to

$$(I_{s0}/V_{i0})_{x=1} = kQ/(1 + k^2 Q^2)R \tag{6.36}$$

The turning points of $(I_{s0}/V_{i0})_{x=1}$ as a function of kQ are given by

$$\frac{d(I_{s0}/V_{i0})_{x=1}}{d(kQ)} = \frac{(1 + k^2 Q^2)R - kQ2kQR}{(1 + k^2 Q^2)^2 R^2} = 0$$

or

$$k^2 Q^2 = 1$$

Since physically both k and Q can only be positive, only the positive root is admissible. Also $(I_{s0}/V_{i0})_{x=1} = 0$ when $kQ = 0$ or $kQ = \infty$, from which it may be deduced that the physically admissible condition for a turning point

$$kQ = 1 \tag{6.37}$$

corresponds to a maximum value of $(I_{s0}/V_{i0})_{x=1}$. This aspect is illustrated in

6.4 Critical coupling of resonant circuits

figure 6.7, where the dependence of I_{s0}/V_{i0} on x is shown for several values of k for the particular case $Q=10$. In the figure $(I_{s0}/V_{i0})_{x=1}$ reaches a maximum when $k=0.1$ in accordance with equation (6.37). The condition $k=1/Q$ is known as *critical coupling*.

The ratio of the amplitude of the primary current to that of the secondary current when $x=1$ is from equation (6.33) simply

$$(I_{p0}/I_{s0})_{x=1} = R/\omega_r kL = 1/kQ \tag{6.38}$$

and therefore from equation (6.36)

$$(I_{p0}/V_{i0})_{x=1} = 1/(1+k^2Q^2)R \tag{6.39}$$

According to equation (6.39), as the coupling increases steadily, the primary current decreases steadily. Physically, as the coupling increases, the primary is increasingly loaded by reflected impedance so that the primary current falls. On the other hand, as the coupling increases, the secondary current increases relative to the primary current as evidenced by equation (6.38). Thus a maximum in $(I_{s0}/V_{i0})_{x=1}$ is feasible as the coupling is varied.

Consider next the frequency response when kQ is small. When $k^2Q^2 \ll 1$, equation (6.35) becomes

$$\left(\frac{I_{s0}}{V_{i0}}\right)_{k^2Q^2 \ll 1} = \frac{kQx}{[1+Q^2(x-1/x)^2]R} \tag{6.40}$$

from which it can be seen that $(I_{s0}/V_{i0})_{k^2Q^2 \ll 1}$ only exhibits one resonant maximum when $x=1/x$, that is, when $\omega=\omega_r$, since negative frequencies have no physical significance. This is the expected reversion towards the

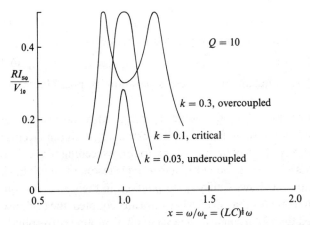

6.7 Frequency response of the amplitude of the secondary current, I_{s0}, as a function of the coefficient of coupling, k, between identical, series resonant circuits. Parameters relate to the circuit shown in figure 6.6.

resonant behaviour of a single series resonant circuit as the coupling becomes weak. Again figure 6.7 illustrates this aspect for the special case $Q = 10$.

Lastly, consider the situation when $k^2Q^2 \gg 1$. When x is far from unity, both squared terms on the denominator of equation (6.35) are large and so I_{s0}/V_{i0} is small. When $x = 1$, the terms in $(x - 1/x)$ vanish but $k^2Q^2x^2 \gg 1$ and so I_{s0}/V_{i0} is small again. The vital point to appreciate is, however, that the first term in the denominator of equation (6.35),

$$[1 - Q^2(x - 1/x)^2 + k^2Q^2x^2]^2$$

goes to zero at a value of x either slightly up or slightly down from unity when the second term

$$[2Q(x - 1/x)]^2$$

is quite modest and therefore resonant maxima occur in I_{s0}/V_{i0} at the two values of x near unity that make the first term zero. The condition $x = 1$ turns out to be that for a minimum when $k^2Q^2 \gg 1$ and the behaviour for the special case $Q = 10$ with $k = 0.3$ so that $k^2Q^2 = 9$, shown in figure 6.7, illustrates the foregoing discussion.

To sum up, a single resonant maximum persists in the secondary current up to the critical coupling $k = 1/Q$ but above this coupling the response breaks up into two resonant peaks. For $k^2 \gg 1/Q^2$, the two peaks clearly occur when, to a good approximation,

$$(x - 1/x)^2 = k^2x^2$$

that is

$$(x - 1/x) = \pm kx$$

or

$$\omega/\omega_r = (1 \pm k)^{-\frac{1}{2}} \tag{6.41}$$

The two peaks are consequently separated in frequency by

$$\Delta f \approx [(1 - k)^{-\frac{1}{2}} - (1 + k)^{-\frac{1}{2}}] f_r \tag{6.42}$$

where $f_r = \omega_r/2\pi$. Adjustment of their separation is possible through variation of k.

Transformer-coupled, high-Q, resonant circuits with the primary and secondary tuned to the same frequency are extremely useful in electronic circuit applications. For coupling equal to or exceeding critical, the response exhibits a greater bandwidth than a single resonant circuit having the same Q-factor yet shows similarly steep fall-off in the wings. It is quite noticeable from figure 6.7 that, even when critically coupled, the response is more flat-topped than for a single resonant circuit. Coupling corresponding to kQ in the range from just over 1.0 to 1.5 is particularly advantageous, giving twin-peaked response but with only a shallow dip in between. Use of

6.4 Critical coupling of resonant circuits

such coupled resonant circuits allows, for example, band-pass amplifiers to be constructed with appreciable bandwidth yet high selectivity and fidelity (the latter meaning constancy of amplification within the bandwidth). Intermediate-frequency amplifiers employed in communications receivers are important amplifiers of this type with sufficient bandwidth to avoid distorting the information carried by the signals.

7

Alternating-current instruments and bridges

7.1 Alternating-current meters

Any system that measures direct current or potential difference can be adapted to measure the corresponding alternating quantity by inserting a *rectifying* circuit in front of it. The term rectification refers to rendering the alternating current or potential difference unidirectional through removing or reversing it whenever it is one of its two possible polarities. It should be clear that, with sufficient damping, a direct measuring system will respond to the mean level of a rectified alternating input. Removal of alternating half-cycles is termed *half-wave* rectification. Reversal of alternate half-cycles is described as *full-wave* rectification and is illustrated in figure 7.1(a) for a sinewave. The neatest and most popular way of implementing full-wave rectification is by means of four diodes arranged in a Wheatstone bridge formation as shown in figure 7.1(b). Understandably, this form of circuit is called a *bridge rectifier*. As explained in section 2.3, a diode is a device that presents a very low resistance to current flow when appreciable potential difference of one sign, known as *forward*, is applied while it presents a very high resistance to current flow when appreciable potential difference of the opposite sign, known as *reverse*, is applied. In the case of modern silicon P–N junction diodes, appreciable here means >0.6 V. The direction of the arrow head in the circuit symbol for the diode indicates the direction of easy current flow. In the bridge rectifier circuit of figure 7.1(b), diodes D_2 and D_3 conduct well during those half-cycles over which the upper input terminal is positive with respect to the lower, diodes D_1 and D_4 being reverse biased and almost open circuit at such times. During the other half-cycles, when the upper input terminal is negative with respect to the lower, it is the diodes D_1 and D_4 that conduct well, diodes D_2 and D_3 then being reverse biased. Evidently, the current or potential difference applied to the direct measuring system is always right to left in the figure,

7.1 A.c. meters

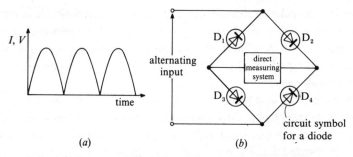

7.1 (a) Full-wave rectified sinewave and (b) bridge rectifier.

the alternating input being full-wave rectified by the arrangement of diodes.

While the type of measuring system just discussed responds to the average rectified value of an alternating quantity, it is customary to describe the magnitudes of alternating quantities in terms of their r.m.s. values. Now the average of an ideally full-wave rectified sinewave $x_0 \sin \omega t$ is

$$\frac{1}{\pi}\int_0^\pi x_0 \sin \omega t \, \mathrm{d}(\omega t) = \frac{x_0}{\pi}(-\cos \omega t)_0^\pi = \frac{2x_0}{\pi}$$

whereas its r.m.s. value is $x_0/\sqrt{2}$. Simple modification of the output scale of the direct measuring system that follows full-wave rectification by a factor $\pi/2\sqrt{2}$ consequently leads to the r.m.s. value of a sinusoidal input being displayed at the output. However, it is important to realise that, where this is done, the system will not display the r.m.s. value at its output for any other input waveform. Another difficulty arises because, although the semiconducting diodes used in this type of circuit pass quite negligible current under reverse bias, they do drop an appreciable potential difference of a few tenths of a volt in typical operation under forward bias. Clearly, the r.m.s. value will only be displayed by the type of system under discussion, even for a sinusoidal input waveform, as long as the amplitude of the input remains sufficiently large in relationship to the forward potential drop across the diodes. To obtain the correct r.m.s. value of a small sinusoidal signal through use of a circuit such as that shown in figure 7.1(b), the signal must first be electronically amplified by a known factor up to a suitable level. The output reading obtained with the enlarged input is then correspondingly scaled down. Small signals can also be measured properly if the simple bridge rectifier is replaced by a more complicated electronic circuit that very much more closely approaches the ideal of 100% pass during forward half-cycles and 100% rejection or reversal during reverse half-cycles of the input. The rectifier circuit of figure 7.1(b) also deteriorates in performance at high frequencies, typically above $\sim 10\,\mathrm{kHz}$, due to capacitive effects. Current spikes at changeover from reverse to forward bias are a particular source of trouble. Naturally, remarks made in section 3.8 with respect to direct measuring systems, about the relevance of their resistance and the possibility of analogue or digital representation of the output, apply equally well to systems that measure alternating currents or potential differences.

While a moving-coil meter does not exhibit any steady deflection when sinusoidal current passes through it and must be preceded by some form of rectifying circuit before it can register such current, other forms of meter exist that respond usefully by themselves to sinusoidal current. In a *moving-iron* meter, the current passes through a fixed coil to create a magnetic induction that magnetises a pivoted, shaped piece of ferromagnetic metal. Since both the induction and magnetisation are proportional to the current, the force exerted on the pivoted element is essentially proportional to the square of the current. Mechanical damping is provided and a steady deflection occurs when the mean deflecting force is balanced by gravity or a restoring spring. The square law of force means that both alternating and direct currents are registered but there is an essentially square-law scale which is cramped at one end. Actually the scale distribution can be modified somewhat by shaping the pivoted element. Problems with the meter are magnetic hysteresis, magnetic shielding, low sensitivity and appreciable inductance of the fixed coil which restricts usage to frequencies below ~ 300 Hz. Moving-iron meters are mainly employed as robust instruments for measuring mains (50 Hz) current. Since the moving element does not carry current, it is not susceptible to overload damage.

In *thermocouple* and *hot-wire* meters the current to be measured passes through a straight piece of resistance wire and heats it up. The temperature rise is registered in the former case by an attached or adjacent thermocouple that delivers its thermoelectric e.m.f. to a moving-coil voltmeter of suitable sensitivity. In the latter case, thermal expansion of the wire proportional to the heating is sensed by mechanical means. Because the heating effect and therefore the temperature rise is proportional to the square of the current, both meters have a square-law output scale and, following calibration with a known direct input, give the true r.m.s. value irrespective of waveform. They also work up to very high frequencies (~ 50 MHz), because the heated wire exhibits negligible inductance, and are often used in the measurement of radio-frequency signals. More sensitive versions of thermocouple meters have the heater element and thermocouple housed in an evacuated enclosure to reduce heat losses and enhance the temperature rise.

Yet another type of meter capable of registering either alternating or direct current is the *electrodynamometer*. The current to be sensed and measured passes through two fixed coils in series with a movable coil. An

7.1 A.c. meters

approximately uniform magnetic induction B is produced by the fixed coils in which the movable coil is suspended as illustrated in figure 7.2(a). The magnetic induction exerts a torque on the suspended coil according to equations (4.8) and (4.9) which rotates it until there is a balancing restoring torque provided by controlling springs. Mechanical damping is again active. The magnetic deflecting torque is proportional to both the current I in the suspended coil and the magnetic induction delivered by the fixed coils which in turn is also proportional to the same current I. Because the deflecting torque is proportional to I^2, both alternating and direct currents cause a steady deflection of the suspended coil, the natural frequency of the suspended system being too low and the damping too great for any fluctuations in I^2 to be followed. A deflection more linearly dependent on the current I than the square law can be achieved by making good use of the fact that the torque also depends on $\cos \theta$ where θ defines the orientation of the suspended coil as shown in figure 7.2(a). It is generally arranged that θ is zero at some suitable deflection such that the fall in $\cos \theta$ at larger deflections significantly counteracts the scale spreading that would otherwise arise from the square-law dependence on current. Appropriate shaping of the fixed coils can also help to linearise the scale.

Although the electrodynamometer is free of hysteresis, it is insensitive unless the coils have a large number of turns which makes both the resistance and inductance high, the latter severely restricting the frequency range of operation. For these reasons it is seldom used for current measurement nowadays but it does find application in a modified form of operation that enables electrical power to be measured in both direct and alternating-current circuits at frequencies up to a few hundred hertz. To measure the power in a load, whichever of the movable or pair of fixed coils has the lower resistance is connected essentially or actually in series with the load as indicated in figure 7.2(b), so that the load current is essentially or

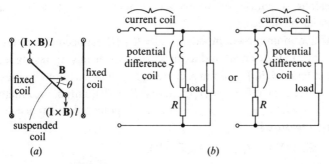

7.2 (a) Arrangement of coils in an electrodynamometer and (b) connections of an electrodynamometer for power measurement.

actually carried. The other is connected essentially or actually in parallel with the load through a large enough resistance R to swamp the inductive reactance, also as indicated in figure 7.2(b). It therefore carries current essentially proportional to the potential difference across the load and the instantaneous couple acting on the suspended coil is proportional to the product of the instantaneous current and potential difference. Connected in this fashion, the electrodynamometer once calibrated gives the mean power in the load.

An instrument of paramount importance that enables the precise nature of any potential difference to be determined is the *cathode-ray oscillograph*. The screen of this complicated electronic instrument presents a stationary picture of the potential difference between its input terminals as a function of time. Every facet of the signal can be studied at leisure, including amplitude, frequency, phase and detailed time dependence. In particular, with respect to this section, it makes a very good alternating voltmeter, possessing calibrated potential difference and time scales and, as mentioned in section 5.2, an input impedance corresponding to resistance as high as $\sim 1\,\mathrm{M}\Omega$ in parallel with capacitance as low as $\sim 30\,\mathrm{pF}$.

7.2   Measurement of impedance by a.c. meters

In the title of this section, *a.c.* means alternating current while impedance implies complex impedance. Such abbreviated language is standard practice and will be widely adopted in the remainder of this book.

Just as an unknown resistance can be determined using direct meters as described in section 3.8, so too can an unknown impedance be determined using a.c. meters. The accuracy of determination again depends on the factors discussed in that section besides the calibration and reading accuracy. Ideally, any alternating voltmeter should possess infinite impedance and any a.c. meter negligible impedance. An alternating voltmeter is, of course, often just an a.c. meter in series with a suitable substantial resistance.

Consider an unknown impedance $Z = R + jX$ connected in series with a sinusoidal e.m.f. $\mathscr{E}_0 \sin \omega t$ and a standard resistor having resistance R_s as shown in figure 7.3(a). The r.m.s. current, I_{rms}, may be found by measuring the r.m.s. potential difference $(V_s)_{\mathrm{rms}}$ across R_s with a suitable voltmeter from which $I_{\mathrm{rms}} = (V_s)_{\mathrm{rms}}/R_s$. If $(V_Z)_{\mathrm{rms}}$ is the r.m.s. potential difference across Z similarly measured

$$|Z| = (V_Z)_{\mathrm{rms}}/I_{\mathrm{rms}} = (V_Z)_{\mathrm{rms}} R_s/(V_s)_{\mathrm{rms}} \qquad (7.1)$$

To find the resistance R and reactance X of the unknown impedance Z, the r.m.s. e.m.f., $\mathscr{E}_{\mathrm{rms}}$, must also be measured with the voltmeter. Armed with

7.2 Impedance measurement by a.c. meters

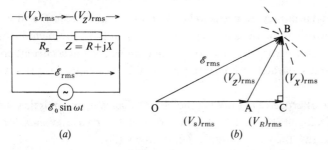

7.3 (a) Circuit for determining impedance from a.c. meter measurements and (b) phasor diagram for circuit (a) deduced from meter measurements of the r.m.s. magnitudes of potential differences across components of it.

$(V_s)_{rms}$, $(V_Z)_{rms}$ and $\mathscr{E}_{rms}$, the phasor diagram for the series circuit can be completed as indicated in figure 7.3(b). First the reference phasor OA, representing the potential difference across the standard resistor, is drawn with length in convenient proportion to $(V_s)_{rms}$. The phasors representing the e.m.f. and potential difference across impedance Z then have lengths in the same proportion to $\mathscr{E}_{rms}$ and $(V_Z)_{rms}$ respectively. To satisfy Kirchhoff's voltage law in the series circuit, the three phasors must complete a triangle as shown in figure 7.3(b). This triangle is established by finding the intersection B of circles centred on O and A with radii in the same proportion to $\mathscr{E}_{rms}$ and $(V_Z)_{rms}$ respectively as OA is to $(V_s)_{rms}$. To summarise up to this point, the r.m.s. magnitudes of phasors $\mathbf{V_s}$, $\mathscr{E}$ and $\mathbf{V_Z}$ have been measured by the voltmeter but not their phases. Drawing the phasor diagram that satisfies the magnitudes of these three potential differences, however, establishes the relative phases. Now because the potential difference across the resistance R of the unknown impedance Z must be in phase with the potential difference across the standard series resistance R_s, the phasor representing it is the projection AC of AB along the direction of OA. Similarly, because the potential difference across the reactance X of the unknown impedance Z must be in phase quadrature (90° out of phase) with the potential difference across the standard series resistance R_s, the phasor representing it is the projection CB of AB perpendicular to the direction of OA. The resistance R and reactance X therefore follow from application of the relations

$$R = \frac{(V_R)_{rms}}{I_{rms}} = \frac{(V_R)_{rms}}{(V_s)_{rms}} R_s = \frac{AC}{OA} R_s \qquad (7.2)$$

$$X = \frac{(V_X)_{rms}}{I_{rms}} = \frac{(V_X)_{rms}}{(V_s)_{rms}} R_s = \frac{CB}{OA} R_s \qquad (7.3)$$

For good accuracy, R_s needs to be similar in magnitude to R and X. If not, small errors in the phasor diagram due to small measurement errors may lead to large errors in R and/or X. Of course, the r.m.s. current could be measured separately by inserting a suitable a.c. meter in series with the circuit.

As an alternative to drawing the phasor diagram, an analytic solution is possible for $(V_R)_{rms}$ and $(V_X)_{rms}$ in terms of the measurements. Applying Pythagorus' theorem to figure 7.3(b) shows that

$$\mathscr{E}_{rms}^2 = (V_X)_{rms}^2 + [(V_s)_{rms} + (V_R)_{rms}]^2 \tag{7.4}$$

$$(V_Z)_{rms}^2 = (V_X)_{rms}^2 + (V_R)_{rms}^2 \tag{7.5}$$

and subtracting these equations gives

$$\mathscr{E}_{rms}^2 - (V_Z)_{rms}^2 = (V_s)_{rms}^2 + 2(V_s)_{rms}(V_R)_{rms} \tag{7.6}$$

in which all but $(V_R)_{rms}$ have been measured so that $(V_R)_{rms}$ can be calculated readily. $(V_X)_{rms}$ then follows from equation (7.5) and R and X from equations (7.2) and (7.3).

Where an unknown impedance is known to approximate closely to a pure reactance or pure resistance, it is only necessary to measure the r.m.s. current I_{rms} through it and r.m.s. potential difference V_{rms} across it with meters to determine its value as V_{rms}/I_{rms}. For a single inductor or capacitor of negligible loss, the latter being more common, the inductance or capacitance can be deduced from the measured impedance provided that the frequency of the current is known. A useful alternative method of finding the capacitance of a low-loss capacitor is to compare the alternating potential drop across it with that across a standard capacitor connected in series, using a very-high-impedance voltmeter. For the particular capacitance meter circuit of figure 7.4(a), the unknown capacitance C is given in terms of the standard capacitance C_s by the relation

$$\frac{(V_s)_{rms}}{\mathscr{E}_{rms}} = \frac{C}{C_s + C} \tag{7.7}$$

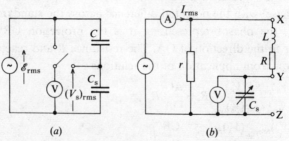

7.4 (a) A capacitance meter and (b) a Q meter.

7.3 Impedance measurement by Wheatstone a.c. bridge

Determination of inductance through comparison of inductors is not viable because there is usually a significant loss and also there may be a problem of mutual inductance between the two inductors. The Q-factor meter, a version of which is shown in figure 7.4(b), is better for determining the inductance L and resistance R of an inductor. With regard to the circuit of figure 7.4(b), the unknown inductor is connected between the terminals X and Y, V being a very-high-impedance voltmeter. The capacitance C_s of the variable standard capacitor is adjusted until the voltmeter V indicates resonance, when the inductance can be found from the theoretical expression (5.42) of section 5.6 for the resonant frequency. Because the resistance r is very small compared with R, most of the current I_{rms} goes through r and, in accordance with the theory developed in section 5.6, the resonant r.m.s. potential difference recorded by the voltmeter V is QrI_{rms}, where Q is the Q-factor. Normally I_{rms} is set to a preset level and the voltmeter V is calibrated to give Q directly, from which the loss resistance R of the inductor can be calculated if required. An unknown capacitance can be determined by connecting it across YZ and finding the change in C_s needed to return to resonance. The power factor corresponding to the connected capacitance is obtained from the change in Q.

7.3 Measurement of impedance by the Wheatstone form of a.c. bridge

Most a.c. bridges are similar in form to the Wheatstone bridge for measuring resistance described in section 3.9. However, in general, each arm of an a.c. bridge constitutes an impedance and an a.c. bridge is energised by an a.c. source while its balance is sensed by a detector that responds to the type of signal delivered by the a.c. source. These essential features are illustrated in figure 7.5(a) where the impedances of the arms are labelled Z_1–Z_4. Normally the source delivers a sinusoidal signal.

The vitally important balance condition of absolutely no current flowing through the detector corresponds to the potential difference across it being zero *at all times*. With reference to figure 7.5(a), such *perfect balance* demands that the potentials of nodes B and D are identical at all times. This only happens when both the amplitudes and phases of the potentials at B and D are equal. Seemingly, two separate conditions must be satisfied to achieve proper balance. Now if $\mathbf{I}_1, \mathbf{I}_2, \mathbf{I}_3$ and $\mathbf{I}_4$ denote the phasor representations of the currents in Z_1, Z_2, Z_3 and Z_4 respectively, then at balance

$$Z_1\mathbf{I}_1 = Z_3\mathbf{I}_3 \quad \text{and} \quad Z_2\mathbf{I}_2 = Z_4\mathbf{I}_4$$

where

$$\mathbf{I}_1 = \mathbf{I}_2 \quad \text{and} \quad \mathbf{I}_3 = \mathbf{I}_4$$

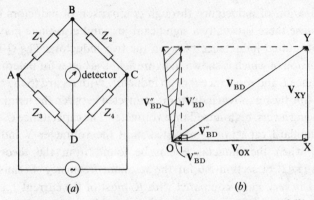

7.5 (*a*) The general Wheatstone form of a.c. bridge and (*b*) the approach to its double balance condition through successive alternate attempts at nulling the in-phase and quadrature components of potential difference across the detector.

because no current is flowing through the detector. Hence at balance

$$Z_1/Z_2 = Z_3/Z_4 \tag{7.8}$$

Although at first sight this may appear to be a single balance condition, in order to be satisfied, the real and imaginary parts of the two sides of the equation must be separately equal so that it is in fact a *double balance* condition as anticipated. Considering the potentials at B and D to have components in phase and in quadrature with the supply, the double balance condition (7.8) is seen to correspond to these in-phase and quadrature components being separately equal which is equivalent to the amplitudes and phases of the potentials at B and D being the same.

Normally several successive alternate balancings of the in-phase and quadrature components are needed before a sufficiently fine approximation to the perfect double balance condition is reached. The reason for this may be understood with reference to figure 7.5(*b*) in which $\mathbf{V}_{BD}$ is the phasor representation of the potential difference between nodes B and D, that is, across the detector, before balancing is commenced. Now the usual type of detector only indicates the magnitude of the potential difference between B and D. Thus, if balance is approached by reducing the in-phase component $\mathbf{V}_{OX}$ of $\mathbf{V}_{BD}$, there will be a range of phasor potential differences across the detector, say, $\mathbf{V}'_{BD}$ to $\mathbf{V}''_{BD}$, that corresponds to magnitudes of potential differences indistinguishable from the minimum V_{XY}. Once this condition is reached, say with potential difference $\mathbf{V}'_{BD}$, balance is approached much better by adjusting the quadrature component $\mathbf{V}_{XY}$ to reach a new range $\mathbf{V}'''_{BD}$ to $\mathbf{V}''''_{BD}$ that corresponds to magnitudes of potential differences indistinguishable from some new much lower minimum. A return to adjusting the in-phase component will now permit a better balance still to be achieved, especially if the sensitivity of the detector can be increased.

If the balance conditions are independent of the frequency of the supply then the use of a nonsinusoidal source poses no balancing problems, the harmonic components of potential being balanced whenever the fundamental components are (see section 11.1). However, a sinusoidal source is always advisable in practice because the electrical parameter being measured might well be significantly dependent on frequency. The most convenient form of source is a tunable electronic oscillator but the mains, through a suitable step-down transformer, permits measurements to be made at the frequency of the mains. The detector can be a pair of headphones at audio frequencies, an a.c. meter or a cathode-ray oscilloscope . In each case, the sensitivity of detection of the balance condition can be enhanced by preceding the detector with a suitable electronic amplifier of adjustable gain. Sometimes while approximate balancing is carried out it is necessary to provide for a reduction in sensitivity of detection through incorporation of a suitable attenuation network. Transformer coupling of the source and/or detector to the Wheatstone network is often adopted to match impedance levels or for isolation purposes. In analogy with the direct Wheatstone bridge circuit, high sensitivity is obtained when the arm impedances Z_1, Z_2, Z_3 and Z_4 are all equal and the source and detector also present this same impedance. Although it is impossible to maintain this condition all the time, it is important to keep the impedances involved similar in magnitude.

Components for use in a.c. bridges present problems. As discussed in section 2.3, wirewound resistors possess considerable inductance and capacitance and, where resistors are required in a.c. bridges, it is best to use modern thin-film types. At very high frequencies difficulties arise through the skin effect which restricts current flow to a region near the surface of a conducting medium. Inductors exhibit inherent resistance and capacitance and are less convenient as standards than capacitors, particularly when continuous variation is required, as discussed in sections 4.1 and 4.2. In variable resistors of the decade box type of construction, the varying capacitance and possibly inductance of the switching mechanism can be a nuisance. Sometimes helpful in determining the values of unknown circuit components are *substitution* and *difference* techniques in which the value indicated from the balance of the bridge with a standard component is compared with that when the unknown is connected on its own or in parallel or series with the standard, whichever is more suitable.

At other than low frequencies (typically above ~ 100 Hz), screening precautions must be taken to avoid unintentional stray coupling, inductive or capacitive (see sections 4.2 and 4.1 respectively), between various parts of the bridge circuit. In implementing screening through incorporation of earthed metallic enclosures, either in the form of boxes round components or the braided outers of coaxial cables, care must be exercised to ensure that not more than one point of the circuit is earthed otherwise part of it will be shorted out. Although the screening increases capacitances to earth, these capacitances are definite, and extraneous potential differences are excluded from the arms of the bridge. What is more, the technique of a *Wagner earth* can be employed to eliminate the effect of the capacitances to earth on the balance point of the bridge. In figure 7.6 the capacitances to earth are represented as lumped together from nodes A, B, C and D. Notice that if nothing is done about these capacitances, they act in pairs across the arms of the bridge; for example, the reactance of C_A in series with that of C_B is in parallel with the impedance Z_1. If large enough, these capacitive reactances would significantly reduce the impedances of the arms of the bridge.

To implement a Wagner earth, an additional point E between extra arms of the bridge having impedances Z_5 and Z_6 is earthed as shown in figure 7.6. A rough balance of the original bridge is obtained first with the detector connected between nodes B and D. The detector is then connected between B and E and another balance obtained through adjustment of Z_5 and/or Z_6. Alternate adjustments in this way bring points B, D and E to earth

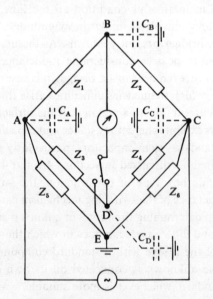

7.6 Introduction of a Wagner earth into the Wheatstone form of a.c. bridge.

potential although only E is actually connected to earth. Capacitances C_B and C_D cannot now affect the bridge because they carry no current. Since node D is not actually connected to earth, capacitances C_A and C_C simply act in parallel with the supply and also do not affect the balance of the main bridge comprising arms Z_1, Z_2, Z_3 and Z_4. That C_C acts in parallel with Z_6 and C_A in parallel with Z_5 is irrelevant because the impedances of these arms do not need to be known when measuring a component through balancing the main bridge.

A number of particularly important a.c. bridges will now be described and their circuits analysed to find their individual double balance conditions. Practical points concerning their use will also be made. For a fuller treatment of a.c. bridges the reader should consult a specialist book such as *Alternating Current Bridge Methods*, Sixth Edition, by B. Hague and T. R. Foord, Pitman, London (1971).

7.4 A.c. bridges for determining inductance

The bridge arrangement depicted in figure 7.7(*a*) and attributed to *Maxwell* allows the inductance of an inductor to be measured in terms of the capacitance of a calibrated variable capacitor and the resistances of two fixed resistors. In the circuit diagram, the inductor to be determined is represented as an inductance L in series with resistance R. When the bridge is balanced, equation (7.8) applies so that

$$R_1(1+j\omega R_2 C_2)/R_2 = (R+j\omega L)/R_4$$

Equating real and imaginary parts reveals that the double balance conditions are

$$R = R_1 R_4 / R_2 \qquad (7.9)$$

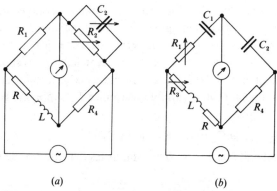

7.7 (*a*) Maxwell's L–C bridge and (*b*) Owen's bridge.

and
$$L = R_1 R_4 \not{C_2} \tag{7.10}$$

Since R_1 and R_4 appear in both equations, independent achievement of the two balance conditions is only possible through variation of just R_2 and C_2. Notice that it is customary to indicate the parameters best varied to achieve balance by drawing an arrow through them where they appear in the double balance equations, as done here. Equation (7.10) shows that the inductance may be calculated from the resistances R_1 and R_4 and the capacitance C_2 of the variable capacitor at balance, in accordance with the initial claim made for the bridge. Equation (7.9) shows that the loss of the inductor is given by the same resistances R_1 and R_4 and the resistance R_2 of the variable resistor at balance. Equations (7.9) and (7.10) combine to show that, when the time constant L/R of the inductor is very large, the time constant $C_2 R_2$ must be just as large at balance so that it may not be feasible to reach the value of R_2 needed to achieve balance. In these circumstances successful balancing can be achieved through a modification due to *Hay* in which the parallel combination of variable resistance and capacitance is replaced by a series combination of variable resistance and capacitance. The parallel equivalent of the balancing series combination is then easily calculated to obtain L and R from equations (7.9) and (7.10). Alternatively, the balance conditions of the Hay bridge can be derived to allow L and R to be calculated. Attention is drawn to the fact that, while the balance conditions (7.9) and (7.10) for the Maxwell L–C bridge are independent of frequency, those for the Hay version are frequency dependent, so that for good results with this version an extremely pure sinusoidal source is needed and, ideally, a detection system that only responds to the same frequency. A sound arrangement is an electronic oscillator and detection system employing ganged selective tuning.

It is more convenient to balance a bridge by adjustment of a calibrated variable resistor than by adjustment of a calibrated variable capacitor. *Owen's bridge*, which is shown in figure 7.7(b) and is balanced by means of two variable resistors, is therefore very attractive for determining inductance. Here the inductance L of an inductor having resistance R is found in terms of a standard fixed capacitance C_2, a standard fixed resistance R_4 and a calibrated variable resistance R_1. Applying equation (7.8) to the bridge gives

$$(R_1 + 1/j\omega C_1)j\omega C_2 = (R + R_3 + j\omega L)/R_4$$

and equating real and imaginary parts establishes that the double balance conditions are

$$(R + \not{R_3})/R_4 = C_2/C_1 \tag{7.11}$$

7.5 The Schering bridge

and
$$L = R_4 R_1' C_2 \tag{7.12}$$

Notice that these balance conditions are independent of frequency and independently achievable through adjustment of R_1 and R_3. Residual inductance of leads, resistors and terminals can be eliminated from the measurements by balancing the bridge first with the inductor included and then with it short-circuited. If R_1 and R_1' are the respective balance values then

$$L = R_4 C_2 (R_1' - R_1'') \tag{7.13}$$

The various bridges considered can be modified to allow the inductance of an inductor to be measured as a function of the direct current carried. This is important for inductors with cores of such as ferromagnetic materials where the incremental or small-signal inductance is strongly dependent on any direct current. It turns out that Hay's bridge is highly convenient for such investigations.

7.5 The Schering bridge for determining capacitance

The particular bridge shown in figure 7.8(a) that was first suggested by *Schering* has proved to be extremely versatile and capable of high accuracy with respect to measuring capacitance. In the bridge, capacitance C_2 is provided by a calibrated, variable, air capacitor because such a device exhibits an absolutely negligible power factor, while resistance R_2 is supplied by a calibrated, variable resistance box. The capacitor undergoing measurement is represented conveniently by capacitance C in series with a small resistance loss R_s. From equation (7.8), balance occurs in the bridge

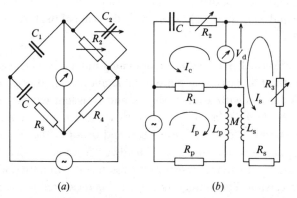

7.8 (a) The Schering bridge and (b) the Heydweiller bridge.

when

$$\left(\frac{1}{j\omega C_1}\right)\left(\frac{1+j\omega R_2 C_2}{R_2}\right) = \left(R_s + \frac{1}{j\omega C}\right)\bigg/R_4$$

and equating real and imaginary parts reveals that the double balance conditions are

$$R_s = R_4 C_2/C_1 \tag{7.14}$$

and

$$C = R_2 C_1/R_4 \tag{7.15}$$

Since both conditions feature R_4 and C_1, their independent satisfaction is only possible through variation of R_2 and C_2. Notice that the capacitance C is determined in terms of the fixed standard capacitance C_1, the fixed standard resistance R_4 and the calibrated variable resistance R_2. Although capacitance C_2 must be varied to reach balance, its value need not be known unless it is also required to determine the loss resistance R_s. Because C is proportional to R_2, the bridge can easily be made direct reading in capacitance through suitable marking of the resistance scale of R_2. Unfortunately, despite the fact that frequency does not appear explicitly in equation (7.14) or (7.15), these balance conditions are not entirely independent of frequency because R_s depends on frequency to some extent. Once more a good sinusoidal source and ganged, tuned detector are desirable.

For the majority of capacitors measured, the loss R_s is small enough to express the power factor as $\omega C R_s$ (refer back to section 5.8 if necessary). Equations (7.14) and (7.15) show that, in such situations, the power factor is given by

$$\cos\phi = \omega R_2 C_2 \tag{7.16}$$

Consequently, if balance is achieved through adjustment of C_2 and R_4 rather than C_2 and R_2, the dials of C_2 and R_4 can be made direct reading in power factor and capacitance although only the scale of the former will be linear.

7.6 The Heydweiller bridge for determining mutual inductance

A method of measuring mutual inductance, devised originally by Carey Foster for use with a direct source and transient techniques, was adapted by *Heydweiller* for operation with an alternating source. The bridge in question is shown in figure 7.8(b) and is not of the Wheatstone form. In adapting it to a.c. operation, Heydweiller found it necessary to incorporate the extra variable resistor of resistance R_2 compared with the Carey Foster version. The mutual inductor being measured is assumed to exhibit mutual inductance M, primary and secondary self inductances L_p

7.6 The Heydweiller bridge

and L_s and corresponding resistances R_p and R_s. Applying Kirchhoff's voltage law to the minimal meshes that include the detector gives

$$j\omega M \mathbf{I}_p + j\omega L_s \mathbf{I}_s + (R_3 + R_s)\mathbf{I}_s + \mathbf{V}_d = 0$$

and

$$(R_2 + 1/j\omega C)\mathbf{I}_c + R_1(\mathbf{I}_c + \mathbf{I}_p) - \mathbf{V}_d = 0$$

where $\mathbf{I}_p$, $\mathbf{I}_s$ and $\mathbf{I}_c$ denote phasor representations of the mesh currents and $\mathbf{V}_d$ the phasor representation of the potential difference across the detector as indicated in the figure. Now at balance $\mathbf{I}_c = \mathbf{I}_s$ and $\mathbf{V}_d = 0$, so that

$$\frac{\mathbf{I}_p}{\mathbf{I}_s} = -\left(\frac{R_3 + R_s + j\omega L_s}{j\omega M}\right) = -\left(\frac{R_1 + R_2 + 1/j\omega C}{R_1}\right)$$

Equating real and imaginary parts, the double balance conditions are seen to be

$$M = R_1(R_3^7 + R_s)C \tag{7.17}$$

and

$$L_s = \left(1 + \frac{R_2^7}{R_1}\right)M \tag{7.18}$$

First of all, note that the mutual inductance must be connected in the correct sense otherwise achievement of balance will be impossible. Secondly, $L_s > M$ is required to allow the balance condition (7.18) to be achieved. Any problem here can be overcome by swopping the primary and secondary windings. Alternatively extra inductance can be added to the secondary circuit, taking care to avoid further mutually inductive couplings.

Once the foregoing points have been taken care of, independent achievement of the balance conditions is possible through variation of R_2 and R_3. Equation (7.17) shows that the mutual inductance is determined in terms of standard fixed capacitance C, standard resistance R_1, and calibrated variable resistance R_3. Because of the finite resistance R_s of the secondary winding it will be necessary to balance the bridge for more than one value of R_1 in order to find M. Balancing for several values of R_1 permits a graph to be drawn of the balance value of R_3 versus $1/R_1$. From equation (7.17)

$$R_3 = M/CR_1 - R_s$$

so that the slope of this graph is M/C and the intercept on the R_3 axis is $-R_s$, from which both M and R_s may be obtained. The secondary inductance can also be deduced from the balance condition (7.18).

One of many interesting alternative ways of measuring mutual inductance uses any bridge that measures self inductance together with the following technique. When the primary and secondary coils of a mutual inductor are connected in series, the self inductance exhibited by the combination is easily shown to be

$$L = L_s + L_p \pm 2M$$

the sign depending on the sense of the series connection of the two coils. Measurement of the self inductance for the two series connections therefore yields the mutual inductance as a quarter of the difference between the self inductances.

7.7 A.c. bridges for determining the frequency of a source

Of the various bridges considered so far, only the Hay bridge mentioned in section 7.4 features balance conditions that exhibit an explicit dependence on frequency so that it can be used to measure the frequency of the source in terms of appropriate components. A simple and popular bridge that exhibits frequency-dependent balance conditions and is widely used to measure the frequencies of sources is the *Wien* bridge shown in figure 7.9(a) comprising just resistors and capacitors. Its essential circuitry is much used for the purpose of frequency selection in electronic equipment, for example, in controlling the output frequency of an electronic oscillator.

At balance of the bridge

$$\frac{R_1/j\omega C_1}{(R_1 + 1/j\omega C_1)R_2} = \frac{R_3 + 1/j\omega C_3}{R_4}$$

or

$$(1/R_1 + j\omega C_1)(R_3 + 1/j\omega C_3) = R_4/R_2$$

Equating real and imaginary parts reveals that the double balance conditions are

$$\frac{R_3}{R_1} + \frac{C_1}{C_3} = \frac{R_4}{R_2} \tag{7.19}$$

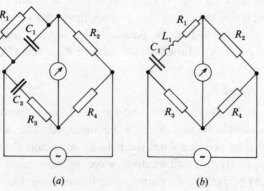

7.9 (a) The Wien bridge and (b) a series resonant bridge.

and
$$\omega^2 = 1/R_1C_1R_3C_3 \tag{7.20}$$
Independent attainment of these balance conditions is not possible. However, a neat way of operating the bridge was devised by *Robinson*. He arranged for C_1 and C_3 to be fixed and equal, say $C_1 = C_3 = C$, and R_1 and R_3 to be variable but equal, say $R_1' = R_3' = R'$, by employing ganged variable resistors to provide these resistances. In these circumstances, equations (7.19) and (7.20) reduce to
$$R_4/R_2 = 2 \tag{7.21}$$
and
$$\omega = 1/RC \tag{7.22}$$
Thus balance can be achieved by making the ratio R_4/R_2 permanently equal to two and then adjusting the ganged resistances $R_1' = R_3' = R'$ for balance. In Robinson's version, resistances R_1 and R_3 were special ganged conductances so that their dials carried a direct frequency reading, according to equation (7.22).

When an inductor and capacitor are arranged in a single arm, the bridge is best regarded as a *resonance* or *tuned-arm* bridge. A simple bridge based on series resonance (see section 5.6) is shown in figure 7.9(b). Resistance R_1 represents the total effective series resistance of the capacitor and inductor. When
$$\omega L_1 = 1/\omega C_1 \tag{7.23}$$
the branch containing the reactances is nonreactive and if also
$$R_1/R_2 = R_3/R_4 \tag{7.24}$$
the bridge is balanced. Clearly the frequency of the source driving this bridge can be determined from the values of L_1 and C_1 at balance, the two balance conditions being easily achieved independently through variation of C_1 and one resistance. The bridge is also sensitive and highly accurate for finding the loss of a capacitor or inductor from the value of R_1 indicated by R_2R_3/R_4 at balance. It is particularly good for determining inductive loss since a capacitor of negligible loss is easily provided.

7.8 Transformer ratio-arm bridges

In sections 6.2 and 6.3 it was established that time-dependent potential differences across the secondary and primary windings of a unity-coupled, lossless transformer are in the same ratio as the turns ratio between these windings. Thus if a close approximation to such a transformer features in the bridge circuit of figure 7.10(a) with its secondary winding tapped so that it divides into two portions having turns N_1 and N_2,

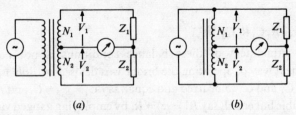

7.10 (*a*) Simple, transformer, ratio-arm bridge and (*b*) the same except that an autotransformer is used.

the ratio V_1/V_2 of the potential differences across these secondary portions will be near enough N_1/N_2. In particular, for sinusoidal excitation, the ratio between the secondary phasor potential differences will be

$$\mathbf{V}_1/\mathbf{V}_2 = N_1/N_2 \tag{7.25}$$

But when the bridge is balanced as judged by null reading of the detector, say by varying Z_2,

$$\mathbf{V}_1/Z_1 = \mathbf{V}_2/Z_2 \tag{7.26}$$

so that

$$Z_1 = (N_1/N_2)Z_2 \tag{7.27}$$

Clearly, balancing the bridge circuit allows an unknown impedance Z_1 to be found in terms of a known impedance Z_2 and known turns ratio N_1/N_2. An important advantage of transformer ratio-arm bridges is that the turns ratio may be known to an accuracy as high as 1 part in 10^7. What is more, by including a number of suitable tapping points, the turns ratio N_1/N_2 may be varied over an enormous range to permit the comparison of widely differing impedances. Just how separate balances can be obtained for the real and imaginary parts of an unknown impedance in a transformer ratio-arm bridge is considered later (see figure 7.13).

Figure 7.10(*b*) presents a variation of the bridge circuit of figure 7.10(*a*) that employs an autotransformer to achieve the potential difference ratio N_1/N_2. It has the advantage that current drawn by the arms of the bridge is supplied by the source and so loading of the ratio windings and possible consequential disturbance of the potential difference ratio from N_1/N_2 is avoided. Actually, the potential difference ratio created by a winding is exactly equal to the turns ratio even when there is magnetic leakage, provided the winding consists of identical sections, each of which is identically coupled to any other, and the tapping point is taken between these sections. This last fact is made use of in achieving highly accurate potential-difference ratios. How such decade section windings can be interconnected to give fine subdivision of an alternating potential difference is illustrated in figure 7.11 for a three-decade divider. Note that the load

7.8 Transformer ratio-arm bridges

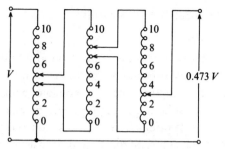

7.11 Three-decade, transformer, potential divider.

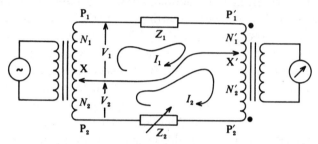

7.12 Essential circuit of a double-transformer, ratio-arm bridge.

each decade places on the previous decade is not serious because the output impedance of a section is very low and essentially resistive while the input impedance of a decade is large and inductive.

A development of the single-transformer ratio-arm bridge, naturally known as the *double-transformer ratio-arm* bridge, is shown in its essential form in figure 7.12. Here the detector is connected through a second closely coupled low-loss transformer, the primary of which has an adjustable tapping X' that divides it into sections with turns N'_1 and N'_2. With reference to figure 7.12, this arrangement acts as a current comparator for the upper and lower meshes. The senses of winding of the primary turns N'_1 and N'_2 are such that when

$$N'_1 \mathbf{I}_1 = N'_2 \mathbf{I}_2 \tag{7.28}$$

there is no magnetic flux in the core of this second transformer and therefore no signal registered by the detector. However, absence of magnetic flux in the core of the second transformer also means that points P'_1, X' and P'_2 are virtually at the same potential. Consequently, at balance, in addition to equation (7.28) applying, the mesh currents are given by

$$\mathbf{I}_1 = \mathbf{V}_1/Z_1, \quad \mathbf{I}_2 = \mathbf{V}_2/Z_2 \tag{7.29}$$

Now the potential differences $\mathbf{V}_1$ and $\mathbf{V}_2$ applied to the two meshes by the source transformer are again in the ratio given by equation (7.25) and combining equations (7.25), (7.28) and (7.29) yields

$$Z_1 = (N_1/N_2)(N'_1/N'_2)Z_2 \tag{7.30}$$

as the balance condition for the double-transformer ratio-arm bridge. The availability of two variable transformer ratios allows even more widely differing impedances to be compared than is possible with single-transformer bridges. Crucially, although the impedances between P'_1, X' and P'_2 are extremely tiny when the bridge is balanced, as soon as it goes out of balance the large primary inductance of the detector transformer comes into play so that the balance condition is very critical and the bridge consequently very sensitive. Yet another attractive feature is that earthing XX' prevents stray capacitances to earth from interfering with component measurement. Strays from P_1 and P_2 simply shunt the transformed supplies and, if the losses in the source transformer are low enough, merely reduce the sensitivity of measurement. Stray capacitances to earth from P'_1 and P'_2 are shorted out by the primary of the detector transformer at balance. The ease with which the effects of stray capacitances are dealt with renders the bridge suitable for making measurements at high frequencies and commercial versions are available that operate up to frequencies ~ 100 MHz.

The connections for one practical double-ratio bridge are outlined in figure 7.13. Balance is reached by adjustment of the decade switches operating over the tappings of the secondary of the source transformer T_1.

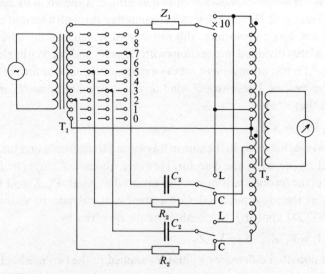

7.13 A practical, double-transformer, ratio-arm bridge.

7.8 Transformer ratio-arm bridges

These control the amplitudes of potential difference applied to a bank of identical fixed standard capacitors and resistors in steps of one-tenth from zero to nine-tenths of a maximum V_s, say. The first resistor is connected to the extremity of the lower primary of the detector transformer T_2, the second to the one-tenth tapping of the same winding and so on. The first capacitor is connected through a switch to the extremity of the lower or upper primary of the detector transformer depending on whether the impedance Z being measured is capacitive or inductive. The second capacitor is connected through a ganged switch to the one-tenth tappings of the same windings and so on. Potential difference of amplitude V_s appearing across the entire secondary of the source transformer T_1 is connected through the unknown impedance Z_1 to the whole, one-tenth and so on tappings of the upper primary of the detector transformer T_2 through a range switch. Thus the tappings of the secondary of the source transformer connected to successive capacitors and resistors give successive decades of the unknown capacitance or inductance and resistance. Only two decades are shown for reasons of labour and clarity. Note that when an inductance is responsible for the reactive part of the unknown impedance Z_1, it is given by $1/\omega^2 C$, where C is the effective value of all the standard capacitances taking into account the tappings of the source and detector transformers at balance and the position of the range switch.

A transformer can also serve as a current comparator in a conventional Wheatstone bridge type of circuit as shown in figure 7.14. It replaces arms that provide a simple resistance ratio, with much advantage. At balance, the currents through the unity-ratio arms BC and CD of figure 7.14 are equal and the sense of the windings is such that the magnetic fluxes created in the core of the transformer by them cancel. The potential differences between B, C and D are again very small, arising only from small losses in

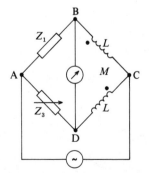

7.14 Wheatstone bridge employing a unity-ratio transformer.

the transformer. Effects of stray capacitance to earth are removed by earthing B, C or D. Off balance, the currents in the coils are no longer equal and a large inductive reactance appears between B and C and between C and D so that the balance condition of the bridge is critical and the bridge consequently sensitive.

8
Attenuators and single-section filters

8.1 Attenuators

The term *attenuation* may describe any reduction in magnitude of an electrical signal but an electrical network is only called an *attenuator* if it reduces the magnitude of a signal without changing its time dependence. Since Fourier analysis shows (see chapter 11) that all signals comprise certain combinations of pure sinusoidal signals, to perform as an attenuator, a circuit must reduce the amplitudes of all sinusoidal signals by the same factor irrespective of frequency and without changing their phases. Attenuators that reduce the potential difference by a given factor are vastly more common than current attenuators and are often appropriately referred to as *potential dividers*.

Purely resistive circuits respond identically to sinusoidal signals of all frequencies and, since they only affect the amplitude, act as attenuators. The simplest form of potential divider that gives a fixed attenuation between input and output is shown in figure 8.1(*a*). It provides attenuation of potential difference represented by

$$V_o/V_i = R_1/(R_1 + R_2) \tag{8.1}$$

when it is negligibly loaded. Its design to avoid significant loading at the output or input has already been considered in section 3.7 for the special case of a direct input, which can be thought of as a sinusoidal input of zero frequency, but the discussion and conclusions reached apply equally well whatever the time dependence of the input signal. The extended version of the basic potential-divider circuit shown in figure 8.1(*b*) provides a range of selectable stepped attenuations. Designs giving decade or binary steps are popular, the latter enabling an indicating instrument connected to the output always to operate at a substantial fraction of full scale as the input signal changes, with attendant reading accuracy advantage. A resistance

168   *Attenuators and single-section filters*

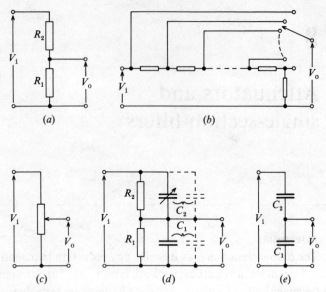

8.1 Potential-divider circuits.

potentiometer connected as shown in figure 8.1(c) delivers a continuous range of potential division from unity to infinite. Potentiometers having linear or logarithmic variation of resistance with position of the contact are available, the latter being, for example, well suited to the control of volume in audio systems since the response of the ear to sound is logarithmic.

Unfortunately it is not possible to construct an absolutely purely resistive circuit in practice. With reference to the circuit of figure 8.1(a), there will always be some capacitance in parallel with R_1 and R_2. The capacitance may be just stray or it may be associated with further circuitry or equipment connected to the potential divider. Such capacitance has a detrimental effect on the attenuating performance, both changing the reduction in amplitude between input and output and introducing a phase shift at sufficiently high frequency. Clearly, the range of frequencies over which a potential divider constructed solely from resistors acts as a satisfactory attenuator is restricted. The detrimental effect on fast pulses is particularly serious, distortion of the types shown in figures 4.12(b) and 4.13(b) occurring on account of the transient response of the resistance–capacitance combinations.

To overcome the capacitive shorting problem, potential dividers for operation at high frequencies or for handling pulses are usually constructed as indicated in figure 8.1(d). Perhaps surprisingly at first sight, additional variable capacitance is introduced in parallel with R_2 and additional fixed capacitance in parallel with R_1. Now if C_1 and C_2 represent total

8.1 Attenuators

capacitances in parallel with R_1 and R_2 respectively, the ratio between the output and input potential difference phasors of this circuit as a function of the pulsatance ω is readily shown to be

$$\frac{V_o}{V_i} = \left(\frac{R_1}{1+j\omega R_1 C_1}\right) \bigg/ \left[\left(\frac{R_1}{1+j\omega R_1 C_1}\right) + \left(\frac{R_2}{1+j\omega R_2 C_2}\right)\right] \quad (8.2)$$

Consequently, if C_2 is adjusted such that

$$R_1 C_1 = R_2 C_2 \quad (8.3)$$

which means that the time constants of the two parallel R–C combinations are the same, the potential difference division avoids phase shift, is given by equation (8.1) again and, in particular, is independent of frequency! Adjustment of C_2 to procure division of potential that is independent of frequency may be carried out directly working with sinusoidal inputs over a wide range of frequencies or more conveniently through monitoring the distortion of fast pulses. Some deliberate fixed capacitance is incorporated in parallel with R_1, firstly to make the performance sufficiently independent of connections to the output and, secondly, to render the value of C_2 that satisfies equation (8.3) large enough to implement in practice, especially when the attenuation is large so that $R_2 \gg R_1$ and $C_2 \ll C_1$. Notice that the technique is extremely difficult to apply to the potentiometer divider of figure 8.1(c) because the resistance ratio can be varied continuously to alter the division of potential, and any introduced ratio of parallel capacitance would need to be capable of being varied in sympathy. However, the technique is applicable to the step divider of figure 8.1(b) to provide variable if not continuous potential division.

When inserting an attenuator between a source and load in order to implement attenuation, it is often highly convenient if the loading of the source is unaltered by the insertion. The advantage of this approach is that, whatever the impedance of the source, the signal fed to the input of the attenuator following its insertion is the same as that fed to the load prior to insertion. Figures 8.2(a) and (b) illustrate the point, the potential divider exhibiting attenuation represented by $V'/V = A < 1$ and input resistance equal to the load resistance R_L. In the case of the simple potential divider depicted in figure 8.2(c), the requirement of unaltered loading on insertion between a source and load resistance R_L demands that

$$R_L = R_2 + \frac{R_1 R_L}{(R_1 + R_L)} \quad (8.4)$$

But the attenuation is

$$A = V'/V = \left(\frac{R_1 R_L}{R_1 + R_L}\right)\left(R_2 + \frac{R_1 R_L}{R_1 + R_L}\right)^{-1}$$

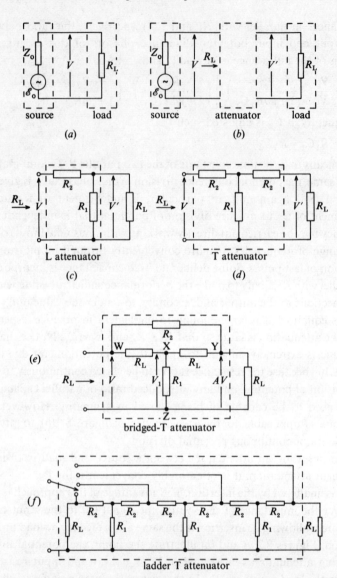

8.2 (a) and (b) Illustration of the introduction of attenuation between a source and load without altering the loading of the source. (c), (d) and (e) Attenuators designed (see text) such that their input resistance equals the load resistance. (f) A switchable ladder attenuator designed (see text) such that its input resistance equals half the load resistance.

8.1 Attenuators

or through condition (8.4)

$$A = R_1/(R_1 + R_L) \tag{8.5}$$

From equations (8.4) and (8.5) it follows that to implement attenuation A through a simple potential divider without altering the loading of the source, the values of resistances R_1 and R_2 must be

$$R_1 = AR_L/(1-A); \quad R_2 = (1-A)R_L \tag{8.6}$$

Notice that to provide variable attenuation yet maintain unaltered loading, that is, input resistance equal to load resistance, both resistances R_1 and R_2 must be adjusted in such a way as to comply with equations (8.6). When $A \ll 1$, equations (8.6) simplify to

$$R_1 = AR_L; \quad R_2 = R_L \tag{8.7}$$

In general, all that is needed to provide a given attenuation and an input resistance equal to the load resistance is a resistive network with two independently selectable resistances. The symmetric T network of figure 8.2(d) comprising two equal resistances R_2 and a different resistance R_1 is therefore also suitable for the purpose. For its input resistance to equal the load resistance R_L

$$R_L = R_2 + \frac{R_1(R_2 + R_L)}{R_1 + R_2 + R_L}$$

while the attenuation is

$$A = \left[\frac{R_1(R_2 + R_L)}{(R_1 + R_2 + R_L)R_L}\right]\left[\frac{R_L}{R_2 + R_L}\right]$$

Combining these last two equations shows that the values of R_1 and R_2 needed for the circuit to perform as required are

$$R_1 = 2AR_L/(1-A^2); \quad R_2 = (1-A)R_L/(1+A) \tag{8.8}$$

The T network is actually the star network considered in the context of direct currents at the end of section 3.2 where it was shown that it transforms into an equivalent delta network. One terminal of this equivalent is again common to the input and output and the delta network can be redrawn with a common line between its input and output and with its components arranged in a Π shape. In such circumstances it is more appropriate to describe the delta network as a Π network and there is clearly a Π version of the attenuator just treated.

A snag with the T attenuator is that three resistors need to be adjusted to vary the attenuation yet maintain the input resistance equal to the load resistance. This difficulty is overcome in the interesting bridged-T attenuator shown in figure 8.2(e). The circuit plus load resistance R_L is of the Wheatstone bridge type with arms WY, YZ, ZX and XW. Insight into its operation is gained by considering the balanced condition in which node-pair potential difference V_1 equals the attenuated output potential difference AV so that no current flows through the connecting branch XY. Balance occurs when R_2/R_L equals R_L/R_1 or

$$R_1 R_2 = R_L^2 \tag{8.9}$$

Thus the input resistance at balance comprising $(R_2 + R_L)$ in parallel with $(R_1 + R_L)$ amounts to

$$\frac{(R_1 + R_L)(R_2 + R_L)}{R_1 + R_2 + 2R_L} = \frac{(R_1 + R_2)R_L + 2R_L^2}{R_1 + R_2 + 2R_L} = R_L$$

as required. To achieve balance with attenuation A

$$R_1/(R_1 + R_L) = A = R_L/(R_2 + R_L)$$

or

$$R_1 = AR_L/(1-A); \quad R_2 = (1-A)R_L/A \tag{8.10}$$

in accordance, of course, with equation (8.9). Notice that when A is very small, equation (8.10) reveals that $R_2 \gg R_L$ and $R_1 \ll R_L$ so that branch WX essentially provides the input resistance R_L. On the other hand, when $A = 1$, equation (8.10) reveals that $R_2 = 0$ and $R_1 = \infty$ so that the load provides the input resistance R_L.

An extremely attractive feature of attenuators having input resistance equal to load resistance is that identical ones may be cascaded to give compound attenuation yet still maintain the overall input resistance equal to the load resistance. A switchable compound or *ladder* attenuator formed from three T sections is shown in figure 8.2(f). The input resistance in this case is, of course, constant at $R_L/2$ as the input switch is changed. Ladder attenuators incorporating Π sections turn out to be more economical in components than those formed from T sections because adjacent pairs of shunt resistors can be merged into single resistors.

A basic high-frequency signal potential divider may be constructed from two low-loss capacitors as shown in figure 8.1(e). The capacitive reactances are designed to be low over the operating range of frequencies compared with any incidental shunting capacitive reactance or resistance. Stray capacitance now simply alters the high-frequency potential division obtained. Whereas resistive potential dividers dissipate electrical power, capacitive dividers do not.

In chapters 6 and 7 it has been established that accurate reduction in the amplitude of alternating potential difference or current can be achieved using a transformer. Here again there is negligible waste of power, but unfortunately the range of frequencies over which a transformer will act as a potential or current divider is restricted by the behaviour of the core, there being both a lower and upper limit.

8.2 Simple single-section filters

Whenever the potential difference across a resistance is reduced to a fraction A of a previous value, the current through it is also reduced to the same fraction A and the attenuation of power is given by A^2. Because power can vary over such enormous ranges, it is customary to describe the attenuation of power in *decibel* units, the ratio of two powers P_1 and P_2 being described in terms of log (P_1/P_2) *bels* or 10 log (P_1/P_2) *decibels*. Thus, making use of the abbreviation dB for decibel, when $A = 0.1$, for example, the power is said to be modified by

$$10 \log (0.01) \text{ decibels} = -20 \text{ dB}$$

and when $A = 0.5$, the modification of power is described as

$$10 \log (0.25) \text{ dB} = 10(\bar{1}.3979) \text{ dB} = -6.021 \text{ dB} \approx -6 \text{ dB}$$

An alternative expression of the power change in these two cases would be that the power is reduced or attenuated by $+20$ dB and $+6$ dB respectively.

Common load resistances that attenuators are designed to operate into are $600\,\Omega$, $75\,\Omega$ and $50\,\Omega$. This is because certain instrumentation and connecting cables (see section 9.5) are arranged to match these impedances; $600\,\Omega$ is the adopted standard in professional audio-frequency systems, $75\,\Omega$ the standard in domestic television and $50\,\Omega$ the standard in radio-frequency instrumentation.

8.2 Simple single-section filters

Selective reduction of the amplitudes of sinusoidal electrical signals as a function of frequency is described as *filtering*. Naturally, networks that achieve such frequency-dependent reduction are termed *filters*. Since any nonsinusoidal signal is the sum of a frequency spectrum of sinusoidal signals (see chapter 11), filtering generally alters the time dependence of a nonsinusoidal signal. In electronics, filters are used, for example, to suppress an undesirable signal that occurs at some frequency, to extract sinusoidal signals over some particular frequency band from a wider range of sinusoidal signals and to convert a nonsinusoidal signal into a sinusoidal signal of the same period.

It should be clear from the basic theory of chapter 5 that, to implement filtering, a circuit must include at least one reactive component. Inevitably, filtering action is accompanied by a frequency-dependent phase shift. Filters are mostly of the *low-pass, high-pass, band-pass* or *band-stop* varieties. Ideally, as the names imply, a low-pass filter passes signals up to some limiting frequency but not above it, a high-pass filter passes signals down to some limiting frequency but not below it, a band-pass filter passes signals over a range of frequencies but not outside it and a band-stop filter only passes signals outside a range of frequencies. Practical filters fall short of these ideals of course.

Two very simple C–R filters are shown in figure 8.3. They have already been encountered in section 4.4 in connection with electrical transients but are reproduced here (in reverse order) for easy reference. Consider their response to a steady sinusoidal input signal when a load resistance R_L is connected across the output terminals. The ratio of the phasor output to input potential difference, V_o/V_i, is known as the *transfer function* and denoting this function by $\mathcal{T}$, for the circuit of figure 8.3(a),

$$\mathcal{T} = \left(\frac{R_L}{1+j\omega CR_L}\right) \bigg/ \left(R + \frac{R_L}{1+j\omega CR_L}\right)$$

$$= R_L/(R_L + R + j\omega CR_L R)$$

$$= \left(\frac{R_L}{R_L + R}\right) \bigg/ (1+j\omega CR')$$

where $R' = R_L R/(R_L + R)$ represents the resistance of R in parallel with R_L. At low-enough frequencies to satisfy $\omega CR' \ll 1$, the transfer function becomes simply $R_L/(R_L + R)$. At higher frequencies the transmission clearly falls below this value so that the circuit behaves as a low-pass filter. To approach the ideal of nearly 100% pass at low frequencies, the circuit must be designed such that $R \ll R_L$. Within this approximation

$$\mathcal{T} = (1+j\omega CR)^{-1} \tag{8.11}$$

$$|\mathcal{T}| = (1+\omega^2 C^2 R^2)^{-\frac{1}{2}} \tag{8.12}$$

$$\tan \phi = -\omega CR \tag{8.13}$$

where ϕ is the phase shift between the input and output potential differences.

The performance of the simple low-pass C–R filter is best displayed over a large range of frequency in a plot of $\log |\mathcal{T}|$ versus $\log \omega$. Such a plot is

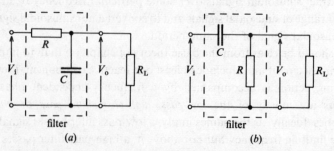

8.3 Basic C–R filters loaded with resistance R_L; (a) low pass and (b) high pass.

8.2 Simple single-section filters

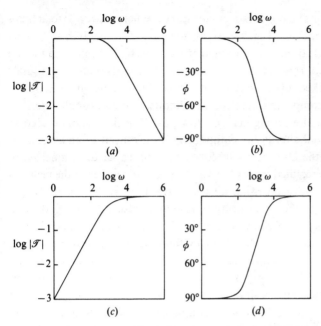

8.4 Responses of simple C–R filters showing the behaviour of the modulus of the transfer function, $|\mathcal{T}|$, and the phase shift, ϕ, as a function of the pulsatance ω; (a) and (b) for the low-pass filter of figure 8.3(a) and (c) and (d) for the high-pass filter of figure 8.3(b), when $RC = 1$ ms.

presented in figure 8.4(a) for the case where the time constant RC of the filter is 1 ms, as might be provided by $R = 100 \, \Omega$, $C = 10 \, \mu\text{F}$, for example. It is of course taken for granted that the loading is such that $R \ll R_L$ always so that the response is given by equations (8.11)–(8.13). The transmission begins to cut off when ω reaches $\sim 1/RC$ and, as depicted in figure 8.4(b), an accompanying change in phase shift ϕ from about zero to a lag of about 90° takes place within a decade of pulsatance either side of $1/RC$. When $\omega CR \gg 1$, $|\mathcal{T}| \approx 1/\omega CR$ and, for constant amplitude of input signal, the amplitude of the output signal falls linearly with frequency, that is, at a rate of 10 dB per decade. The corresponding rate of fall of the output signal power is 20 dB per decade or very close to 6 dB per octave (an octave is the musical term for two notes, one of which is double the frequency of the other). Actually when $\omega CR \gg 1$, the capacitive reactance $1/\omega C$ is very small compared with R and R_L so that the input impedance of the filter is almost constant and equal to R. Consequently the amplitude of input signal delivered by a sinusoidal source may well remain constant as the frequency changes.

The simple form of C–R filter just discussed is widely used in conjunction with an operational amplifier, in circumstances where $\omega CR \gg 1$, to achieve electronic integration of a signal as mentioned in section 4.4. This important practical circuit is treated in section 10.5. A more mundane application is to the reduction of interfering mains-frequency signals accidentally picked up in circuits designed to operate at even lower frequency, for example in direct-current circuits. In direct supplies derived by rectifying the alternating mains, a low-pass C–R filter is often used to reduce the residual alternating component of the output to an acceptable low level. Here the filter normally follows a tank capacitor connected in parallel with the rectified mains. This first capacitor charges to the peaks of the rectified sinewave potential during forward intervals but discharges somewhat through the load during reverse intervals as the rectified e.m.f. first falls from its peak value and then rises back to the potential difference retained on the capacitor. The following filter incorporated to reduce the residual potential fluctuation may be C–R in type provided that the series resistance R introduces only an insignificant drop in the direct output potential difference. Such a situation arises when the supply only delivers a small direct load current. Yet another application of the simple low-pass C–R filter is in *amplitude demodulation*. The information carried by an amplitude-modulated signal (see section 5.9) is recovered by rectifying it and filtering out the carrier wave to leave the amplitude-modulating wave. In this case, the RC time constant of the filter must be long compared with the period of the carrier but short compared with the briefest period involved in the modulation.

Turning to the simple C–R circuit of figure 8.3(b)

$$\mathcal{T} = R'/(R' - \mathrm{j}/\omega C) = (1 - \mathrm{j}/\omega CR')^{-1} \tag{8.14}$$

$$|\mathcal{T}| = [1 + 1/\omega^2 C^2 (R')^2]^{-\frac{1}{2}} \tag{8.15}$$

$$\tan \phi = 1/\omega CR' \tag{8.16}$$

where again R' represents the resistance of R in parallel with R_L and ϕ the shift in phase of the potential difference between input and output. To illustrate the behaviour of this type of filter, figures 8.4(c) and (d) respectively show plots of $\log |\mathcal{T}|$ and ϕ versus $\log \omega$ for the case where the time constant $R'C$ is 1 ms. At high-enough frequencies to satisfy $\omega CR' \gg 1$, there is virtually 100% transmission and negligible phase shift. When the pulsatance falls to $\sim 1/R'C$, the transmission begins to fall and reaches a cut-off rate of 6 dB per octave in power when $\omega CR' \ll 1$. This time the phase shift changes over from about zero to a lead of about 90° within a decade of pulsatance either side of $1/R'C$. The response of the C–R circuit of figure 8.3(b) is clearly complementary to that of figure 8.3(a); it acts as a high-pass filter. At low-enough frequencies to satisfy $\omega CR' \ll 1$, the input impedance is

8.2 Simple single-section filters

virtually $1/\omega C$ which is becoming very large. Thus the amplitude of input signal delivered to the filter by a sinusoidal source may well remain constant as the frequency changes in this range. Very important electronic applications of the simple high-pass form of C–R circuit are to coupling a signal while blocking a direct potential difference and to differentiation of a signal as discussed in section 4.4.

Simple L–R filters that correspond to the simple C–R filters of figure 8.3 are shown loaded with resistance R_L in figure 8.5. The transfer function of the circuit of figure 8.5(a) is

$$\mathcal{T} = R'/(R' + j\omega L) = (1 + j\omega L/R')^{-1} \quad (8.17)$$

where R' represents the resistance of R in parallel with R_L, while the transfer function of the circuit of figure 8.5(b) is

$$\mathcal{T} = \left(\frac{j\omega L R_L}{R_L + j\omega L}\right) \bigg/ \left(R + \frac{j\omega L R_L}{R_L + j\omega L}\right)$$

$$= R_L \bigg/ \left(R_L + R + \frac{R R_L}{j\omega L}\right)$$

or

$$\mathcal{T} = \left(\frac{R_L}{R_L + R}\right) \bigg/ \left(1 - j\frac{R'}{\omega L}\right) \quad (8.18)$$

Respective comparison of equations (8.11) and (8.14) with equations (8.17) and (8.18) reveals that the circuit of figure 8.5(a) responds similarly to that of figure 8.3(a) and acts as a low-pass filter while the circuit of figure 8.5(b) responds similarly to that of figure 8.3(b) and acts as a high-pass filter. Notice that the frequency responses of the inductive circuits are characterised by the inductive time constant L/R'. Because of the greater size and expense of an L–R filter compared with its C–R counterpart, not to mention the less-ideal behaviour of inductors compared with capacitors, the C–R version of a filter is usually preferred to the L–R version.

Replacement of the inductor of a simple L–R filter by a series or parallel

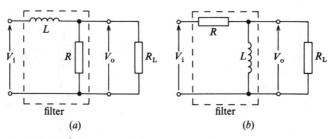

8.5 Basic L–R filters loaded with resistance R_L; (a) low pass and (b) high pass.

combination of an inductor and capacitor creates a band-pass or band-stop filter on account of the resonant response. Series resonant versions of such filters together with sketches of their frequency responses are presented in figure 8.6. The central frequency of the pass or stop band is, of course, given by $1/2\pi(LC)^{\frac{1}{2}}$ and the width of response by the Q-factor of the resonant combination.

To procure a steeper cut-off than is exhibited by the frequency responses of the simple C–R or L–R low and high-pass filters, further reactive components must be added to the network. Actually, the band-pass and band-stop filters just treated illustrate this point nicely. Consider next the unloaded low-pass L–C filter drawn in figure 8.7(a). Enhanced performance stems from the capacitive reactance falling simultaneously with the inductive reactance increasing as the frequency increases. The unloaded transfer function is

$$\mathcal{T} = \left(\frac{1}{j\omega C}\right) \bigg/ \left(j\omega L + \frac{1}{j\omega C}\right) = 1/(1 - \omega^2 LC) \qquad (8.19)$$

At sufficiently high frequencies to satisfy $1/\omega C \ll \omega L$, $|\mathcal{T}|$ falls off as $1/\omega^2$ compared with the fall off as $1/\omega$ for the simple C–R and L–R filters. Figure 8.7(b) shows $\log |\mathcal{T}|$ plotted against $\log \omega$ for the case $LC = 10^{-6}$ s^2. The infinite singularity in the response would not occur in a practical circuit because of inevitable resistive loss. The fall in $|\mathcal{T}|$ of 20 dB per decade when

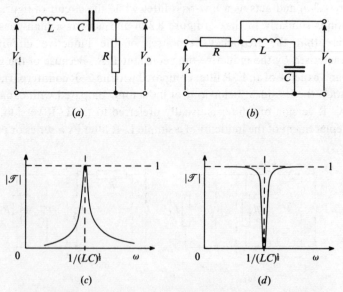

8.6 Series resonant filters; (a) band-pass, (b) band-stop, (c) and (d) sketches of the frequency responses of (a) and (b) respectively.

8.3 Wien, bridged-T and twin-T rejection filters

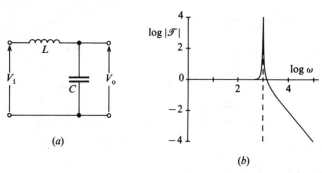

8.7 (a) Low-pass, L–C filter (unloaded) and (b) its frequency response when $LC = 10^{-6}\,\text{s}^2$.

$\omega \gg 1/(LC)^{\frac{1}{2}}$ is clear. A particularly appropriate application of this form of circuit is to the filtering of rectified mains in mains-derived direct supplies. The practical transfer function is very close to unity for the required direct component of the potential difference if low-loss reactors are incorporated yet is very tiny at the ripple frequency if $1/(LC)^{\frac{1}{2}}$ is made very small compared with the ripple pulsatance (200π for full-wave rectification, since the ripple to be smoothed is at twice the mains frequency). Note that the input impedance of this filter is extremely dependent on frequency with a sharp minimum at series resonance.

8.3 Wien, bridged-T and twin-T rejection filters

The creation of highly selective band-pass or band-stop filters based on appropriate resonant branches has been alluded to in the previous section. A problem arises with the design of such filters for passing or stopping low frequencies. To obtain a low resonant frequency, the inductance has to be very large since it is difficult to achieve very high capacitance. Even with capacitance of as much as 100 μF, inductance of 0.1 H is needed to procure resonance at 50 Hz. Components that provide such high inductance are inconveniently big and rather expensive. Fortunately band-pass and band-stop filters can be constructed from just capacitors and resistors, thereby avoiding the inductive problem. Band filters that can be tuned down to low frequencies are useful in a host of applications including electronic oscillators. As already mentioned, they are useful for strong rejection of low-frequency interfering signals originating from the mains supply, which is absolutely essential in many instances. Often band-stop filters are described alternatively as *rejection filters*.

One well-known band filter that is formed from resistors and capacitors

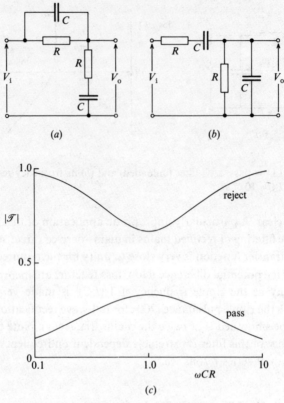

8.8 (a) Wien band-stop filter, (b) Wien band-pass filter and (c) the frequency responses of these two filters.

only is the Wien network shown in figure 8.8(a). Its transfer function in the unloaded condition is

$$\mathcal{T} = \left(R + \frac{1}{j\omega C}\right) \bigg/ \left(R + \frac{1}{j\omega C} + \frac{R/j\omega C}{R + 1/j\omega C}\right)$$

$$= (1 + j\omega CR) \bigg/ \left(1 + j\omega CR + \frac{j\omega CR}{1 + j\omega CR}\right)$$

or
$$\mathcal{T} = (1 - \omega^2 C^2 R^2 + j\omega 2CR)/(1 - \omega^2 C^2 R^2 + j\omega 3CR) \tag{8.20}$$

Hence its unloaded transmission is given by

$$|\mathcal{T}| = \left[\frac{(1 - \omega^2 C^2 R^2)^2 + 4\omega^2 C^2 R^2}{(1 - \omega^2 C^2 R^2)^2 + 9\omega^2 C^2 R^2}\right]^{\frac{1}{2}} \tag{8.21}$$

from which it can be seen that the transmission approaches 100% as the frequency tends to zero or infinity, but reaches a minimum value of $\frac{2}{3}$ when

8.3 Wien, bridged-T and twin-T rejection filters

$\omega = 1/RC$. Inspection of the circuit diagram reveals that the tendency to perfect transmission at high-enough frequencies is due to that capacitor which shorts the output to input in this range. Almost perfect transmission at low-enough frequencies stems from both capacitors tending to become open circuit. Clearly, the circuit behaves as a rejection filter and figure 8.8(c) gives its response over a range of frequencies either side of the rejection frequency $1/2\pi RC$. Provided any loading impedance is high compared with resistance R, the response will stay close to that of figure 8.8(c). Making the capacitance $C = 1\,\mu F$ and resistance $R = 1\,k\Omega$ yields a time constant $RC = 1$ ms and leads to rejection at a frequency of 160 Hz, for example. Notice that, according to equation (8.20), the output and input are in phase at the rejection frequency. Most importantly, the plot of $|\mathcal{T}|$ in figure 8.8(c) reveals that the rejection provided by a Wien network is neither sharp nor strong. The reason that a Wien bridge based on this network and already described in section 7.7 performs well is because the bridge arrangement achieves null potential difference across the detector at the rejection frequency, thereby enhancing the effect of rejection.

The Wien band-pass filter complementary to the rejection filter just considered is shown in figure 8.8(b). This time the transfer function is

$$\mathcal{T} = \left(\frac{R/j\omega C}{R + 1/j\omega C}\right) \bigg/ \left(R + 1/j\omega C + \frac{R/j\omega C}{R + 1/j\omega C}\right)$$

$$= 1 \bigg/ \left[1 + \frac{(R + 1/j\omega C)^2}{R/j\omega C}\right]$$

$$= 1 \bigg/ \left[1 + \frac{(1 + j\omega CR)^2}{j\omega CR}\right]$$

or

$$\mathcal{T} = j\omega CR/(1 - \omega^2 C^2 R^2 + j\omega 3CR) \tag{8.22}$$

Hence

$$|\mathcal{T}| = 1 \bigg/ \left[9 + \frac{(1 - \omega^2 C^2 R^2)^2}{\omega^2 C^2 R^2}\right]^{\frac{1}{2}} \tag{8.23}$$

Now there is zero transmission when $\omega = 0$ and $\omega = \infty$ with maximum transmission amounting to $|\mathcal{T}| = \frac{1}{3}$ when $\omega = 1/RC$. The asymptotic approach to zero transmission at high frequencies is associated with the capacitance in parallel with the output while the similar behaviour at low frequencies comes about because of the series capacitance between input and output. Again the filtering action is far from sharp, as can be seen from the response plotted in figure 8.8(c). Despite the somewhat diffuse action, satisfactory sinusoidal oscillators can be formed based on Wien filters as explained in section 10.6. An irritating feature of Wien filters is that to vary the pass or rejection frequency, ideally both capacitances or, as is rather easier, both resistances should be varied in sympathy.

A simple *bridged-T* form of rejection filter is shown in figure 8.9(a). Applying Kirchhoff's current law to the unloaded network, the phasor node-pair potentials V_i, V_o and V_k are found to be related by

$$j\omega kCV_k = \frac{V_o - V_k}{R} + \frac{V_i - V_k}{R}$$

$$(V_o - V_k)/R + j\omega C(V_o - V_i) = 0$$

Substituting for V_k in the second equation in terms of V_o and V_i from the first yields

$$\left(\frac{1}{R} + j\omega C\right)V_o - j\omega CV_i - \frac{V_o + V_i}{R(2 + j\omega kCR)} = 0$$

from which the transfer function is

$$\mathcal{T} = \frac{1 - \omega^2 kC^2R^2 + j\omega 2CR}{1 - \omega^2 kC^2R^2 + j\omega(k+2)CR} \qquad (8.24)$$

Consequently the ratio of potential-difference amplitude between output and input is

$$|\mathcal{T}| = \left[\frac{(1 - \omega^2 kC^2R^2)^2 + 4\omega^2 C^2R^2}{(1 - \omega^2 kC^2R^2)^2 + (k+2)^2\omega^2 C^2R^2}\right]^{\frac{1}{2}} \qquad (8.25)$$

When $k = 1$ this is the same transmission as provided by the Wien network of figure 8.8(a). In general, equation (8.25) shows that the bridged-T filter of figure 8.9(a) exhibits minimum transmission when

$$\omega = 1/k^{\frac{1}{2}}RC \qquad (8.26)$$

amounting to

$$|\mathcal{T}| = 2/(k+2) \qquad (8.27)$$

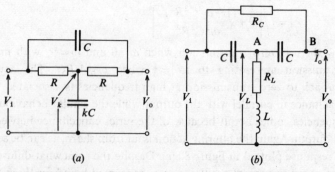

8.9 (a) A bridged-T, R–C, rejection filter and (b) a bridged-T, L–C–R, rejection filter.

8.3 Wien, bridged-T and twin-T rejection filters

Making k larger gives better rejection. Changing the rejection frequency by tuning only capacitance kC simultaneously varies the degree of rejection.

An extremely popular R–C filter, that in its ideal form provides total rejection at the designed rejection frequency, is the *twin-T* filter shown in figure 8.10(a). In practice the maximum attenuation is finite and depends on the quality of the components used to construct an approximation to the theoretical circuit of figure 8.10(a). For a high degree of rejection, the capacitors must be very low loss and the resistors must exhibit very little capacitance. Behaviour of the twin-T as a rejection filter is easily understood qualitatively on appreciating that it comprises a low-pass filter $(R, R, 2C)$ in parallel with a high-pass filter $(C, C, R/2)$. Treating the twin-T of figure 8.10(a) quantitatively by the method of node-pair analysis, Kirchhoff's current law applied at nodes X and Y respectively gives

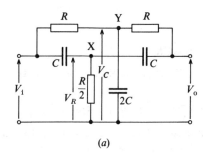

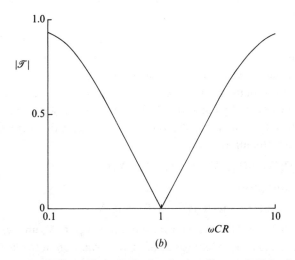

8.10 (a) Twin-T, R–C, rejection filter and (b) its frequency response.

$$2\mathbf{V}_R/R = j\omega C(\mathbf{V}_o - \mathbf{V}_R) + j\omega C(\mathbf{V}_i - \mathbf{V}_R)$$
$$j\omega 2C\mathbf{V}_C = (\mathbf{V}_o - \mathbf{V}_C)/R + (\mathbf{V}_i - \mathbf{V}_C)/R$$

On collecting terms these equations become

$$j\omega C\mathbf{V}_o + j\omega C\mathbf{V}_i - (2/R + j\omega 2C)\mathbf{V}_R = 0 \tag{8.28}$$
$$(1/R)\mathbf{V}_o + (1/R)\mathbf{V}_i - (2/R + j\omega 2C)\mathbf{V}_C = 0 \tag{8.29}$$

Now, provided that the twin-T is unloaded,

$$(\mathbf{V}_o - \mathbf{V}_C)/R + j\omega C(\mathbf{V}_o - \mathbf{V}_R) = 0$$

or

$$\mathbf{V}_C = (1 + j\omega CR)\mathbf{V}_o - j\omega CR\mathbf{V}_R \tag{8.30}$$

Substituting for $\mathbf{V}_C$ from equation (8.30) in equation (8.29)

$$(2\omega^2 C^2 R - 1/R - j\omega 4C)\mathbf{V}_o + (1/R)\mathbf{V}_i - (\omega^2 2C^2 R - j\omega 2C)\mathbf{V}_R = 0$$

and substituting for $\mathbf{V}_R$ from equation (8.28)

$$(2\omega^2 C^2 R - 1/R - j\omega 4C)\mathbf{V}_o + (1/R)\mathbf{V}_i + j\omega CRj\omega C(\mathbf{V}_o + \mathbf{V}_i) = 0$$

Hence the transfer function for the unloaded twin-T is

$$\mathcal{T} = \frac{1 - \omega^2 C^2 R^2}{1 - \omega^2 C^2 R^2 + j\omega 4CR} \tag{8.31}$$

and the ratio of potential-difference amplitude between the output and input is

$$|\mathcal{T}| = (1 - \omega^2 C^2 R^2)/[(1 - \omega^2 C^2 R^2)^2 + 16\omega^2 C^2 R^2]^{\frac{1}{2}} \tag{8.32}$$

The behaviour of $|\mathcal{T}|$ as a function of frequency according to equation (8.32) is plotted in figure 8.10(b). There is much sharper rejection than obtained with the other filters described in this section and, very significantly, there is *total* rejection when

$$\omega = 1/RC \tag{8.33}$$

While the basic circuit of figure 8.10(a) is only really suitable for rejection at a fixed frequency, variants exist which are amenable to tuning.

Before leaving the topic of the twin-T filter, it is worth pointing out that it rejects at the frequency given by equation (8.33) no matter what the load. Whenever $\mathbf{V}_o = 0$, the output current is also zero and so, because the current is continuous between the input terminals,

$$2\mathbf{V}_R/R + j\omega 2C\mathbf{V}_C = (\mathbf{V}_i - \mathbf{V}_C)/R + j\omega C(\mathbf{V}_i - \mathbf{V}_R)$$

Rearranging terms, this gives

$$(2/R + j\omega C)\mathbf{V}_R + (1/R + j\omega 2C)\mathbf{V}_C = (1/R + j\omega C)\mathbf{V}_i$$

Through equations (8.28) and (8.29) under the condition $\mathbf{V}_o = 0$, $\mathbf{V}_R$ and $\mathbf{V}_C$ can be expressed in terms of $\mathbf{V}_i$. Making use of this information in the last equation yields

8.3 Wien, bridged-T and twin-T rejection filters

or

$$\frac{(2/R+j\omega C)j\omega C}{(2/R+j\omega 2C)} + \frac{(1/R+j\omega 2C)(1/R)}{(2/R+j\omega 2C)} = \frac{1}{R} + j\omega C$$

$$\frac{j\omega 2C}{R} - \omega^2 C^2 + \frac{1}{R^2} + \frac{j\omega 2C}{R} = \frac{2}{R^2} + \frac{j\omega 4C}{R} - 2\omega^2 C^2$$

While the imaginary terms of this equation balance, the real terms give equation (8.33) again.

Another interesting network that totally rejects signals of a certain frequency is the *L–C–R bridged-T* of figure 8.9(b). Since $\mathbf{I}_o = \mathbf{V}_o = 0$ when the circuit is totally rejecting a signal, application of Kirchhoff's current law to nodes A and B reveals that total rejection occurs when the simultaneous equations

$$\mathbf{V}_L/(R_L + j\omega L) + j\omega C \mathbf{V}_L + j\omega C (\mathbf{V}_L - \mathbf{V}_i) = 0$$
$$\mathbf{V}_i/R_C + j\omega C \mathbf{V}_L = 0$$

are satisfied. Eliminating $\mathbf{V}_i$ between these equations gives

$$1/(R_L + j\omega L) + j\omega 2C - \omega^2 C^2 R_C = 0$$

or on separately equating the real and imaginary parts

$$1 - 2\omega^2 LC - \omega^2 C^2 R_C R_L = 0$$
$$2\omega C R_L - \omega^3 LC^2 R_C = 0$$

The second of these two relations simplifies to

$$\omega^2 LC = 2R_L/R_C \tag{8.34}$$

and substitution of this condition into the first yields

$$R_L = R_C/(4 + \omega^2 C^2 R_C^2) \tag{8.35}$$

Equations (8.34) and (8.35), representing the conditions that must be satisfied to procure total rejection, become much simpler if it is assumed that $\omega^2 C^2 R_C^2 \ll 4$, for they then reduce to

$$R_L \approx R_C/4 \tag{8.36}$$
$$\omega \approx 1/(2LC)^{\frac{1}{2}} \tag{8.37}$$

Evidently, if $\omega^2 C^2 R_C^2 \ll 4$ and $R_L = R_C/4$, total rejection occurs at a frequency close to $1/2\pi(2LC)^{\frac{1}{2}}$. Making use of equations (8.36) and (8.37), the simplifying condition on $\omega C R_C$ can be seen to be equivalent to

$$\omega^2 L^2 \approx 1/4\omega^2 C^2 \gg R_C^2/16 \approx R_L^2 \tag{8.38}$$

which in practice simply means that the coil providing impedance $R_L + j\omega L$ must have a high Q-factor.

Apart from the applications already referred to in this section, filters that provide total rejection at some frequency are particularly useful for measuring *distortion*. A periodic signal with a waveform distorted from sinusoidal is equivalent to a Fourier spectrum of sinusoidal signals (see section 11.1). Total rejection of the fundamental just leaves the harmonic content that represents the distortion and can be measured readily.

Although the single-section filters covered in this chapter are quite important, their treatment merely serves as an introduction to the vast subject of filters. Multiple-section filters are analysed in the following chapter while the topic of active filters is broached in chapter 10. The modern approach of filter synthesis features in chapter 12. Before concluding the present chapter, a brief discussion of *phase-shift* networks is appropriate.

8.4 Phase-shift networks

Although all the filter networks treated so far in this chapter cause phase shift, in each case it is accompanied by attenuation. What is more, should a component be varied to alter the phase shift, the attenuation also changes. Many practical situations require the introduction of a variable phase shift, ideally with no attenuation but at least with fixed attenuation.

A fixed phase shift of π radians may be obtained without attenuation from a unity-ratio, close-coupled, low-loss transformer as explained in chapter 6. Figure 8.11(*a*) depicts the derivation of signals of equal amplitude but separated in phase by π radians through the action of a centre-tapped transformer. Provision of such phase-related signals is described as *phase splitting* and is widely used in electronics for various purposes. One area of application is in null balancing methods of measurement such as the transformer ratio-arm bridges described in section 7.8. Phase splitting also plays a vital role in the important variable phase-shift network to be considered in a moment. Two alternatives to a transformer for splitting the phase of a signal are the potential-divider arrangement of figure 8.11(*b*), which attenuates the input signal by a factor two while splitting the phase, and the transistor phase splitter of figure 8.11(*c*), which has virtually unity gain provided R is large enough.

A very simple network, that while preserving a constant amplitude introduces a variable phase shift through adjustment of a single resistor, is shown in figure 8.12(*a*). How this network operates is most easily explained by means of the phasor diagram presented in figure 8.12(*b*). A phase-split signal $\pm V$ derived from any input, for example by one of the circuits appearing in figure 8.11, is applied between the terminal pairs BO and AO. Thus the phasor potential difference between B and A is $2V$. The phasor potential difference V_C across the capacitance C lags $90°$ behind the phasor potential difference V_R across the resistance R, assuming negligible loading

8.4 Phase-shift networks

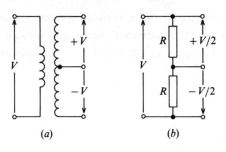

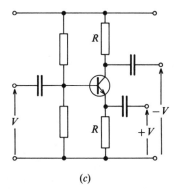

8.11 Phase splitting by (a) a centre-tapped transformer, (b) a resistive potential divider and (c) a transistor amplifier.

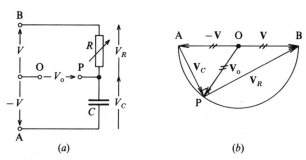

8.12 (a) Network that introduces a variable phase shift but preserves a constant output amplitude and (b) its analysis by a phasor diagram.

of the output taken between terminals P and O, while

$$V_C + V_R = 2V$$

All of these aspects are maintained in the phasor diagram of figure 8.12(b) and its geometry is seen to be such that the point P moves over a circle, centre O, as the resistance R is varied. But OP in the diagram represents the phasor output potential difference $\mathbf{V}_o$ taken between terminals P and O. Consequently, through altering the resistance R, the output may be varied in phase by π radians with respect to the input while maintaining its amplitude constant.

9

Multiple-section filters and transmission lines

9.1 Ladder filters

The sharpness of filtering may be vastly improved by cascading individual filter sections to create what is known, rather appropriately, as a *ladder* filter because of its appearance. One obvious approach is simply to cascade a number of identical sections. The problem with such ladder filters is that, in general, each section loads the preceding one with a different impedance making it difficult to predict the overall performance of the ladder or to design a ladder to meet a given specification. Avoidance of this difficulty is only possible by following the approach adopted in the design of ladder attenuators and arranging that the input impedance of any section is equal to its load impedance. Cascaded identical sections that meet this criterion are all identically loaded and, therefore, behave identically. In particular, if $\mathcal{T}$ is the transfer function of any such section, the overall transfer function of the ladder filter is $\mathcal{T}^n$ where n is the number of sections. In the common case of a symmetric section, the load impedance that renders the input impedance equal to it is called the *characteristic* impedance. While the identical symmetric sections of an *infinite* ladder would obviously be loaded with the characteristic impedance, this is an impractical arrangement. A *finite* ladder of identical symmetric sections, which has the last section loaded with the characteristic impedance in order that all sections are so loaded, is said to be *correctly terminated*.

Consider now the form of ladder filter shown in figure 9.1(*a*), which has repeated series and parallel impedances Z_1 and Z_2. It can be thought of as comprising cascaded identical symmetric T or Π-sections, the circuit diagrams of which are presented in figures 9.1(*b*) and (*c*). The characteristic impedances Z_{kT} and $Z_{k\Pi}$ of these T and Π-sections are respectively given by

$$Z_{kT} = \tfrac{1}{2}Z_1 + \frac{(\tfrac{1}{2}Z_1 + Z_{kT})Z_2}{\tfrac{1}{2}Z_1 + Z_{kT} + Z_2}$$

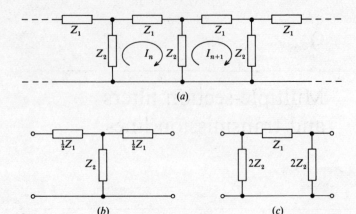

9.1 (*a*) A ladder network which may be considered as comprising cascaded identical T-sections, each of which is as shown in (*b*) or cascaded identical Π-sections, each of which is as shown in (*c*).

or

$$\tfrac{1}{2}Z_1 Z_{kT} + Z_{kT}^2 + Z_2 Z_{kT} = \tfrac{1}{4}Z_1^2 + \tfrac{1}{2}Z_1 Z_{kT} + \tfrac{1}{2}Z_1 Z_2 + \tfrac{1}{2}Z_1 Z_2 + Z_2 Z_{kT}$$

which reduces to

$$Z_{kT} = [Z_1 Z_2 (1 + Z_1/4Z_2)]^{\tfrac{1}{2}} \tag{9.1}$$

and

$$Z_{k\Pi} = 2Z_2 \left(Z_1 + \frac{2Z_2 Z_{k\Pi}}{2Z_2 + Z_{k\Pi}} \right) \bigg/ \left(2Z_2 + Z_1 + \frac{2Z_2 Z_{k\Pi}}{2Z_2 + Z_{k\Pi}} \right)$$

$$= 2Z_2 (2Z_1 Z_2 + Z_1 Z_{k\Pi} + 2Z_2 Z_{k\Pi}) / (4Z_2^2 + 4Z_2 Z_{k\Pi} + 2Z_1 Z_2 + Z_1 \ldots$$

which reduces to

$$Z_{k\Pi} = [Z_1 Z_2 / (1 + Z_1/4Z_2)]^{\tfrac{1}{2}} \tag{9.2}$$

If the mesh currents in the *n*th and (*n*+1)th meshes of the ladder network are denoted by I_n and I_{n+1} as indicated in figure 9.1(*a*), then, provided the ladder is correctly terminated, the transfer functions of its T and Π-sections can be expressed as

$$\mathcal{T}_T = Z_{kT} I_{n+1} / Z_{kT} I_n = I_{n+1}/I_n \tag{9.3}$$

and

$$\mathcal{T}_\Pi = [2Z_2 Z_{k\Pi}/(2Z_2 + Z_{k\Pi})] I_{n+1} / [2Z_2 Z_{k\Pi}/(2Z_2 + Z_{k\Pi})] I_n$$
$$= (\tfrac{1}{2}Z_1 + Z_{kT}) I_{n+1} / (\tfrac{1}{2}Z_1 + Z_{kT}) I_n = I_{n+1}/I_n \tag{9.4}$$

respectively, which are identical. The result $\mathcal{T}_T = \mathcal{T}_\Pi$ is, of course, to be expected, since T and Π-sections are different ways of breaking the *same* network into repeated sections. To find I_{n+1} in terms of I_n, the network must be analysed. Application of Kirchhoff's voltage law in the (*n*+1)th mesh yields

9.1 Ladder filters

$$(Z_2 + \tfrac{1}{2}Z_1 + Z_{kT})\mathbf{I}_{n+1} = Z_2\mathbf{I}_n \qquad (9.5)$$

in terms of Z_{kT} or

$$\left(Z_2 + Z_1 + \frac{2Z_2 Z_{k\Pi}}{2Z_2 + Z_{k\Pi}}\right)\mathbf{I}_{n+1} = Z_2\mathbf{I}_n \qquad (9.6)$$

in terms of $Z_{k\Pi}$. That these two expressions relating $\mathbf{I}_{n+1}$ to $\mathbf{I}_n$ are equivalent is easily verified since it requires

$$\tfrac{1}{2}Z_1 + \frac{2Z_2 Z_{k\Pi}}{2Z_2 + Z_{k\Pi}} = Z_{kT}$$

or

$$Z_1 Z_2 + \tfrac{1}{2}Z_1 Z_{k\Pi} + 2Z_2 Z_{k\Pi} = 2Z_2 Z_{kT} + Z_{k\Pi} Z_{kT}$$

The validity of this last relation is best seen by appreciating that, according to equations (9.1) and (9.2), the characteristic impedances obey

$$Z_{kT} Z_{k\Pi} = Z_1 Z_2 \qquad (9.7)$$

and

$$Z_{kT}/Z_{k\Pi} = 1 + Z_1/4Z_2 \qquad (9.8)$$

Making use of equations (9.1) and (9.5) and writing the ratio $Z_1/4Z_2$ as u, it follows from equations (9.3) and (9.4) that the transfer function of a T or Π-section of the form of ladder filter under consideration is given by

$$\mathcal{T} = \mathcal{T}_T = \mathcal{T}_\Pi = \{1 + 2u + [4u(1+u)]^{\frac{1}{2}}\}^{-1} = \{1 + 2u - [4u(1+u)]^{\frac{1}{2}}\} \qquad (9.9)$$

In general u and hence $\mathcal{T}$ will be complex and it is helpful at this juncture to put

$$\mathcal{T} = \exp{-\gamma} = \exp{-(\alpha + j\beta)} \qquad (9.10)$$

where γ is known as the *propagation constant*. The quantity $\exp{-\alpha}$ is the ratio between the amplitudes of currents in successive meshes or the ratio between the amplitudes of potential differences at the inputs of successive sections. Accordingly, the parameter α is known as the *attenuation constant*. The parameter β represents the phase shift introduced by a section. From equations (9.9) and (9.10)

$$\cosh\gamma = \tfrac{1}{2}(\{1 + 2u + [4u(1+u)]^{\frac{1}{2}}\} + \{1 + 2u - [4u(1+u)]^{\frac{1}{2}}\}) = 1 + 2u \qquad (9.11)$$

that is, the propagation constant is given by the simple relation

$$\cosh\gamma = 1 + Z_1/2Z_2 \qquad (9.12)$$

Attention will now be focussed on correctly terminated ladder filters in which Z_1 and Z_2 are pure reactances, for practical approximations to these networks are widely used. When Z_1 and Z_2 are pure reactances, $\cosh\gamma$ is

real and since
$$\cosh \gamma = \cosh(\alpha + j\beta) = \cosh \alpha \cos \beta + j \sinh \alpha \sin \beta$$
it follows that
$$\sinh \alpha \sin \beta = 0 \tag{9.13}$$
$$\cosh \alpha \cos \beta = 1 + Z_1/2Z_2 = 1 + 2u \tag{9.14}$$

The solution $\sinh \alpha = 0$ to equation (9.13) implies that $\alpha = 0$ or $\exp -\alpha = 1$ which means that there is no reduction in amplitude through a section of the filter. The solution $\sinh \alpha = 0$ also implies that $\cosh \alpha = 1$ and so from equation (9.14), there is a phase shift β given by

$$\cos \beta = 1 + \frac{Z_1}{2Z_2} = 1 + 2u \tag{9.15}$$

However, $\cos \beta$ has to be in the range -1 to $+1$. Consequently the solution $\sinh \alpha = 0$ and the associated absence of attenuation corresponds to u being in the range -1 to 0, that is,

$$-1 \leqslant u \leqslant 0 \tag{9.16}$$

For particular reactive impedances, this last condition is satisfied over a certain range of frequency. Thus any correctly terminated filter section with purely reactive elements perfectly passes signals over a certain range of frequency defined by the inequality (9.16) and appropriately known as the *pass* or *transmission* band but introduces a phase shift given by equation (9.15) in this frequency band. The twin solutions to equation (9.15) of equal positive and negative phase shifts correspond to the possibility of feeding a signal in at either end of the symmetric section and loading it with its characteristic impedance at the other end.

The alternative solution $\sin \beta = 0$ to equation (9.13) implies that $\beta = 0$ or $\beta = \pm \pi$. In the first case, $\cos \beta = 1$ and so from equation (9.14) there is a reduction in amplitude given by

$$\cosh \alpha = 1 + Z_1/2Z_2 = 1 + 2u \tag{9.17}$$

Since $\cosh \alpha \geqslant 1$, this in turn implies that

$$u \geqslant 0 \tag{9.18}$$

The case $\beta = \pm \pi$ implies that $\cos \beta = -1$, and so from equation (9.14) there is a reduction in amplitude given by

$$\cosh \alpha = -1 - Z_1/2Z_2 = -1 - 2u \tag{9.19}$$

which in turn implies that

$$u \leqslant -1 \tag{9.20}$$

For given reactive impedances Z_1 and Z_2, inequalities (9.18) and (9.20) define ranges of frequency over which the signal is attenuated, there being a phase shift of zero in one range and $\pm \pi$ in the other. This time, the term

9.2 Constant-k filters

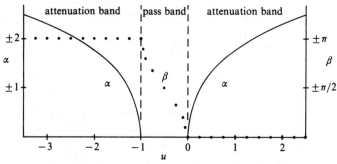

9.2 Variation of the attenuation constant α and phase shift β with the parameter $u = Z_1/4Z_2$ for a correctly terminated section of a purely reactive ladder filter of the type shown in figure 9.1.

attenuation band is an apt description of each range of frequency. Twin solutions to equations (9.17) and (9.19) of equal positive and negative values of α again correspond to the possibility of feeding the symmetric section at either end and terminating it at the other end.

Figure 9.2 shows the variation of the attenuation constant α and phase shift β with the parameter u according to the theory just presented. For a filter having n sections the total phase shift is, of course, $n\beta$ and the total attenuation $|\mathcal{T}^n| = \exp(-n\alpha)$.

9.2 Constant-k filters

The theory of the foregoing section will now be applied to particular purely reactive filters. As it is thereby illustrated and developed its implications should clarify. Consider first a ladder filter of the type studied in the previous section in which each series impedance Z_1 is due to inductance L and each parallel impedance Z_2 to capacitance C as shown in figure 9.3(a). Since the series impedances are small and the parallel impedances are high at low frequencies, while the opposite is the case at high frequencies, this network behaves as a low-pass filter. Its quantitative response, when correctly terminated, is governed by the parameter $u = Z_1/4Z_2$ introduced in the last section and substitution of $Z_1 = j\omega L$ and $Z_2 = 1/j\omega C$ establishes that

$$u = -\omega^2 LC/4 \tag{9.21}$$

Because u cannot be positive at any frequency, there is no range of frequency that corresponds to inequality (9.18) and equation (9.17) is inapplicable. In the range $-1 \leq u \leq 0$ which corresponds to

$$0 \leq \omega \leq 2/(LC)^{\frac{1}{2}} \tag{9.22}$$

the attenuation constant α is zero and the phase shift β per section is given by equation (9.15) which becomes

$$\cos \beta = 1 - \omega^2 LC/2 \tag{9.23}$$

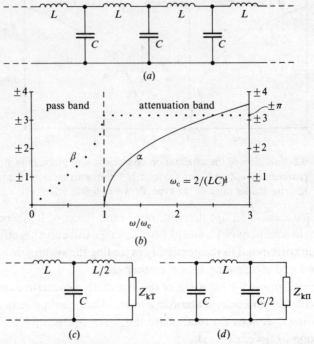

9.3 (a) Low-pass, symmetric, L–C, ladder filter, (b) its attenuation constant α and phase shift β per section as a function of the pulsatance ω when correctly terminated, (c) correct termination where Z_{kT} is given by equation (9.28) and (d) correct termination where $Z_{k\Pi}$ is given by equation (9.29).

In the range $u \leqslant -1$ which corresponds to

$$\omega \geqslant 2/(LC)^{\frac{1}{2}} \tag{9.24}$$

the phase shift is $\pm \pi$ and the attenuation is given by equation (9.19) which becomes

$$\cosh \alpha = \omega^2 LC/2 - 1 \tag{9.25}$$

Figure 9.3(b) presents plots of the attenuation constant α and phase shift β per section according to the relations just deduced. The most significant feature of the response is the existence of a *critical pulsatance*

$$\omega_c = 2/(LC)^{\frac{1}{2}} \tag{9.26}$$

which marks the end of an ideal pass band and the beginning of attenuation. In terms of this critical pulsatance, at pulsatances above and below it respectively, the attenuation constant and phase shift per section are given by

$$\cosh \alpha = -\cos \beta = 2(\omega/\omega_c)^2 - 1 \tag{9.27}$$

The fact that the transfer function $\mathcal{T}$ becomes $[1 - j\omega(LC)^{\frac{1}{2}}]$ when $\omega \ll \omega_c$

9.2 Constant-k filters

establishes that, in the pass band, the output signal of a section *lags* in phase behind the input signal to that section.

According to equations (9.1), (9.2) and (9.26), the characteristic impedances that correctly terminate T and Π-sections of this low-pass filter are

$$Z_{kT} = \left(\frac{L}{C}\right)^{\frac{1}{2}} \left(1 - \frac{\omega^2}{\omega_c^2}\right)^{\frac{1}{2}} \tag{9.28}$$

and

$$Z_{k\Pi} = \left(\frac{L}{C}\right)^{\frac{1}{2}} \Bigg/ \left(1 - \frac{\omega^2}{\omega_c^2}\right)^{\frac{1}{2}} \tag{9.29}$$

respectively. Figures 9.3(c) and (d) show the circuit arrangements needed for correct termination by Z_{kT} or $Z_{k\Pi}$. Unfortunately, no combination of circuit components can produce an impedance with the required terminating frequency dependence of equation (9.28) for Z_{kT} or that of equation (9.29) for $Z_{k\Pi}$. At best these dependences can only be approximated by complicated networks and the performance of any real filter must fall short of that represented by figure 9.3(b) for a correctly terminated filter. The usual procedure adopted is to terminate with a fixed resistance

$$R = (L/C)^{\frac{1}{2}} \tag{9.30}$$

in place of Z_{kT} or $Z_{k\Pi}$. Such termination is very close to correct in the pass band until ω approaches ω_c, say until ω reaches $\sim 0.3\omega_c$. Further, although incorrectly terminated, the input impedance of a section terminated with resistance $(L/C)^{\frac{1}{2}}$ is much closer to the characteristic impedance than resistance $(L/C)^{\frac{1}{2}}$. Thus a ladder filter comprising several sections and terminated in resistance $(L/C)^{\frac{1}{2}}$ tends to correct termination of sections rapidly along the ladder and responds overall quite closely to how it would respond if it were possible to correctly terminate the end section. Figure 9.4 illustrates the point by comparing the input impedance of a low-pass, symmetric, L–C T-section terminated in resistance $(L/C)^{\frac{1}{2}}$, which is

$$Z_{(L/C)^{\frac{1}{2}}} = \left(\frac{L}{C}\right)^{\frac{1}{2}} \left[\frac{1 - j2x^3(1 - 2x^2)}{1 + 4x^4}\right] \tag{9.31}$$

where $x = \omega/\omega_c$, with its characteristic impedance given by equation (9.28).

An interesting application of the low-pass, L–C, ladder filter, apart from harnessing its filtering action, is its use to delay an electrical signal by a known time interval without attenuation or distortion. To achieve this end, the filter is arranged to operate entirely in its pass band by designing it so that its critical frequency is well above the highest frequency present in the Fourier frequency spectrum of the signal being handled. This being so, there is virtually 100% transmission of each Fourier component of the signal with

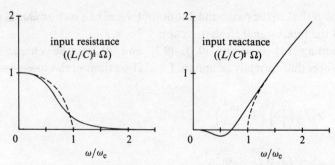

9.4 The input impedance of a terminated T-section of a symmetric, low-pass, L–C, ladder filter. Solid lines give values of the input resistance and reactance when terminated by fixed resistance $(L/C)^{\frac{1}{2}}$ while dashed lines show values of these quantities when terminated by the characteristic impedance.

each component suffering a phase shift of β per section, where β is given by equation (9.23) or (9.27). Because the frequencies of all the components are well below the critical frequency, the phase shift β is always small enough to make the approximation

$$\beta^2/2 \approx 1 - \cos\beta = 2(\omega/\omega_c)^2$$

or

$$\beta \approx 2\omega/\omega_c \tag{9.32}$$

Corresponding to the phase shift there is a time delay t_d per section given by

$$t_d = \beta/\omega \approx 2/\omega_c = (LC)^{\frac{1}{2}} \tag{9.33}$$

The crucial point to emerge from equation (9.33) is that the delay is almost independent of frequency so that all Fourier components of the signal experience virtually the same delay and the signal is transmitted, delayed but virtually undistorted, as well as virtually unattenuated. Any network that behaves in this way is called a *delay line*. In practice, a single-section, low-pass, L–C, delay line works reasonably well provided that the highest frequency present in the signal is less than one-half of the critical frequency. Above this frequency, the attenuation ceases to be negligible and the phase shift ceases to be sufficiently proportional to frequency. If individual sections are cascaded to form a delay line, the critical frequency must be correspondingly higher in relation to the frequencies present in the signal, otherwise unacceptable attenuation and distortion will again occur on account of the compound transfer function. Thus from equation (9.33) the delay per cascaded section is correspondingly shorter and the longest delay that can be achieved by means of a low-pass, L–C, ladder filter is rather a small fraction of the fundamental period or pulse duration of the signal being handled.

9.2 Constant-k filters

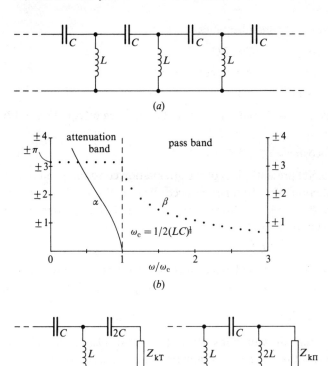

9.5 (a) High-pass, symmetric, L–C, ladder filter, (b) its attenuation constant α and phase shift β per section as a function of the pulsatance ω when correctly terminated, (c) correct termination where Z_{kT} is given by equation (9.41) and (d) correct termination where $Z_{k\Pi}$ is given by equation (9.42).

Interchanging capacitance C with inductance L in the low-pass, L–C, ladder filter creates the complementary, high-pass, L–C, ladder filter shown in figure 9.5(a). Qualitatively, the high-pass filtering action stems from the series and parallel impedances being respectively low and high at high frequencies while the opposite is true at low frequencies. Quantitatively, the series and parallel impedances are respectively $Z_1 = 1/j\omega C$ and $Z_2 = j\omega L$ so that

$$u = -1/4\omega^2 LC \tag{9.34}$$

and equation (9.17) is again inapplicable because inequality (9.18) cannot be satisfied at any frequency. In the range $-1 \leqslant u \leqslant 0$ which corresponds to

$$\omega \geqslant 1/2(LC)^{\frac{1}{2}} \tag{9.35}$$

the attenuation constant is zero and the phase shift β per section is given by equation (9.15) which becomes

$$\cos \beta = 1 - 1/2\omega^2 LC \tag{9.36}$$

In the range $u \leqslant -1$ which corresponds to

$$0 \leqslant \omega \leqslant 1/2(LC)^{\frac{1}{2}} \tag{9.37}$$

the phase shift is $\pm \pi$ and the attenuation is given by equation (9.19) which becomes

$$\cosh \alpha = 1/2\omega^2 LC - 1 \tag{9.38}$$

Figure 9.5(b) presents plots of the attenuation constant α and phase shift β as a function of pulsatance ω according to the results represented by relations (9.35)–(9.38). This time the critical pulsatance separating the pass and attenuation bands is

$$\omega_c = 1/2(LC)^{\frac{1}{2}} \tag{9.39}$$

In terms of the critical pulsatance, at frequencies below and above it respectively, the attenuation constant and phase shift per section are given by

$$\cosh \alpha = -\cos \beta = 2(\omega_c/\omega)^2 - 1 \tag{9.40}$$

According to equations (9.1), (9.2) and (9.39), the characteristic impedances that correctly terminate T and Π-sections of this high-pass filter are

$$Z_{kT} = \left(\frac{L}{C}\right)^{\frac{1}{2}} \left(1 - \frac{\omega_c^2}{\omega^2}\right)^{\frac{1}{2}} \tag{9.41}$$

and

$$Z_{k\Pi} = \left(\frac{L}{C}\right)^{\frac{1}{2}} \bigg/ \left(1 - \frac{\omega_c^2}{\omega^2}\right)^{\frac{1}{2}} \tag{9.42}$$

respectively. Figures 9.5(c) and (d) show the circuit arrangements needed for correct termination by Z_{kT} or $Z_{k\Pi}$. In the limit of sufficiently high frequencies to satisfy $\omega \gg \omega_c$, the transfer function $\mathcal{T}$ reduces to $[1 + j/\omega(LC)^{\frac{1}{2}}]$ establishing that, in the pass band, the output signal of a section *leads* the input signal to that section in phase. Once again it is impossible to correctly terminate the filter for all frequencies and the usual practice is to terminate with resistance $(L/C)^{\frac{1}{2}}$ which is correct for most of the transmission band. Thus terminated, a high-pass ladder filter tends to correct termination of sections rapidly along the ladder so that the overall response is quite close to that of a correctly terminated ladder. The high-pass filter is unsuitable for use as a delay line because the delay in the pass band depends markedly on frequency. For example, when $\omega \gg \omega_c$ the phase shift β is always small enough to make the approximation

$$\beta^2/2 \approx 1 - \cos \beta = 2(\omega_c/\omega)^2$$

9.2 Constant-k filters

or

$$\beta \approx 2\omega_c/\omega \qquad (9.43)$$

and the corresponding time delay per section is inversely proportional to ω^2.

The low and high-pass, L–C, ladder filters already discussed in this section are said to be *constant-k* filters because the product of the series and parallel impedances is independent of frequency, which fact can be expressed in terms of a constant k as

$$Z_1 Z_2 = k^2 \qquad (9.44)$$

For the low and high-pass filters considered, k is just $(L/C)^{\frac{1}{2}}$. In general, two impedances satisfy equation (9.44) if the networks providing them are *dual* which means that the admittance of one network exhibits the same frequency dependence as the impedance of the other. Writing the impedance of one network as $Z_1 = (R_1 + jX_1)$ and the admittance of the other as $Y_2 = 1/Z_2 = G_2 - jB_2$, it is clear that equation (9.44) is satisfied if

$$G_2 - jB_2 = (R_1 + jX_1)/k^2 \qquad (9.45)$$

For equation (9.45) to be applicable, the conductance G_2 and susceptance B_2 must exhibit the same frequency dependences as the resistance R_1 and negative reactance $-X_1$ respectively. That such behaviour occurs for series and parallel resonant circuits has already been pointed out in section 5.7. The concern in the present section is with purely reactive circuits and it is worth noting that in dual versions the frequencies at which the reactance becomes zero or infinite is the same, one circuit being resonant ($X_1 = 0$) whenever the other is antiresonant ($X_2 = \infty$) and vice versa. In addition, purely reactive dual networks exhibit opposite signs of imaginary impedance at all frequencies.

Figure 9.6(a) shows a T-section of a constant-k, band-pass, ladder filter. The series and parallel arms of the ladder are dual, resonant, reactive networks, their impedances being

$$Z_1 = j(\omega L_1 - 1/\omega C_1); \quad Z_2 = j\omega L_2/(1 - \omega^2 L_2 C_2) \qquad (9.46)$$

To make the product $Z_1 Z_2$ independent of frequency requires

$$L_1 C_1 = L_2 C_2 \qquad (9.47)$$

under which condition

$$Z_1 Z_2 = L_2/C_1 = L_1/C_2 = k^2 \qquad (9.48)$$

Satisfaction of condition (9.47) causes the series and parallel arms to resonate at the same frequency and there is 100% transmission at that frequency. At low frequencies the impedances of the series arms are high on account of series capacitance while the impedances of the parallel arms are

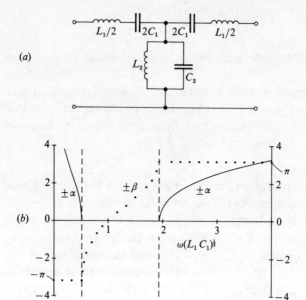

9.6 (a) T-section of a constant-k, band-pass, ladder filter and (b) its attenuation constant α and phase shift β per section as a function of the pulsatance ω when correctly terminated and $L_2 = L_1/2$, $C_2 = 2C_1$.

low on account of parallel inductance. Consequently the transmission is low, with the circuit behaving like a high-pass filter. At high frequencies, the impedances of the series and parallel arms are again high and low respectively but on account of series inductance and parallel capacitance. Once more, the transmission is low, but now the circuit behaves like a low-pass filter. Evidently, the overall response of this particular filter may be described as band-pass. When correctly terminated, its quantitative response is governed by the parameter u, which from equations (9.46) and (9.48) is

$$u = Z_1/4Z_2 = Z_1^2/4k^2 = -(\omega^2 L_1 C_1 - 1)^2/4k^2\omega^2 C_1^2 \tag{9.49}$$

Once more u cannot be positive at any frequency and there is no range of frequency that corresponds to inequality (9.18) so that equation (9.17) is inapplicable. The condition $-1 \leqslant u \leqslant 0$ is conveniently expressed by

$$(\omega^2 L_1 C_1 - 1)^2/4k^2\omega^2 C_1^2 = \zeta^2 \tag{9.50}$$

where

$$0 \leqslant \zeta^2 \leqslant 1 \tag{9.51}$$

Taking the square root of both sides of equation (9.50), the quadratic equation

$$L_1 C_1 \omega^2 \pm 2k\zeta C_1 \omega - 1 = 0$$

9.2 Constant-k filters

is obtained which has roots

$$\omega = \mp(k\zeta/L_1) \pm (k^2\zeta^2/L_1^2 + 1/L_1C_1)^{\frac{1}{2}}$$

or, since only positive pulsatances have physical meaning

$$\omega = \mp(k\zeta/L_1) + (k^2\zeta^2/L_1^2 + 1/L_1C_1)^{\frac{1}{2}}$$

Substituting for k from equation (9.48), these pulsatances are

$$\omega = +(\zeta^2/L_1C_2 + 1/L_1C_1)^{\frac{1}{2}} \pm (\zeta^2/L_1C_2)^{\frac{1}{2}}$$

and imposing condition (9.51) establishes that the pulsatance range corresponding to $-1 \leq u \leq 0$ is

$$(1/L_1C_2 + 1/L_1C_1)^{\frac{1}{2}} - (1/L_1C_2)^{\frac{1}{2}} = \omega_1 \leq \omega \leq \omega_2$$
$$= (1/L_1C_2 + 1/L_1C_1)^{\frac{1}{2}} + (1/L_1C_2)^{\frac{1}{2}} \quad (9.52)$$

In this range of pulsatance, the attenuation constant is zero, but from equation (9.15) there is a phase shift per section given by

$$\cos\beta = 1 - (\omega^2 L_1 C_1 - 1)^2 / 2\omega^2 L_2 C_1 \quad (9.53)$$

In the range $u \leq -1$ which corresponds to pulsatances satisfying $\omega \geq \omega_2$ and $\omega \leq \omega_1$, where ω_1 and ω_2 are the critical pulsatances defined in condition (9.52), the phase shift is $\pm\pi$ but there is attenuation characterised by an attenuation constant α where

$$\cosh\alpha = (\omega^2 L_1 C_1 - 1)^2 / 2\omega^2 L_2 C_1 - 1 \quad (9.54)$$

Perhaps surprisingly, condition (9.52) yields the simple relationship

$$\omega_1 \omega_2 = 1/L_1 C_1 = 1/L_2 C_2 \quad (9.55)$$

Figure 9.6(b) shows the attenuation constant α and phase shift β per section, according to the theoretical expressions just deduced, for a correctly terminated, constant-k, band-pass, ladder filter in which $L_2 = L_1/2$ and $C_2 = 2C_1$. The characteristic impedance of a T-section of a constant-k, band-pass, ladder filter is from equations (9.1), (9.48) and (9.49)

$$Z_{kT} = k\left[1 - \frac{(\omega^2 L_1 C_1 - 1)^2}{4k^2 \omega^2 C_1^2}\right]^{\frac{1}{2}} \quad (9.56)$$

However, the usual procedure is to terminate such a ladder with resistance k given by equation (9.48), in which case the performances of its later sections depart somewhat from that indicated by figure 9.6(b).

Interchanging the series and parallel forms of resonant reactive circuit between the series and parallel arms of the band-pass filter just discussed, naturally creates a band-stop filter. It is left as a useful exercise for the reader to show that if L_1 and C_1 represent the inductances and capacitances in the parallel resonant series arms, while L_2 and C_2 represent the corresponding quantities in the series resonant parallel arms,

$$u = -\omega^2 L_1^2 / 4k^2 (1 - \omega^2 L_1 C_1)^2 \tag{9.57}$$

and the critical pulsatances between which attenuation occurs are given by

$$\omega = \frac{1}{4}\left[+ \left(\frac{1}{L_2 C_1} + \frac{16}{L_1 C_1}\right)^{\frac{1}{2}} \mp \left(\frac{1}{L_2 C_1}\right)^{\frac{1}{2}} \right] \tag{9.58}$$

It is worthwhile recognising that in the attenuation bands of the purely reactive ladder filters considered, the characteristic impedances

$$Z_{kT} = k(1+u)^{\frac{1}{2}} \quad \text{and} \quad Z_{k\Pi} = k/(1+u)^{\frac{1}{2}}$$

are pure reactances because $u \leqslant -1$. Thus, under correct termination, no electrical power is fed into these filters over the attenuation bands of frequencies. The current and potential difference are everywhere 90° out of phase. There is also, of course, no power in the reactive terminating loads. In the pass bands, by contrast, u is in the range 0 to -1 so that Z_{kT} and $Z_{k\Pi}$ are real resistances and, under correct termination, power is fed into the filters. Obviously, since the filters are purely reactive, none of this power can be absorbed in them and it all reaches the terminating load in each case.

Purely reactive filters are, of course, impossible to achieve in practice and, in any approximation to them, the components inevitably exhibit resistive losses. Such resistances raise the attenuation constant to a finite value in the pass bands and give rise to finite power dissipation in the filter. They also adversely affect the rate at which the attenuation changes with frequency near critical frequencies.

It will be appreciated that ladder filters with configurations other than the one considered so far are possible. Sometimes *lattice* filters are used in which each section is configured as shown in figure 9.7. In this case, the characteristic impedance turns out to be simply

$$Z_k = (Z_1 Z_2)^{\frac{1}{2}} \tag{9.59}$$

Particularly interesting behaviour arises when the choice $Z_1 = j\omega L$, $Z_2 = 1/j\omega C$ is made, for then Z_k equals the fixed resistance $(L/C)^{\frac{1}{2}}$ at all frequencies and correct termination is easy. Such a correctly terminated low-pass lattice filter constitutes an extremely good delay line, the time delay per section being constant and equal to $(LC)^{\frac{1}{2}}$ at pulsatances well below $1/(LC)^{\frac{1}{2}}$.

9.3 *m*-Derived filters

Individual constant-*k* filter sections suffer from insufficient attenuation just outside the pass band and, although several identical constant-*k* sections may be cascaded to overcome this drawback, the resulting filter is rather unwieldy. A much better way of achieving a

9.3 m-Derived filters

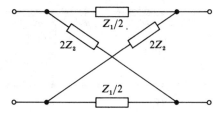

9.7 Section of a lattice filter.

composite filter with strong rejection outside the pass band is to combine what is known as an *m-derived* section with a constant-*k* section. As its name implies, the *m*-derived section is based on the constant-*k* section with which is cascaded and it is customary to refer to the undeveloped constant-*k* section as the *prototype*. The *m*-derived section is arranged to exhibit the same characteristic impedance and critical frequency or frequencies as the prototype. Enhanced attenuation over a suitable range of frequency outside the pass band is attained by making the shunt arm series resonant in the case of a T-type *m*-derived section, or the series arm parallel resonant in the case of a Π-type *m*-derived section. Usually, the *m*-derived section is designed to resonate at a frequency just outside the pass band of the prototype so that its strong resonant attenuation coincides with the range of weakest attenuation of the prototype.

How a T-type *m*-derived section is developed from a prototype T-section is shown in figure 9.8. The impedances of the series components are multiplied by a factor *m* compared with the prototype where *m* lies between zero and unity. To maintain the same characteristic impedance as the prototype, the shunt impedance Z'_2 of the *m*-derived section is arranged to satisfy

$$Z_1 Z_2 + \frac{Z_1^2}{4} = m Z_1 Z'_2 + \frac{m^2 Z_1^2}{4}$$

or

$$Z'_2 = \frac{Z_2}{m} + \frac{(1-m^2) Z_1}{4m} \qquad (9.60)$$

Thus the shunt arm of the *m*-derived section must be formed from the shunt impedance Z_2 of the prototype divided by *m* in series with extra impedance $(1-m^2)Z_1/4m$ as illustrated in figure 9.8(c). The extra impedance, being of the kind Z_1 rather than Z_2, has to be provided by extra components.

The design of a Π-type *m*-derived section based on a prototype Π-section is *not* the Π-section equivalent of the *m*-derived T-section just deduced, for this does not have the same characteristic impedance as the Π-prototype. To form a Π-type *m*-derived section, the procedure is as illustrated in figure 9.9.

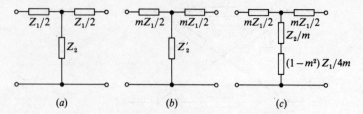

9.8 Development of an m-derived section from a T-section; (a) prototype, (b) series arms multiplied by m where $0 \leqslant m \leqslant 1$, leaving the shunt arm to be determined and (c) complete T-type m-derived section with two series impedances in the shunt arm so as to achieve the same characteristic impedance as the prototype.

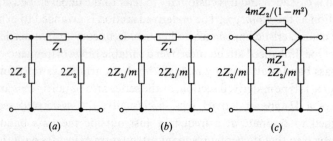

9.9 Development of an m-derived section from a Π-section; (a) prototype, (b) shunt arms divided by m where $0 \leqslant m \leqslant 1$, leaving the series arm to be determined and (c) complete Π-type m-derived section with two parallel impedances in the series arm so as to achieve the same characteristic impedance as the prototype.

First the impedances of the shunt components are divided by a factor m compared with the prototype where m lies between zero and unity. To maintain the same characteristic impedance as the prototype, the series impedance Z_1' of the m-derived section is chosen so as to satisfy

$$\frac{Z_1 Z_2}{(Z_1 Z_2 + Z_1^2/4)^{\frac{1}{2}}} = \frac{Z_1' Z_2/m}{(Z_1' Z_2/m + (Z_1')^2/4)^{\frac{1}{2}}}$$

or

$$\frac{Z_1}{Z_2 + Z_1/4} = \frac{Z_1'/m^2}{Z_2/m + Z_1'/4}$$

Thus

$$Z_1' = \frac{4m Z_1 Z_2}{4Z_2 + (1-m^2)Z_1} \qquad (9.61)$$

and the series arm of the m-derived section must be formed from the series impedance Z_1 of the prototype multiplied by m in parallel with extra impedance $4mZ_2/(1-m^2)$ as illustrated in figure 9.9(c). Again, the extra

9.3 m-Derived filters

impedance, being of the kind Z_2 rather than Z_1, has to be provided by extra components.

Making the characteristic impedances of the constant-k and m-derived filter sections identical also ensures that their critical frequencies are the same. Inspection of the theory of sections 9.1 and 9.2 reveals the reason for this. In particular, notice that the critical frequencies of m-derived T and Π-sections are determined by the general condition $u = -1$ which renders their characteristic impedance zero or infinite respectively. Thus an m-derived section having the same characteristic impedance as a constant-k prototype involving series and parallel impedances Z_1 and Z_2, that is, characteristic impedance $k[1+(Z_1/4Z_2)]^{\pm\frac{1}{2}}$ depending on whether the sections are T or Π-type, exhibits critical frequencies determined by $Z_1/4Z_2 = -1$. This is precisely the condition that determines the critical frequencies of the prototype and so the critical frequencies of the m-derived and constant-k prototype sections are bound to be the same. Checking this point, the critical frequencies of m-derived T and Π-sections are determined by $u = -1$, that is, by

$$mZ_1 \bigg/ 4\left[\frac{Z_2}{m} + \frac{(1-m^2)Z_1}{4m}\right] = -1$$

and

$$(mZ_1)\left(\frac{4mZ_2}{1-m^2}\right) \bigg/ \left(mZ_1 + \frac{4mZ_2}{1-m^2}\right)\left(\frac{4Z_2}{m}\right) = -1$$

(9.62)

respectively. It is easily seen that these conditions are identical and reduce to just $Z_1/4Z_2 = -1$, the condition governing the critical frequencies of the constant-k prototype.

When the constant-k prototype is purely reactive, so that Z_1 and Z_2 are simply opposing imaginary quantities, it is easily appreciated that series resonance occurs in the shunt arm of the T-type m-derived section and parallel resonance in the series arm of the Π-type m-derived section. Putting $Z'_2 = 0$ in equation (9.60) and $Z'_1 = \infty$ in equation (9.61) further reveals that the resonant frequencies of both such derived sections are given by

$$Z_2/Z_1 = -(1-m^2)/4 \tag{9.63}$$

In particular, for given Z_1, Z_2 and m, the resonances of the T and Π-type sections are coincident in frequency. Notice, however, that, while the attenuation of an m-derived section based on a purely reactive prototype would be infinite at resonance, any practical version of such a section actually exhibits strong but finite attenuation at resonance.

According to equation (9.63), a low-pass m-derived section with $Z_1 = j\omega L$, $Z_2 = 1/j\omega C$ resonates at pulsatance

$$\omega_\infty = \omega_c/(1-m^2)^{\frac{1}{2}} \tag{9.64}$$

where ω_c is the critical pulsatance $2/(LC)^{\frac{1}{2}}$. In the case of a complementary high-pass m-derived section with $Z_1 = 1/j\omega C$, $Z_2 = j\omega L$, equation (9.63) shows that the resonant pulsatance ω_∞ is related to the critical pulsatance ω_c by

$$\omega_\infty = (1 - m^2)^{\frac{1}{2}} \omega_c \qquad (9.65)$$

the critical pulsatance being $1/2(LC)^{\frac{1}{2}}$. Since $0 \leqslant m \leqslant 1$, resonance occurs in the attenuation band in either case. Making m small sets the resonant frequency very close to the critical frequency so that the attenuation of the prototype is enhanced where it is weakest. The parameter m is often chosen to be 0.3 which separates the resonant frequency by about 5% from the critical frequency. Figure 9.10 shows the behaviour of the attenuation constant of a correctly terminated, low-pass, m-derived, filter section in which $m = 0.3$. Also shown is the dependence of the attenuation constant of the corresponding prototype on frequency and the behaviour of the prototype and m-derived sections when cascaded. Note that the low-pass, m-derived T-section corresponding to $m = 0.3$ has inductance $0.15L$ in each series arm compared with $0.5L$ in each series arm of the prototype and capacitance $0.3C$ in series with inductance of approximately $0.76L$ in the shunt arm compared with just capacitance C in the shunt arm of the prototype.

The problems of correct termination and providing constant input resistance in the pass band may both be eased by incorporating suitable m-derived *half-sections* at the input and output. Consider first the T-type half-section shown in figure 9.11(a). When this particular half-section is terminated in the characteristic impedance Z_{kT} of the prototype, its input

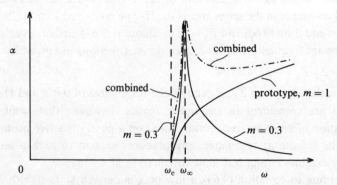

9.10 Attenuation constant α as a function of the pulsatance ω for a constant-k, low-pass, prototype, filter section, for the corresponding m-derived section having $m = 0.3$ and for these two sections cascaded.

9.3 m-Derived filters

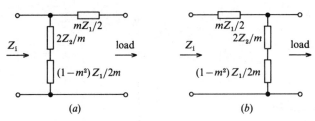

9.11 (a) An m-derived T-type half-section and (b) the same with input and output connections interchanged.

impedance is

$$Z_i = \left(\frac{mZ_1}{2} + Z_{kT}\right)\left[\frac{2Z_2}{m} + \frac{(1-m^2)Z_1}{2m}\right] \Big/ \left[\frac{mZ_1}{2} + Z_{kT} + \frac{2Z_2}{m} + \frac{(1-m^2)Z_1}{2m}\right]$$

$$= Z_1 Z_2 \left(1 + \frac{2Z_{kT}}{mZ_1}\right)\left[1 + \frac{(1-m^2)Z_1}{4Z_2}\right] \Big/ \left[Z_{kT} + \frac{2Z_2}{m}\left(1 + \frac{Z_1}{4Z_2}\right)\right]$$

which reduces to

$$Z_i = \frac{Z_1 Z_2}{Z_{kT}}\left[1 + \frac{(1-m^2)Z_1}{4Z_2}\right] \tag{9.66}$$

on making use of equation (9.1). Notice that if $m=0$, equation (9.66) further simplifies to $Z_i = Z_{kT}$ which is just as expected, since the series and parallel arms of the half-section are then short and open circuit respectively. At the opposite extreme of $m=1$, the half-section is half of a constant-k T-section and equation (9.66) shows that in this case the load Z_{kT} gives rise to input impedance $Z_i = Z_1 Z_2 / Z_{kT} = Z_{k\Pi}$, the characteristic impedance of the corresponding Π-section. Here the half-section is said to convert the load Z_{kT} to impedance $Z_{k\Pi}$ at the input. For values of m between 0 and 1, the input impedance lies between Z_{kT} and $Z_{k\Pi}$ and is a function of ω and m. The dependence on ω and m for a low-pass m-derived half-section of the form of figure 9.11(a) is shown in figure 9.12. When $m=0.6$, the input resistance remains close to $(L/C)^{\frac{1}{2}}$ for frequencies up to 85% of the critical frequency. Thus such a section placed in front of any number of correctly terminated constant-k or m-derived T-sections presents an almost constant input resistance of $(L/C)^{\frac{1}{2}}$ over 85% of the pass band.

Now consider the other half of the T-type m-derived section shown in figure 9.11(b), which is of course just the same network as the first half already considered and shown in figure 9.11(a) but with the input and

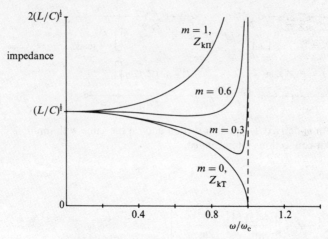

9.12 Input impedance of an m-derived low-pass half-section of the form of figure 9.11(a) with $Z_1 = j\omega L$, $Z_2 = 1/j\omega C$, as a function of the pulsatance ω and parameter m, when loaded with the characteristic impedance Z_{kT} of the prototype.

output terminals interchanged. Its input impedance when terminated in the impedance given by equation (9.66) is

$$Z_i = \frac{mZ_1}{2} + \frac{\left[\dfrac{2Z_2}{m} + \dfrac{(1-m^2)Z_1}{2m}\right]\dfrac{Z_1 Z_2}{Z_{kT}}\left[1 + \dfrac{(1-m^2)Z_1}{4Z_2}\right]}{\dfrac{2Z_2}{m} + \dfrac{(1-m^2)Z_1}{2m} + \dfrac{Z_1 Z_2}{Z_{kT}}\left[1 + \dfrac{(1-m^2)Z_1}{4Z_2}\right]}$$

$$= \frac{Z_1 Z_2 + \dfrac{(1-m^2)Z_1^2}{4} + \left(\dfrac{2Z_2}{m} + \dfrac{Z_1}{2m}\right)\dfrac{Z_1 Z_2}{Z_{kT}}\left[1 + \dfrac{(1-m^2)Z_1}{4Z_2}\right]}{\dfrac{2Z_2}{m} + \dfrac{(1-m^2)Z_1}{2m} + \dfrac{Z_1 Z_2}{Z_{kT}}\left[1 + \dfrac{(1-m^2)Z_1}{4Z_2}\right]}$$

On using equation (9.1) again, this relation reduces to

$$Z_i = \frac{Z_1 Z_2\left[1 + \dfrac{(1-m^2)Z_1}{4Z_2}\right] + \dfrac{2Z_2 Z_{kT}}{m}\left[1 + \dfrac{(1-m^2)Z_1}{4Z_2}\right]}{\dfrac{2Z_2}{m}\left[1 + \dfrac{(1-m^2)Z_1}{4Z_2}\right] + \dfrac{Z_1 Z_2}{Z_{kT}}\left[1 + \dfrac{(1-m^2)Z_1}{4Z_2}\right]} = Z_{kT} \quad (9.67)$$

Thus the network of figure 9.11(b) terminated in the impedance given by equation (9.66) presents impedance Z_{kT} at its input terminals, precisely the required termination for constant-k or m-derived T-sections. Since the impedance given by equation (9.66) is very close to $(L/C)^{\frac{1}{2}}$ over 85% of the pass band when $m = 0.6$, the half-section of figure 9.11(b) with $m = 0.6$ interposed between a fixed resistance $(L/C)^{\frac{1}{2}}$ and the output of a constant-k

or m-derived T-section will provide the T-section with virtually correct termination over most of the pass band. An example of a composite low-pass filter terminated at its input and output with suitable m-derived half-sections is shown in figure 9.13. It is of course possible to design Π-type m-derived half-sections to fulfil corresponding roles to those of the T-type m-derived half-sections. Better performance still than that provided by the m-derived sections that have been described is available from what are known as double m-derived sections.

9.4 Asymmetric sections

The term *iterative* impedance is applied to a load that renders the input impedance of an asymmetric section equal to it. There are of course two differing iterative impedances for any asymmetric section corresponding to the possibility of loading either end. For the asymmetric T-section shown in figure 9.14(*a*), the two iterative impedances are given by

$$Z_{it} = Z_1 + \frac{Z_2(Z_3 + Z_{it})}{Z_2 + Z_3 + Z_{it}}$$

and

$$Z'_{it} = Z_3 + \frac{Z_2(Z_1 + Z'_{it})}{Z_2 + Z_1 + Z'_{it}}$$

which readily rearrange into the quadratic equations

$$Z_{it}^2 + (Z_3 - Z_1)Z_{it} - (Z_1Z_2 + Z_2Z_3 + Z_3Z_1) = 0$$

and

$$(Z'_{it})^2 + (Z_1 - Z_3)Z'_{it} - (Z_1Z_2 + Z_2Z_3 + Z_3Z_1) = 0$$

Thus

$$\left.\begin{array}{l}Z_{it} = \tfrac{1}{2}\{(Z_1 - Z_3) \pm [(Z_3 - Z_1)^2 + 4(Z_1Z_2 + Z_2Z_3 + Z_3Z_1)]^{\frac{1}{2}}\} \\ Z'_{it} = \tfrac{1}{2}\{(Z_3 - Z_1) \pm [(Z_1 - Z_3)^2 + 4(Z_1Z_2 + Z_2Z_3 + Z_3Z_1)]^{\frac{1}{2}}\}\end{array}\right\} \quad (9.68)$$

Of the two solutions to each of the equations (9.68), those having positive components of iterative resistance are appropriate and these normally

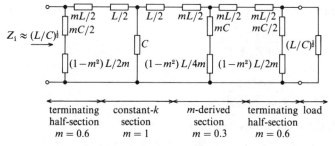

9.13 Multiple-section low-pass filter comprising a constant-k section, an m-derived section and terminating half-sections.

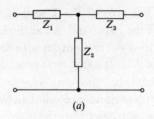

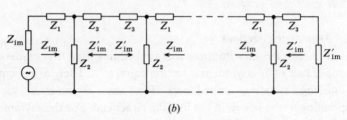

9.14 (a) An asymmetric T-section and (b) a source matched to a load through alternately reversed image sections.

correspond to taking the positive root in each case. For any *symmetric* section, one possibility being a section of the form of figure 9.14(a) with $Z_1 = Z_3$, the two iterative impedances are equal and the common impedance is said to be the characteristic impedance as stated earlier.

Clearly, maximum transfer of power will not be achieved by cascading identical asymmetric sections between a source and correctly terminating iterative impedance. However, any asymmetric section also has two *image* impedances Z_{im} and Z'_{im} such that, if one pair of terminals is terminated in Z'_{im} and the input impedance at the other pair is Z_{im}, then, if that other pair is terminated in Z_{im}, the input impedance at the first pair is Z'_{im}. Maximum transfer of power does occur between a source and load, if their impedances are the image impedances of identical asymmetric sections connected in cascade between them such that alternate sections are reversed as illustrated in figure 9.14(b). For the T-section of figure 9.14(a), the image impedances are given by

$$Z_{im} = Z_1 + \frac{Z_2(Z_3 + Z'_{im})}{Z_2 + Z_3 + Z'_{im}}$$

and

$$Z'_{im} = Z_3 + \frac{Z_2(Z_1 + Z_{im})}{Z_2 + Z_1 + Z_{im}}$$

Rearranged, these equations become

$$(Z_2 + Z_3)Z_{im} + Z_{im}Z'_{im} = Z_1Z_2 + Z_2Z_3 + Z_3Z_1 + (Z_1 + Z_2)Z'_{im}$$
$$(Z_2 + Z_1)Z'_{im} + Z_{im}Z'_{im} = Z_1Z_2 + Z_2Z_3 + Z_3Z_1 + (Z_3 + Z_2)Z_{im}$$

9.4 Asymmetric sections

Subtracting yields
$$Z'_{im}/Z_{im} = (Z_2+Z_3)/(Z_1+Z_2)$$
while adding gives
$$Z'_{im}Z_{im} = Z_1Z_2 + Z_2Z_3 + Z_3Z_1$$
Hence
$$\left.\begin{array}{l}Z'_{im} = [(Z_2+Z_3)(Z_1Z_2+Z_2Z_3+Z_3Z_1)/(Z_1+Z_2)]^{\frac{1}{2}} \\ Z_{im} = [(Z_1+Z_2)(Z_1Z_2+Z_2Z_3+Z_3Z_1)/(Z_2+Z_3)]^{\frac{1}{2}}\end{array}\right\} \quad (9.69)$$

Interestingly, the image impedances of a section can be more neatly expressed in terms of its input impedances under open and short-circuit termination. Representing open and short-circuit terminating conditions by subscripts o and s respectively, these extreme cases of input impedance for the T-section of figure 9.14(a) are

$$Z_o = Z_1 + Z_2; \quad Z'_o = Z_3 + Z_2$$
$$Z_s = Z_1 + \frac{Z_2Z_3}{Z_2+Z_3} = \frac{Z_1Z_2+Z_2Z_3+Z_3Z_1}{Z_2+Z_3} \quad (9.70)$$
$$Z'_s = Z_3 + \frac{Z_1Z_2}{Z_1+Z_2} = \frac{Z_1Z_2+Z_2Z_3+Z_3Z_1}{Z_1+Z_2}$$

Hence from equations (9.69) and (9.70)
$$Z_{im} = (Z_oZ_s)^{\frac{1}{2}}; \quad Z'_{im} = (Z'_oZ'_s)^{\frac{1}{2}} \quad (9.71)$$

In the particular case of a symmetric section, the two image impedances are identical, the common image impedance being, of course, just the characteristic impedance. Putting $Z_1 = Z_3$ in equations (9.68) and (9.69) bears this out, yielding

$$Z_{it} = Z'_{it} = Z_{kT} = (2Z_1Z_2 + Z_1^2)^{\frac{1}{2}} = Z_{im} = Z'_{im}$$

for the section of figure 9.14(a) with $Z_1 = Z_3$. In comparing this particular expression for the characteristic impedance with equation (9.1) relevant to the circuit of figure 9.1(b), do not forget to replace Z_1 by $Z_1/2$.

One interesting asymmetric section is that L-section which back-to-back with itself forms the symmetric T or Π-section of figure 9.1 as illustrated in figure 9.15. From equations (9.71) the image impedances are

$$Z_{im} = (\tfrac{1}{2}Z_1 + 2Z_2)^{\frac{1}{2}}(\tfrac{1}{2}Z_1)^{\frac{1}{2}} = (Z_1Z_2)^{\frac{1}{2}}(1+Z_1/4Z_2)^{\frac{1}{2}} \quad (9.72)$$
$$Z'_{im} = (2Z_2)^{\frac{1}{2}}[Z_1Z_2/(\tfrac{1}{2}Z_1+2Z_2)]^{\frac{1}{2}} = (Z_1Z_2)^{\frac{1}{2}}/(1+Z_1/4Z_2)^{\frac{1}{2}} \quad (9.73)$$

The characteristic impedance Z_{kT} of the symmetric T-section formed by following the L-section of figure 9.15(a) with that of figure 9.15(b) is clearly going to be Z_{im} given by equation (9.72) which is in agreement with equation (9.1). Similarly, the characteristic impedance $Z_{kΠ}$ of the symmetric Π-section formed by following the L-section of figure 9.15(b) with that of figure 9.15(a)

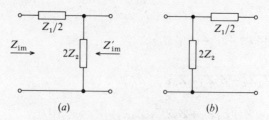

9.15 (a) An L-section which when cascaded with itself with reversed input and output connections as depicted in (b) forms the symmetric T-section of figure 9.1(b). Note that reversing the order of cascading these two sections leads to the symmetric Π-section of figure 9.1(c).

is going to be Z'_{im} given by equation (9.73) which is in agreement with equation (9.2). As noted previously in section 9.3, terminal impedance can be converted between Z_{kT} and $Z_{k\Pi}$ by introducing an appropriate half-section.

9.5 Transmission lines

An arrangement consisting of a delivery and return conductor by means of which an electrical signal can be efficiently conveyed between two points is known as a *transmission line*. Transmission lines vary in length from centimetres to thousands of kilometres, and to achieve satisfactory transmission it is essential for the delivery and return conductors to be of low-enough resistance and sufficiently insulated from each other. The simplest type of transmission line comprises just a pair of parallel wires kept a uniform distance apart by suitably inserted insulating spacers. However, a much more convenient and popular type of transmission line, generally referred to as a *coaxial cable*, essentially consists of a central conducting lead wire insulated from a coaxial outer return conductor. In the usual form of construction both the lead and return conductors are made of copper to minimise resistive loss, the outer return is braided for flexibility and the entire space in between is filled with highly insulating material such as polythene, polythene foam or polytetrafluoroethylene. The whole coaxial arrangement is enclosed within an outer protective insulating jacket. There are two great advantages associated with the coaxial geometry. Because the magnetic fields due to the delivery and return currents cancel outside the braided outer conductor, there is no loss of signal power through electromagnetic radiation as a signal passes along the cable. In addition, if the outer braid is earthed, as is common practice, the inner lead wire is electrostatically screened from interfering signals due to extraneous external sources. Fluctuating fields cannot exist inside a fixed equipotential

9.5 Transmission lines

surface (that of the braid here) on account of fluctuating external charge.

It will be appreciated that the conductors of a transmission line inevitably exhibit some series inductance and some capacitance between each other besides some series resistance and some conductance between each other. Moreover, all the circuit properties of a line are distributed along its length, uniformly so in the ideal situation of a uniformly constructed line. The distributed aspect contrasts sharply with the lumped nature of circuit representation of discrete components that has been adopted in all networks considered so far. In view of the distributed nature of a transmission line, to determine its behaviour it will generally be necessary to consider the response of an infinitesimal element of it. As will be confirmed by such analysis in a moment, the necessity to consider an element strictly depends on the length l of the line compared with the wavelength λ of the signal, for there is a phase difference $2\pi l/\lambda$ between the line's extremities. If this phase difference is negligibly small, say, less than a few degrees, then the line may as well be represented in terms of lumped components corresponding to the total series and parallel impedances. This means that, in the case of a signal of mains frequency, 50 Hz, for which the wavelength is $3 \times 10^8/50$ m $= 6000$ km in air, the line has to exceed around 60 km in length before its distributed nature needs to be taken into account. For signals of ultra-high radio frequency, on the other hand, say, 300 MHz, the air wavelength is 100 cm and the distributed nature of the line needs to be taken into account whenever its length exceeds around 1 cm. Note in passing that the actual wavelength in a transmission line is a little less than that in air on account of the dielectric constant of the plastic insulating material but this does not significantly affect the foregoing order of magnitude estimates. An important corollary of the present discussion is that circuit components of centimetre dimensions can properly be regarded as discrete until the frequency gets as high as about 300 MHz (recall discussion of this topic near the beginning of section 4.3).

Consider now, with reference to figure 9.16(a), an elementary length dx of a transmission line. Let the total series resistance and inductance of the lead and return per unit length be R and L respectively. Similarly, let the total shunt conductance and capacitance between the lead and return per unit length be G and C respectively. The shunt aspect of any element is illustrated in the blow-up of figure 9.16(b) and application of Kirchhoff's current law to this aspect gives

$$\mathbf{I} = G\,dx\mathbf{V} + j\omega C\,dx\mathbf{V} + \mathbf{I} + d\mathbf{I}$$

or

$$d\mathbf{I}/dx = -(G + j\omega C)\mathbf{V} \qquad (9.74)$$

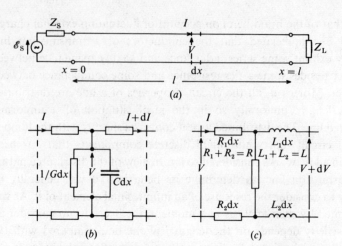

9.16 (a) Illustration of a transmission line connected between a load impedance Z_L and a source represented by e.m.f. $\mathscr{E}_S$ in series with impedance Z_S, (b) blow-up of an element dx showing the shunt aspect and (c) blow-up of an element dx showing the series aspect.

where **I** is the phasor representing series current in the lead or return and **V** the phasor representing potential difference between them, both at distance x along the line. Note that the full derivative is appropriate here because **I** and **V** being phasors are independent of time. Also, the change in the potential difference phasor **V** over the infinitesimal element has been neglected in arriving at equation (9.74) since it only gives rise to terms that are second order in smallness. The series aspect of any element is illustrated in the blow-up of figure 9.16(c) and application of Kirchhoff's voltage law to this aspect gives

$$\mathbf{V} = R\,dx\mathbf{I} + j\omega L\,dx\mathbf{I} + \mathbf{V} + d\mathbf{V}$$

or

$$d\mathbf{V}/dx = -(R+j\omega L)\mathbf{I} \tag{9.75}$$

This time note that it has been possible to neglect the change in the current phasor **I** over the infinitesimal element as it only leads to terms that are second order in smallness. Combining equations (9.74) and (9.75) yields

$$d^2\mathbf{I}/dx^2 = \gamma^2\mathbf{I}; \quad d^2\mathbf{V}/dx^2 = \gamma^2\mathbf{V} \tag{9.76}$$

where the parameter γ is given by

$$\gamma = (R+j\omega L)^{\frac{1}{2}}(G+j\omega C)^{\frac{1}{2}} \tag{9.77}$$

and is known as the propagation constant for reasons that will emerge in a moment.

The solutions of equations (9.76) are

$$\mathbf{I} = \mathbf{I}_1 \exp{-\gamma x} + \mathbf{I}_2 \exp{+\gamma x} \tag{9.78}$$

9.5 Transmission lines

$$V = V_1 \exp -\gamma x + V_2 \exp +\gamma x \qquad (9.79)$$

the physical meaning of which becomes clear on appreciating that the propagation constant is complex, say, $\gamma = \alpha + j\beta$ and that I_1, I_2, V_1 and V_2 are phasors relating to quantities of the form $A \exp[j(\omega t + \phi)]$. The first term in each equation represents a wave travelling in the positive x direction and the second term a similar wave travelling in the negative x direction. Clearly α is the attenuation constant and β the phase constant related to the phase velocity ω/β. Overall, equations (9.78) and (9.79) allow for a signal being fed in at one end of a transmission line, propagating along it and being partially reflected at the other end to give a wave travelling in the opposite direction.

The next important point to notice is that I_1, I_2, V_1 and V_2 are not independent. Inserting equations (9.78) and (9.79) for I and V into equation (9.75) establishes that

$$-\gamma V_1 \exp -\gamma x + \gamma V_2 \exp +\gamma x = -(R + j\omega L)(I_1 \exp -\gamma x + I_2 \exp +\gamma x)$$

Hence

$$V_1/I_1 = -V_2/I_2 = (R + j\omega L)/\gamma = Z_k \qquad (9.80)$$

where from equation (9.77)

$$Z_k = [(R + j\omega L)/(G + j\omega C)]^{\frac{1}{2}} \qquad (9.81)$$

For a transmission line of *infinite length*, I_2 and V_2 are zero in equations (9.78) and (9.79) respectively because it is physically impossible for I or V to be infinite as x goes to infinity. Thus

$$V = V_1 \exp -\gamma x = Z_k I_1 \exp -\gamma x = Z_k I \qquad (9.82)$$

From this it will be seen that the impedance between the lead and return at any point, including the input, along an infinite transmission line is the same, namely Z_k. Apparently Z_k corresponds to the *characteristic impedance* concept already introduced in this chapter in connection with ladder filters. Terminating a finite line of length l with this characteristic impedance forces

$$(V)_{x=l} = Z_k (I)_{x=l} \qquad (9.83)$$

But from equations (9.78), (9.79) and (9.80)

$$(I)_{x=l} = I_1 \exp -\gamma l + I_2 \exp +\gamma l \qquad (9.84)$$

$$(V)_{x=l} = Z_k I_1 \exp -\gamma l - Z_k I_2 \exp +\gamma l \qquad (9.85)$$

To satisfy equations (9.83), (9.84) and (9.85) simultaneously, the current I_2 must be zero and so equation (9.82) applies again. This means that, just as for a ladder filter, a finite transmission line terminated in the characteristic impedance Z_k exhibits impedance Z_k at all elements. Because the elements are now infinitesimal, the impedance is everywhere Z_k.

Resistive losses are often very small in transmission lines in comparison with corresponding reactive effects. This is certainly the case for coaxial cables designed to carry radio-frequency signals (1–300 MHz). Completely neglecting R compared with ωL and G compared with ωC in the interest of simplicity, it follows from equations (9.77) and (9.81) that

$$\gamma = j\omega(LC)^{\frac{1}{2}} = j\beta \tag{9.86}$$
$$Z_k = (L/C)^{\frac{1}{2}} \tag{9.87}$$

Observe that the characteristic impedance is purely resistive in the lossless approximation while the attenuation constant, α, is zero and the phase velocity, ω/β, is $1/(LC)^{\frac{1}{2}}$ which is independent of frequency so that dispersion is absent. When present, dispersion causes unwanted distortion of propagating signals. The properties just deduced are entirely those expected from the theory of section 9.2. A lossless transmission line can be regarded as a multiple-section low-pass filter with sectional inductance and capacitance $L\,dx$ and $C\,dx$ respectively. The multiple-section treatment yields a cut-off pulsatance of $2/(LC)^{\frac{1}{2}}dx$ which tends to infinity as dx tends to zero. Consequently all frequencies are passed without attenuation and, from either equation (9.28) or equation (9.29), the characteristic impedance is $(L/C)^{\frac{1}{2}}$. The delay per section is from equation (9.33) equal to $(LC)^{\frac{1}{2}}\,dx$ so that the phase velocity is confirmed as $1/(LC)^{\frac{1}{2}}$. Insertion of the theoretical expressions for L and C for a coaxial transmission line reveals that the phase velocity is alternatively $1/(\varepsilon\mu)^{\frac{1}{2}}$ where ε and μ are the permittivity and permeability of the medium between the lead and coaxial return. This is precisely the velocity of light in the medium.

In the case of a lossless transmission line, equations (9.78) and (9.79) become, through incorporation of the results of equations (9.80), (9.86) and (9.87),

$$\mathbf{I} = \mathbf{I}_1 \exp{-j\beta x} + \mathbf{I}_2 \exp{+j\beta x} \tag{9.88}$$
$$\mathbf{V} = (L/C)^{\frac{1}{2}}\mathbf{I}_1 \exp{-j\beta x} - (L/C)^{\frac{1}{2}}\mathbf{I}_2 \exp{+j\beta x} \tag{9.89}$$

When such a line is terminated by impedance Z_L at $x = l$ as depicted in figure 9.16(a)

$$\left(\frac{\mathbf{V}}{\mathbf{I}}\right)_{x=l} = \left[\frac{\mathbf{I}_1 \exp{-j\beta l} - \mathbf{I}_2 \exp{+j\beta l}}{\mathbf{I}_1 \exp{-j\beta l} + \mathbf{I}_2 \exp{+j\beta l}}\right](L/C)^{\frac{1}{2}} = Z_L$$

Hence
$$[(L/C)^{\frac{1}{2}} - Z_L](\exp{-j\beta l})\mathbf{I}_1 = [(L/C)^{\frac{1}{2}} + Z_L](\exp{+j\beta l})\mathbf{I}_2$$
or
$$\frac{\mathbf{I}_2}{\mathbf{I}_1} = -\frac{\mathbf{V}_2}{\mathbf{V}_1} = \left[\frac{(L/C)^{\frac{1}{2}} - Z_L}{(L/C)^{\frac{1}{2}} + Z_L}\right]\exp{-j2\beta l} \tag{9.90}$$

It follows that the input impedance of a lossless transmission line of length l

9.5 Transmission lines

terminated by impedance Z_L is

$$Z_i = \left(\frac{V}{I}\right)_{x=0} = \left(\frac{I_1 - I_2}{I_1 + I_2}\right)\left(\frac{L}{C}\right)^{\frac{1}{2}}$$

$$= \left\{\frac{[(L/C)^{\frac{1}{2}} + Z_L] - [(L/C)^{\frac{1}{2}} - Z_L]\exp -j2\beta l}{[(L/C)^{\frac{1}{2}} + Z_L] + [(L/C)^{\frac{1}{2}} - Z_L]\exp -j2\beta l}\right\}\left(\frac{L}{C}\right)^{\frac{1}{2}}$$

$$= \left[\frac{Z_L(1 + \exp -j2\beta l) + (L/C)^{\frac{1}{2}}(1 - \exp -j2\beta l)}{Z_L(1 - \exp -j2\beta l) + (L/C)^{\frac{1}{2}}(1 + \exp -j2\beta l)}\right]\left(\frac{L}{C}\right)^{\frac{1}{2}}$$

or

$$Z_i = \left[\frac{Z_L + j(L/C)^{\frac{1}{2}}\tan \beta l}{(L/C)^{\frac{1}{2}} + jZ_L \tan \beta l}\right]\left(\frac{L}{C}\right)^{\frac{1}{2}} \tag{9.91}$$

In discussing equation (9.91), firstly notice that, when the line is loaded by the characteristic impedance, that is, when $Z_L = (L/C)^{\frac{1}{2}}$, it reduces to just $Z_i = (L/C)^{\frac{1}{2}}$ as expected. When the lossless line is short circuited so that $Z_L = 0$, the input impedance is apparently

$$Z_{isc} = (j \tan \beta l)(L/C)^{\frac{1}{2}} \tag{9.92}$$

Under open-circuit loading, that is, when $Z_L = \infty$, the input impedance becomes

$$Z_{ioc} = (-j \cot \beta l)(L/C)^{\frac{1}{2}} \tag{9.93}$$

Combining equations (9.92) and (9.93) gives

$$j \tan \beta l = (Z_{isc}/Z_{ioc})^{\frac{1}{2}} \tag{9.94}$$

$$(L/C)^{\frac{1}{2}} = Z_k = (Z_{isc} Z_{ioc})^{\frac{1}{2}} \tag{9.95}$$

from which it is apparent that the values of Z_k and β can be determined for a line from measurements of Z_{isc} and Z_{ioc}. Equation (9.95) is in agreement with the earlier equations (9.71) of course. Turning to the particular case of a quarter-wavelength line, $l = \lambda/4$, so that $\beta l = 2\pi l/\lambda = \pi/2$ and equation (9.91) reduces to the simple result

$$Z_i = (L/C)/Z_L \tag{9.96}$$

Thus a quarter-wavelength line can easily be used to transform one impedance into another. In particular, choice of the magnitude of the characteristic impedance $(L/C)^{\frac{1}{2}}$ allows a quarter-wavelength line to match two impedances. Notice that a short-circuited quarter-wavelength line provides open-circuit conditions at its input. Correspondingly an open-circuited quarter-wavelength line provides short-circuit conditions at its input. The former of these last two arrangements permits connections to lines without loading them.

Regarding reflection at the end of a transmission line, naturally the ratio of the potential difference of the immediate reflected wave to that of the incident wave is called the *reflection coefficient*. Equation (9.89) shows that the reflection coefficient of a lossless line is

$$\Gamma = -(\mathbf{I}_2/\mathbf{I}_1)\exp j2\beta l = (\mathbf{V}_2/\mathbf{V}_1)\exp j2\beta l \tag{9.97}$$

or using equation (9.90)

$$\Gamma = \frac{Z_L - (L/C)^{\frac{1}{2}}}{Z_L + (L/C)^{\frac{1}{2}}} \tag{9.98}$$

This result reveals that when the line is terminated in the characteristic impedance $(L/C)^{\frac{1}{2}}$ there is no reflection. Any other termination causes reflection. One consequence of reflection is the existence of standing waves on the line. In terms of the reflection coefficient Γ, equation (9.89) becomes

$$\mathbf{V} = (L/C)^{\frac{1}{2}}\mathbf{I}_1\{\exp -j\beta x + \Gamma \exp j\beta(x-2l)\} \tag{9.99}$$

For a general load impedance Z_L, the reflection coefficient Γ is complex and writing it as $\Gamma_0 \exp j\theta$, the positional dependence of the magnitude of $\mathbf{V}$ is given by

$$|\exp -j\beta x + \Gamma_0 \exp j(\beta x - 2\beta l + \theta)|$$

$$= \{[\cos \beta x + \Gamma_0 \cos(\beta x - 2\beta l + \theta)]^2 + [\Gamma_0 \sin(\beta x - 2\beta l + \theta) - \sin \beta$$

$$= \{1 + \Gamma_0^2 + 2\Gamma_0[\cos \beta x \cos(\beta x - 2\beta l + \theta) - \sin \beta x \sin(\beta x - 2\beta l + \theta)$$

$$= \{1 + \Gamma_0^2 + 2\Gamma_0 \cos(2\beta x - 2\beta l + \theta)\}^{\frac{1}{2}}$$

This expression shows that the magnitude of $\mathbf{V}$ passes through maxima and minima as x varies. The ratio S of the maximum to minimum of these standing waves, known as the *voltage standing wave ratio*, is evidently

$$S = (1+\Gamma_0)/(1-\Gamma_0) \tag{9.100}$$

Nodes, that is minima, of the standing wave occur when

$$2\beta x - 2\beta l + \theta = (2n+1)\pi \tag{9.101}$$

where n is any integer. However, $\beta = 2\pi/\lambda$ and so nodes exist at positions given by

$$l - x = [\theta/\pi - (2n+1)]\lambda/4 \tag{9.102}$$

showing that adjacent nodes are separated by $\lambda/2$.

The impedance of a lossless transmission line may be neatly expressed in terms of the voltage standing wave ratio S at positions where there is a maximum or minimum of the voltage standing wave. In terms of the reflection coefficient Γ, equation (9.88) becomes

$$\mathbf{I} = \mathbf{I}_1\{\exp -j\beta x - \Gamma \exp j\beta(x-2l)\} \tag{9.103}$$

Dividing the corresponding equation (9.99) for $\mathbf{V}$ by this equation and putting $\Gamma = \Gamma_0 \exp j\theta$ as before shows that the impedance of the line at any

9.5 Transmission lines

position x is expressible as

$$Z = \frac{V}{I} = \left\{\frac{1+\Gamma_0 \exp j(2\beta x - 2\beta l + \theta)}{1-\Gamma_0 \exp j(2\beta x - 2\beta l + \theta)}\right\}\left(\frac{L}{C}\right)^{\frac{1}{2}} \tag{9.104}$$

Since voltage nodes or minima occur when $2\beta x - 2\beta l + \theta = (2n+1)\pi$, equation (9.104) reveals that the impedance at such points is simply

$$Z_{min} = \left(\frac{1-\Gamma_0}{1+\Gamma_0}\right)\left(\frac{L}{C}\right)^{\frac{1}{2}} = \left(\frac{1}{S}\right)\left(\frac{L}{C}\right)^{\frac{1}{2}} \tag{9.105}$$

Similarly, voltage peaks or maxima occur when $2\beta x - 2\beta l + \theta = 2n\pi$ so that the impedance at these points is simply

$$Z_{max} = \left(\frac{1+\Gamma_0}{1-\Gamma_0}\right)\left(\frac{L}{C}\right)^{\frac{1}{2}} = S\left(\frac{L}{C}\right)^{\frac{1}{2}} \tag{9.106}$$

Apparently the impedance is purely resistive at voltage maxima or minima and is respectively a maximum and minimum at such positions. At some point between an adjacent maximum and minimum the real part of the impedance is $(L/C)^{\frac{1}{2}}$ but there is an associated reactive component. Now equations (9.92) and (9.93) show that an open or short-circuited line of length less than half a wavelength can provide any desired reactance or susceptance. Thus by connecting a subsidiary section of open or short-circuited line of suitable length ($<\lambda/2$) across a main transmission line at a point where the conductance is $1/(L/C)^{\frac{1}{2}}$, the associated susceptance or reactance can be neutralised. The subsidiary section connected for such purpose is called a *stub* and the procedure renders the main line correctly terminated at the stubbing point. Such impedance transformation is more flexible than that achieved through the use of quarter-wavelength sections.

In practice, signal attenuation in radio-frequency coaxial cables is caused by skin effect losses, which are proportional to the square-root of the frequency, and by dielectric losses, which are proportional to the frequency. Temperature and cable ageing also affect the attenuation. Dimensional or material irregularities incorporated during manufacture and badly assembled connectors all cause deviations in the impedance relative to the nominal characteristic impedance. All such deviations, like incorrect termination, cause a portion of the radio-frequency signal to be reflected and impair the quality of the transmitted signal.

10
Signal analysis of nonlinear and active networks

10.1 Two-terminal nonlinear networks

The signal responses of two and four-terminal *passive linear* networks have been considered extensively in the previous six chapters. Attention is now turned to deducing the signal responses of both *active and passive nonlinear* networks, a topic of great importance in view of the key roles that nonlinear devices play in electronics. The topic has already been briefly broached, of course, in section 5.9, where certain consequences of nonlinearity were established by examining a few illustrative passive circuits. At this juncture the objective is to treat the analysis of nonlinear circuits in a much wider context and in a much more general manner.

Graphical analysis of the response of any nonlinear network can be achieved through its terminal static characteristic or characteristics irrespective of the magnitudes of the signals involved. This approach is, however, clearly most appropriate under *large-signal* conditions. At the opposite extreme, whenever the signals in a nonlinear network are small enough, the network is *effectively linear with respect to the signals* so that the methods developed for linear network analysis may be applied with advantage. Although maintenance of such *small-signal* conditions may appear somewhat restrictive, their occurrence is quite widespread. Electronic systems are very often concerned with processing weak signals and sometimes the nonlinearity involved is sufficiently slight for quite large signals to qualify as small enough for the purpose of linear analysis.

Explanation of the methods of large and small-signal analysis of nonlinear networks is best undertaken initially in terms of two-terminal networks. Graphical determination of the steady-state response of a two-terminal nonlinear component to connection of a direct source has already been treated in section 3.10 where the concepts of *load line* and *operating point* were introduced. With a two-terminal nonlinear network rather than

10.1 Two-terminal nonlinear networks

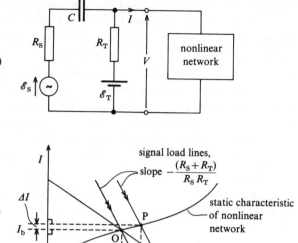

10.1 (a) A circuit comprising a signal source capacitor coupled to a biased nonlinear network and (b) its graphical solution using load lines.

a two-terminal nonlinear component, the same procedure applies except that the relevant static characteristic is that of the network. Referring to figure 3.22(a), if the applied direct e.m.f. $\mathscr{E}_T$ is replaced by a signal e.m.f. $\mathscr{E}_S$, then the intercept of the load line on the potential V axis of figure 3.22(b) varies with time accordingly. Thus the time dependences of the current I and potential difference V are given by the time dependence of the intersection of the load line with the characteristic, provided the frequency of the source is not so high that reactive effects render the static characteristic inappropriate. With a signal e.m.f. included in series with the direct bias e.m.f. $\mathscr{E}_T$, a similar solution follows, the time-dependent intercept of the load line on the potential V axis now being equal to the total e.m.f. Notice that the presence of the signal e.m.f. in both these cases causes the load line to move parallel to itself with time. Another interesting and common circuit arrangement has a signal source $\mathscr{E}_S$, R_S capacitor-coupled to a biased nonlinear network as shown in figure 10.1(a). The capacitor segregates the direct bias circuit $\mathscr{E}_T$, R_T so that the bias load line is as shown in figure 10.1(b) with intercept $\mathscr{E}_T$ on the V axis and slope $-1/R_T$. Intersection of this load line with the characteristic occurs at some point O representing the operating bias I_b, V_b of the nonlinear network. The coupled signal causes the current I and potential difference V to fluctuate about the values I_b and V_b corresponding to the point O. Assuming that the capacitor exhibits negligible reactance at the operating frequency so that it couples the signal properly, Thévenin transformation reveals an effective signal circuit connected to the nonlinear network terminals comprising e.m.f. $R_T \mathscr{E}_S/(R_S + R_T)$ in series with resistance $R_S R_T/(R_S + R_T)$. Consequently the slope and intercept of the signal load line are as shown in the figure and, as $\mathscr{E}_S$ varies, the displacement of the intersection P of the signal load line with the characteristic gives the signal potential difference ΔV and signal current ΔI. In general, the waveform of the signal ΔV will be completely different from that of the signal source e.m.f. $\mathscr{E}_S$. However, if the amplitude of the signal source e.m.f. $\mathscr{E}_S$ becomes small enough for the excursion of the intersection point P to be over an effectively linear region of the characteristic, then the signals ΔV and ΔI will be proportional to $\mathscr{E}_S$. In other words, the nonlinear network will respond linearly to the signal source.

Whatever the time dependences of the signals in a two-terminal nonlinear network, its terminal behaviour can be represented by

$$I = f_1(V) \quad \text{or} \quad V = f_2(I) \tag{10.1}$$

Moreover, the relationship between sufficiently small signals ΔI and ΔV, denoted according to convention by lower-case letters i and v, is

$$i = \left(\frac{dI}{dV}\right)_{V_b} v \quad \text{or} \quad v = \left(\frac{dV}{dI}\right)_{I_b} i \tag{10.2}$$

where (V_b, I_b) is the bias operating point. If the signals only vary slowly, $(dI/dV)_{V_b}$ and $(dV/dI)_{I_b}$ respectively represent the small-signal conductance and resistance at the operating point and are given by the slope of the static characteristic and its inverse at the operating point. When the terminal potential difference varies more rapidly, reactive effects arise. With regard to sinusoidal signals that are small enough in amplitude for the network response to be effectively linear, whatever the frequency the result embodied in equations (10.2) is conveniently expressed in terms of phasors. Thus writing the terminal small-signal current and potential difference phasors as **i** and **v** respectively

$$\mathbf{i} = Y\mathbf{v} \quad \text{or} \quad \mathbf{v} = Z\mathbf{i} \tag{10.3}$$

The terminal, small-signal complex admittance Y or impedance Z introduced here may be found at any particular frequency from measurements of the phasors **i** and **v** with the appropriate bias applied. As equations (10.2) clearly demonstrate, the small-signal impedance of a nonlinear network depends on its operating bias. Armed with the

10.2 Four-terminal nonlinear networks

appropriate value of the small-signal complex impedance, the signal response of a nonlinear network to any small input may be found in just the same way as the response of a linear network to an input of any magnitude.

As pointed out at the beginning of this section, some of the interesting responses that arise on applying somewhat larger sinusoidal e.m.f.s to nonlinear resistive circuits such that the signal behaviour is governed by the nonlinear relation

$$i = av + bv^2$$

have been considered in section 5.9. The reader is referred back to that section on this point but is again reminded that phasor and complex algebraic methods of linear circuit analysis are inapplicable to situations where an effective nonlinearity exists.

10.2 Four-terminal nonlinear networks

Many circuits and electronic systems process a signal between a pair of input terminals and a pair of output terminals as indicated in figure 10.2. Such networks are aptly described as *four-terminal* networks. Numerous four-terminal networks of the passive linear type have been considered in the earlier chapters, for example, transformers, attenuators, filters and phase-shift networks. A prime active example of a nonlinear four-terminal network is an electronic *amplifier* which increases the power of a signal between its input and output terminals. Three-terminal devices and networks often function as four-terminal networks with one terminal common between the input and output. Nonlinear devices such as triode thermionic valves and various kinds of transistor belong to this latter category. Now consider nonlinear four-terminal networks in general.

A four-terminal network responds to external input circuit connections and drives external load circuits through the terminal input and output potential differences and currents V_i, I_i, V_o and I_o and its behaviour is specified by the interdependences of these terminal variables. Once these dependences are determined, the response of the network to input and output connections can be deduced independently of any knowledge of the internal action or circuitry. Interdependences of the terminal variables are

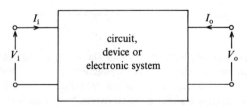

10.2 Four-terminal network.

usually presented in the form of static characteristics. Notice that a convention is normally followed in which the current at a terminal is regarded as positive if it flows into that terminal. The labelling of the terminal variables I_i and I_o in figure 10.2 conforms with this convention.

Of the four terminal variables V_i, I_i, V_o, I_o, any two may be regarded as independent and the other two may then be determined or expressed in terms of them. Plots of I_o versus V_o for sets of values of V_i or I_i are known appropriately as *output* characteristics. Correspondingly, graphs of I_i versus V_i for sets of values of V_o or I_o are described as *input* characteristics. Plots of an output terminal variable as a function of an input terminal variable or vice versa for various values of one of the other two variables are termed *transfer* or *mutual* characteristics.

Figure 10.3 shows examples of each type of static characteristic for the particularly important, four-terminal, nonlinear, active network of a three-terminal, N–P–N, bipolar, junction transistor operating in what is known as the common-emitter configuration. Of the three terminals, collector, emitter and base, the emitter is common to the input and output in this configuration, with the output taken between the collector and emitter and the input applied between the base and emitter. In labelling the axes of the characteristics, subscripts c, e and b have been used to denote the collector, emitter and base. Thus, for example, I_c and V_{ce} respectively represent the

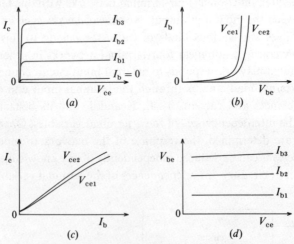

10.3 Examples of static characteristics for the four-terminal active network formed by arranging a bipolar junction transistor in the common-emitter configuration; (*a*) output, (*b*) input, (*c*) forward transfer and (*d*) reverse transfer characteristics. Note that $I_{b3} > I_{b2} > I_{b1}$ and $V_{ce2} > V_{ce1}$.

10.2 Four-terminal nonlinear networks

collector current and potential difference between the collector and emitter. With respect to input and output notation, $V_i = V_{be}$, $I_i = I_b$, $V_o = V_{ce}$ and $I_o = I_c$. For reasons that will emerge shortly, it is customary to present the output and reverse transfer characteristics for various base currents and the input and forward transfer characteristics for various potential differences between the collector and emitter. Notice that the output depends on the input and vice versa. Moreover, although the characteristics are markedly nonlinear overall, certain characteristics are virtually linear over substantial ranges.

Just as for two-terminal networks, the large-signal response of a four-terminal network may be determined by graphical analysis of the static characteristics. Consider, for example, the simple capacitor-coupled common-emitter amplifier circuit of figure 10.4(a) in which the terminals of the standard symbol denoting the N–P–N transistor are labelled. The bias

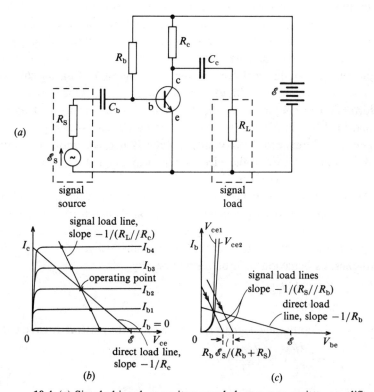

10.4 (a) Simply biased, capacitor-coupled, common-emitter amplifier, (b) output characteristics of amplifying transistor with load lines corresponding to circuit (a) superimposed and (c) input characteristics of amplifying transistor also with load lines corresponding to circuit (a) superimposed.

circuit, segregated by the coupling capacitors C_c and C_b, comprises just the e.m.f. $\mathscr{E}$ and resistors R_b and R_c. Consequently the direct load lines for the output and input are as shown in figures 10.4(b) and (c) respectively. To find the output operating point, the input bias current is needed. Fortunately, the direct e.m.f. $\mathscr{E}$ is usually very large compared with the input potential difference V_{be} (≈ 0.6 V for a silicon transistor) so that to a fair approximation $I_b = \mathscr{E}/R_b$. Whatever the magnitude of $\mathscr{E}$, because I_b only depends slightly on V_{ce}, without proper knowledge of V_{ce} a better approximation to I_b can be obtained from intersections of the input load line with input characteristics for which $0 \leqslant V_{ce} \leqslant \mathscr{E}$. The intersection of the appropriate output characteristic with the output load line then gives a good approximation to V_{ce} which in turn leads to a better value of I_b from the input characteristics and input load line. Such iteration is rapidly convergent and soon yields accurate values of both the input and output operating bias if needed. The coupling capacitors are chosen to be of sufficiently large capacitance to present negligible reactance to signals over the operating bandwidth (frequency range) of the amplifier. Thus, just as for a two-terminal network with capacitor-coupled signal source, the input-signal load line behaves as shown in figure 10.4(c), its changing intersection giving the input signal current ΔI_b. Once more, this information in conjunction with the output characteristics and output-signal load line of figure 10.4(b) gives the output signal current ΔI_c and potential difference ΔV_{ce}. If high accuracy is needed, another iterative process will take care of the effect of the output signal ΔV_{ce} on the input signal current and vice versa.

As with two-terminal networks, when the signals become sufficiently small, the response becomes linear and signal analysis is better conducted in terms of suitable small-signal parameters rather than graphically. Choosing the input and output currents of the four-terminal network as independent variables, its behaviour can be represented by

$$V_i = f_3(I_i, I_o)$$
$$V_o = f_4(I_i, I_o)$$
(10.4)

and so sufficiently small signals v_i, i_i, v_o and i_o in the network are related by

$$v_i = \left(\frac{\partial V_i}{\partial I_i}\right)_{I_o} i_i + \left(\frac{\partial V_i}{\partial I_o}\right)_{I_o} i_o$$

$$v_o = \left(\frac{\partial V_o}{\partial I_i}\right)_{I_o} i_i + \left(\frac{\partial V_o}{\partial I_o}\right)_{I_i} i_o$$
(10.5)

Since the derivatives appearing in equations (10.5) have the dimensions of impedance, it is normal to put

10.2 Four-terminal nonlinear networks

$$Z_{11} = \left(\frac{\partial V_i}{\partial I_i}\right)_{I_o} = \left(\frac{v_i}{i_i}\right)_{i_o=0} ; \quad Z_{12} = \left(\frac{\partial V_i}{\partial I_o}\right)_{I_i} = \left(\frac{v_i}{i_o}\right)_{i_i=0}$$
$$Z_{21} = \left(\frac{\partial V_o}{\partial I_i}\right)_{I_o} = \left(\frac{v_o}{i_i}\right)_{i_o=0} ; \quad Z_{22} = \left(\frac{\partial V_o}{\partial I_o}\right)_{I_i} = \left(\frac{v_o}{i_o}\right)_{i_i=0}$$
(10.6)

for easy reference. In terms of these Z-parameters, equations (10.5) are just

$$\left. \begin{array}{l} v_i = Z_{11}i_i + Z_{12}i_o \\ v_o = Z_{21}i_i + Z_{22}i_o \end{array} \right\}$$
(10.7)

and, because of the linear relationship between small signals, the response to small sinusoidal signals is governed by corresponding phasor equations

$$\left. \begin{array}{l} \mathbf{v}_i = Z_{11}\mathbf{i}_i + Z_{12}\mathbf{i}_o \\ \mathbf{v}_o = Z_{21}\mathbf{i}_i + Z_{22}\mathbf{i}_o \end{array} \right\}$$
(10.8)

When the signals involved are of low-enough frequency for reactive and other frequency-dependent effects to be negligible, the Z-parameters are given by the slopes of appropriate static characteristics at the operating bias levels. From equations (10.6), Z_{11} is the reciprocal of the slope of the input characteristic I_i versus V_i for constant output current I_o while Z_{12} is the slope of the reverse transfer characteristic V_i versus I_o for constant input current I_i. Also, Z_{21} is the slope of the forward transfer characteristic V_o versus I_i for constant output current I_o and Z_{22} is the reciprocal of the slope of the output characteristic I_o versus V_o for constant input current I_i.

While the Z-parameters relevant to low-frequency operation may be obtained from static characteristics, they may be determined at any frequency by measuring certain small signals of the network under suitable open-circuit conditions. In accordance with equations (10.8), Z_{11} is the small-signal, complex, input impedance $\mathbf{v}_i/\mathbf{i}_i$ when the output is open circuit to signals so that $\mathbf{i}_o = 0$. Similarly Z_{12} is the ratio of the small, input, signal voltage to small, output, signal current when the input is open circuit to signals and Z_{21} is the ratio of the small, output, signal voltage to small, input, signal current when the output is open circuit to signals. Finally, Z_{22} is the small-signal, complex, output impedance $\mathbf{v}_o/\mathbf{i}_o$ when the input is open circuit to signals.

An alternative choice of independent variables that is more convenient for certain types of four-terminal network is the input current I_i and output voltage V_o. In terms of these particular variables the behaviour is expressible as

$$\left. \begin{array}{l} V_i = f_5(I_i, V_o) \\ I_o = f_6(I_i, V_o) \end{array} \right.$$
(10.9)

and so sufficiently small signals are related by

$$v_i = h_{11}i_i + h_{12}v_o$$
$$i_o = h_{21}i_i + h_{22}v_o \qquad (10.10)$$

where

$$h_{11} = \left(\frac{\partial V_i}{\partial I_i}\right)_{V_o} = \left(\frac{v_i}{i_i}\right)_{v_o=0} \;;\quad h_{12} = \left(\frac{\partial V_i}{\partial V_o}\right)_{I_i} = \left(\frac{v_i}{v_o}\right)_{i_i=0}$$

$$h_{21} = \left(\frac{\partial I_o}{\partial I_i}\right)_{V_o} = \left(\frac{i_o}{i_i}\right)_{v_o=0} \;;\quad h_{22} = \left(\frac{\partial I_o}{\partial V_o}\right)_{I_i} = \left(\frac{i_o}{v_o}\right)_{i_i=0} \qquad (10.11)$$

Again, the small-signal linearity means that the small-signal sinusoidal response is governed by corresponding phasor equations

$$\left.\begin{array}{l}\mathbf{v}_i = h_{11}\mathbf{i}_i + h_{12}\mathbf{v}_o \\ \mathbf{i}_o = h_{21}\mathbf{i}_i + h_{22}\mathbf{v}_o\end{array}\right\} \qquad (10.12)$$

Symbols h_{ij} are universally adopted to denote the differential parameters involved here because they have hybrid dimensions. Such dimensions are, of course, a direct consequence of selecting hybrid independent variables and the differential parameters are appropriately known as *hybrid* or *h-parameters*. Parameters h_{12} and h_{21} are in fact dimensionless while h_{11} and h_{22} respectively exhibit impedance and admittance dimensions. At low-enough frequencies the parameters are again given by the slopes of characteristics. For example, h_{21} is the slope of the forward transfer characteristic I_o versus I_i at constant V_o and h_{22} is the slope of the output characteristic I_o versus V_o at constant input current I_i. To determine a set of h-parameters at any frequency from small-signal measurements demands both short-circuit and open-circuit signal terminations. For instance, h_{11} is the small-signal, complex, input impedance $\mathbf{v}_i/\mathbf{i}_i$ when the output is *short circuited* for signals so that $\mathbf{v}_o = 0$. On the other hand, h_{12} is the reciprocal of the small-signal voltage gain $\mathbf{v}_o/\mathbf{v}_i$ when the input is *open circuit* to signals so that $\mathbf{i}_i = 0$.

Remember that, in general, the set of Z-parameters to be used in equations (10.7) or (10.8) and the set of h-parameters to be used in equations (10.10) or (10.12) will depend on the operating bias. Just how small the signals must be for the Z and h-parameter equations to apply with constant sets of parameters depends on the particular case. For a bipolar junction transistor, the output and current transfer characteristics are almost linear over substantial parts of the normal operating range but the input characteristics are markedly nonlinear (recall figure 10.3). Thus the output signals can be quite large without violating the approximate applicability of the second of the two h-parameter equations with constant parameters.

10.2 Four-terminal nonlinear networks

However, the input signals must be extremely small for the first of these two equations to apply with constant parameters.

Signal measurements to determine the small-signal parameters of four-terminal networks must be executed under particular bias conditions implemented by direct circuits connected between the input terminals and between the output terminals. The relative merits of measuring Z or h-parameters rests on which of the signal termination conditions is the easier to provide. To effectively present open-circuit signal conditions between terminals requires the external impedance to be high compared with the internal impedance. Thus the open-circuit conditions needed for Z-parameter determination are more easily arranged when the input and output impedance of the four-terminal network concerned are both low. For effectively open-circuit operation the resistance of the bias circuit has to be high enough or a choke of high enough reactance has to be connected in series with it. On the input side, the signal source can be introduced without disturbing the bias conditions or effectively open-circuit signal operation by simply connecting it in series with the bias circuit or by capacitor coupling it to the input through a high series resistance. Measurement of h-parameters demands that the input terminals are effectively open circuit with respect to signals while the output terminals are effectively short circuited as far as signals are concerned. The latter is readily achieved in practice without disturbing the bias arrangements by connecting a capacitor of large capacitance across the output terminals. Four-terminal networks for which the input impedance is low and the output impedance is high are clearly well suited to the achievement of the termination conditions needed for h-parameter measurement. A junction transistor in the common-base or common-emitter configuration is a good example of a network exhibiting such input and output impedances.

Since the small-signal behaviour of any four-terminal network is represented by equations (10.8) in terms of Z-parameters and by equations (10.12) in terms of h-parameters, the Z and h-parameters of a given four-terminal network are always interrelated. Comparison of the second equation of the pair (10.8) with the second equation of the pair (10.12) immediately reveals that for a given network

$$\left. \begin{array}{l} Z_{22} = 1/h_{22} \\ Z_{21} = -h_{21}/h_{22} \end{array} \right\} \tag{10.13}$$

Also, elimination of v_o between equations (10.12) and comparison of the resulting equation with the first equation of the pair (10.8) establishes that for a given network

$$\left. \begin{array}{l} Z_{11} = h_{11} - h_{12}h_{21}/h_{22} \\ Z_{12} = h_{12}/h_{22} \end{array} \right\} \tag{10.14}$$

It is a matter of simple algebraic manipulation to show that the inverse relations to equations (10.13) and (10.14) are

$$\left.\begin{array}{ll}h_{11}=Z_{11}-Z_{12}Z_{21}/Z_{22}; & h_{12}=Z_{12}/Z_{22}\\ h_{21}=-Z_{21}/Z_{22}; & h_{22}=1/Z_{22}\end{array}\right\} \quad (10.15)$$

Now consider the Z-parameters for the very simple passive linear T-network of figure 10.5(a). Applying Kirchhoff's voltage law to the input and output meshes gives

$$\mathbf{v}_i = (Z_1 + Z_3)\mathbf{i}_i + Z_3 \mathbf{i}_o$$
$$\mathbf{v}_o = Z_3 \mathbf{i}_i + (Z_2 + Z_3)\mathbf{i}_o$$

Thus

$$\begin{array}{l}Z_{11}=Z_1+Z_3\\ Z_{12}=Z_{21}=Z_3\\ Z_{22}=Z_2+Z_3\end{array} \quad (10.16)$$

The results obtained for this particular network are illustrative of the fact that $Z_{12}=Z_{21}$ or $h_{12}=-h_{21}$ in all passive networks. In active networks $Z_{12}\neq Z_{21}$. If a passive network is also symmetric, that is, if the same response is obtained with the output and input connections interchanged, $Z_{11}=Z_{22}$ also. The simple network of figure 10.5(a) is symmetric of course if $Z_1=Z_2$ and equations (10.16) confirm that $Z_{11}=Z_{22}$ in this particular case. Notice that from equations (10.13) and (10.14), $Z_{11}=Z_{22}$ is equivalent in terms of h-parameters to $h_{11}h_{22}-h_{12}h_{21}=1$.

10.3 Small-signal equivalent circuits and analysis

Networks that exhibit the same terminal behaviour as some device, system or more complicated network are naturally known as *equivalent circuits*. This section is concerned with the introduction and application of certain particularly useful types of equivalent circuit that display the *same form of linear small-signal response as any nonlinear four-terminal network*. Consider first the four-terminal network shown in figure 10.5(b). Application of Kirchhoff's voltage law to its input and output circuits

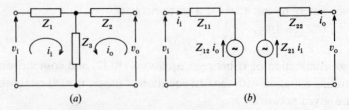

10.5 (*a*) Passive linear T-network and (*b*) Z-parameter, small-signal, equivalent circuit.

10.3 Small-signal equivalent circuits

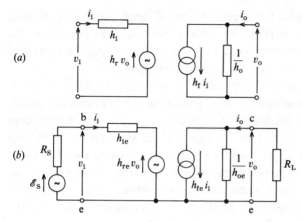

10.6 (a) General, small-signal, hybrid, equivalent circuit and (b) small-signal, low-frequency, equivalent circuit of a bipolar junction transistor connected in the common-emitter configuration to a source of e.m.f. $\mathscr{E}_S$, R_S and load resistance R_L.

generates equations (10.8). It therefore reproduces the linear small-signal response of any four-terminal network and is appropriately referred to as the Z-parameter equivalent circuit. In accordance with Thévenin's theorem both the input and output circuits comprise a signal e.m.f. in series with an impedance. However, beyond that, each signal e.m.f. is related to the signal current in the other circuit. Especially note that the senses in which the signal e.m.f.s act are related to the senses of the input and output signal currents.

The four-terminal small-signal equivalent circuit based on the h-parameter relations (10.12) and therefore called the *hybrid* equivalent circuit is shown in figure 10.6(a). Notice that here the input representation satisfies Thévenin's theorem while the output representation satisfies Norton's theorem. Again the signal sources in the input and output circuits relate in both magnitude and sense of action to the output and input conditions respectively. Labelling the parameters $h_i = h_{11}, h_r = h_{12}, h_f = h_{21}$ and $h_o = h_{22}$ is neater and, compared with matrix style subscripts, perhaps more directly infers the nature of each parameter in that i stands for input, r for reverse, f for forward and o for output. It is left as a simple exercise to show that application of Kirchhoff's laws to the circuit does indeed yield relations (10.12).

Although through appropriate complex parameters the Z and h-parameter equivalent circuits can reproduce the small-signal sinusoidal response of any four-terminal network at any frequency, it is more usual to represent just the low-frequency behaviour by such an equivalent circuit with real parameters. Corresponding small-signal behaviour at high frequencies is then covered by adding reactive components to the equivalent circuit that directly relate to the particular physical mechanisms that cause the differing response at such frequencies.

To illustrate how equivalent circuits may be applied to analyse the small-signal responses of nonlinear systems, the low-frequency response of a basic junction transistor amplifier will now be examined. The hybrid equivalent circuit is especially suited to representing the small-signal amplifying behaviour of a bipolar junction transistor since its parameters are easily determined for the most useful common-emitter configuration by appropriate small-signal measurements. Sometimes the transistor is operated with advantage as an amplifying four-terminal network with the collector or base rather than emitter common between input and output, giving the so-called common-collector and common-base configurations. Hybrid parameters relevant to the three configurations are distinguished by adding a second subscript so that, for example, the common-emitter h-parameters are denoted by h_{ie}, h_{re}, h_{fe} and h_{oe}.

Figure 10.6(b) shows the low-frequency, small-signal, equivalent circuit of a bipolar junction transistor connected in the common-emitter configuration to a source of e.m.f. $\mathscr{E}_S$, R_S and a load resistance R_L. All the h-parameters are real here with h_{ie} and $1/h_{oe}$ denoting resistances and h_{re} and h_{fe} pure numbers. Typical values for a low-power transistor would be $h_{ie} = 3\,\text{k}\Omega$, $h_{re} = 10^{-4}$, $h_{fe} = 200$ and $1/h_{oe} = 50\,\text{k}\Omega$. Note that, due to the common connection of the emitter to input and output, there is now a common rail compared with figure 10.6(a). Applying Kirchhoff's voltage and current laws respectively to the input and output circuits of figure 10.6(b) yields

$$\mathbf{v}_i = h_{ie}\mathbf{i}_i + h_{re}\mathbf{v}_o \tag{10.17}$$

$$\mathbf{i}_o = h_{fe}\mathbf{i}_i + h_{oe}\mathbf{v}_o \tag{10.18}$$

However, $\mathbf{v}_o = -R_L\mathbf{i}_o$ and so from equation (10.18) the small-signal current gain between input and output is

$$A_i = \frac{\mathbf{i}_o}{\mathbf{i}_i} = \frac{h_{fe}}{1 + h_{oe}R_L} \tag{10.19}$$

while the input resistance presented to small signals is from equation (10.17)

$$R_i = \frac{\mathbf{v}_i}{\mathbf{i}_i} = h_{ie} - h_{re}R_L(\mathbf{i}_o/\mathbf{i}_i)$$

Making use of equation (10.19), the latter becomes

$$R_i = h_{ie} - \frac{h_{re}h_{fe}R_L}{1 + h_{oe}R_L} \tag{10.20}$$

10.3 Small-signal equivalent circuits

Proceeding further, the small-signal voltage gain between input and output is

$$A_v = \frac{v_o}{v_i} = \frac{-R_L i_o}{R_i i_i} = -\frac{R_L A_i}{R_i}$$

and, on substituting for A_i and R_i from equations (10.19) and (10.20), it is seen that

$$A_v = \left[h_{re} - \frac{h_{ie}}{h_{fe}} \left(\frac{1 + h_{oe} R_L}{R_L} \right) \right]^{-1} \tag{10.21}$$

To obtain an expression for the output resistance, observe that in the input circuit

$$\mathcal{E}_S - (R_S + h_{ie}) i_i = h_{re} v_o$$

Eliminating i_i through equation (10.18) gives

$$\mathcal{E}_S + [(R_S + h_{ie})/h_{fe}](h_{oe} v_o - i_o) = h_{re} v_o$$

and rearranging this equation in the form

$$v_o = \mathcal{E}_o + R_o i_o$$

establishes that the output resistance is

$$R_o = \left(h_{oe} - \frac{h_{re} h_{fe}}{h_{ie} + R_S} \right)^{-1} \tag{10.22}$$

Notice that when $R_L \ll 1/h_{oe}$, $A_i \approx h_{fe}$ and, since h_{re} is very small, $A_v \approx -h_{fe} R_L / h_{ie}$. The negative sign here simply means that the signal voltage is inverted between the input and output, that is, a small sinusoidal signal voltage is shifted in phase by 180° between the input and output. With a typical load resistance of a few kΩ, the magnitude of the signal voltage gain A_v is around 100. Combined with the simultaneous signal current gain, $A_i \approx h_{fe} \sim 200$, this means that a vast *signal power gain* $A_p = A_i A_v \sim 2 \times 10^4$ is achieved. Attainment of such huge signal power gain is of the greatest electronic significance. Whenever a signal experiences power gain, *amplification* is said to occur and the network achieving it is described as an amplifier. Such networks are also said to be active in the spirit of the application of the term active at the end of chapter 2. In simple language, the signal becomes vastly more powerful as it passes through this type of network. Recall that although either signal voltage gain or signal current gain can be obtained separately with a transformer, there is always attenuation of signal power through one. Signal voltage gain is always accompanied by greater signal current attenuation and vice versa with a transformer.

A casual glance at the equivalent circuit of figure 10.6(b) might suggest that the output resistance of the common-emitter amplifier is $1/h_{oe}$ and that maximum signal power gain arises with this network when the load resistance R_L equals $1/h_{oe}$. Neither of these conclusions is correct. To begin with, the signal current source $h_{fe}i_i$ in the output is only constant if the input signal current i_i is constant. In general, as R_L varies, v_o varies and so i_i changes because of the signal e.m.f. $h_{re}v_o$ in the input circuit. Although with i_i held constant in some way (making R_S infinite in equation (10.22)), the output resistance is indeed $1/h_{oe}$ and maximum signal power is transferred between the output signal source and load when $R_L = 1/h_{oe}$, the input signal power then depends on R_L through v_o because the latter affects the reverse transfer e.m.f. $h_{re}v_o$ and hence v_i. Again it is clear that maximum signal power gain does not correspond to $R_L = 1/h_{oe}$. In practice, because h_{re} is very small, the maximum signal-power-gain condition is not far removed from $R_L = 1/h_{oe}$. Furthermore, the availability of cheap transistors makes obtaining maximum power gain through each transistor rather unimportant in any case.

With respect to the input and output resistance of the common-emitter configuration, equations (10.20) and (10.22) reveal that, no matter what the magnitude of the load resistance R_L or source resistance R_S, $R_i \sim h_{ie}$ and $R_o \sim 1/h_{oe}$. Thus for a low-power transistor, R_i is typically 3 kΩ and R_o is typically 50 kΩ.

A common cause of deterioration in the performance of amplifiers at high frequencies is the presence of capacitance between the output and input. Such capacitance is, of course, kept to a minimum, but some is inevitable through the device involved in the amplification and due to the external wiring of the circuit. The situation is depicted schematically in figure 10.7 where the triangular symbol represents the amplifier of gain, say $-A$, between the input and output terminals. Extra input current drawn through the troublesome capacitance C amounts to

$$\mathbf{i}_C = j\omega C(\mathbf{v}_i - \mathbf{v}_o) = j\omega C(A+1)\mathbf{v}_i$$

From the point of view of the input terminals, the presence of capacitance C

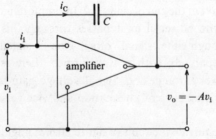

10.7 The origin of the Miller effect.

makes it appear that there is additional reactive impedance
$$v_i/i_C = 1/j\omega(A+1)C \tag{10.23}$$
bridging the input terminals. In other words, an amplified version $(A+1)C$ of the capacitance C appears to exist across the input terminals. This interesting phenomenon is known as the *Miller* effect. At sufficiently high frequencies, the capacitive reactance becomes small enough to reduce the magnitude of the input signal v_i delivered from any practical source of finite internal impedance.

10.4 Feedback

An extremely valuable electronic technique is to return a *fraction* β of the output signal s_o of an amplifier of gain A to the input in such a way that the net input signal to the amplifier becomes
$$s = s_i - \beta s_o \tag{10.24}$$
where s_i is the input signal intended for amplification. Since
$$s_o = As \tag{10.25}$$
it follows that the overall gain under such *feedback* conditions is
$$\frac{s_o}{s_i} = \frac{A}{1+\beta A} \tag{10.26}$$
Notice that, in this broad introductory description of the feedback technique, the symbol **s** has been deliberately introduced to signify any signal current or voltage. The fedback gain represented by expression (10.26) is aptly referred to as the *closed-loop* gain and quite naturally corresponding descriptions *open-loop* gain and *loop* gain are often applied to the quantities A and βA respectively. While both β and A may be complex quantities in general, it will be assumed for the purposes of the present section that β and A are just positive or negative real quantities. Positive or negative signs simply imply, of course, phase shifts of zero or 180° for sinusoidal signals through the feedback network or open-loop amplifier.

If the fedback signal *opposes* the input signal then the feedback is said to be *negative*. In a similar way, if the fedback signal *augments* the input signal then the feedback is said to be *positive*. Observe, however, that negative and positive feedback respectively correspond to the loop gain βA being positive and negative in equation (10.26). Of enormous significance is the fact that when the feedback is negative such that $\beta A \gg 1$ then
$$s_o/s_i \approx 1/\beta \tag{10.27}$$
and the overall gain is virtually independent of A! In other words, through the introduction of suitable negative feedback, amplification that is

virtually independent of the open-loop amplifier can be achieved. In particular, the negative feedback technique enables very linear amplification to be attained despite substantial nonlinearity in the open-loop amplifier. Moreover, through it, the same predictable behaviour can be obtained irrespective of the particular active device incorporated in the open-loop amplifier even when the characteristics of the devices concerned vary greatly. Other important effects of negative feedback are changes in the input and output impedances and improvement of the frequency response (gain constant over a greater range of frequency). It is, of course, vital to appreciate that the changes described occur at the expense of gain; the closed-loop gain under negative feedback is always less than the open-loop gain. Provided that the negative feedback is implemented in a manner suited to a particular application, all the changes will normally be beneficial and the reduction in magnitude of gain worthwhile. Deliberate, controlled positive feedback also has some application in connection with amplifiers but accidental, uncontrolled positive feedback is often troublesome in amplifiers leading to instability. The main application of positive feedback is in the attainment of oscillation and switching. Instability and oscillation will be discussed in sections 10.6 and 10.7.

Consider now the four basic circuit connections by means of which feedback is implemented. These are shown in figure 10.8. In each case the input and output impedances of the open-loop amplifier are represented by Z_i and Z_o respectively while its gain is denoted by a suitable parameter A. Observe that while the fedback signal is derived in parallel with the output in the circuits of figures 10.8(a) and (c), it is derived in series with the output in the circuits of figures 10.8(b) and (d). In a similar way, the fedback signal is inserted in series with the input in the circuits of figures 10.8(a) and (b) but in parallel with the input in the circuits of figures 10.8(c) and (d). Thus convenient descriptions of the basic arrangements of circuits 10.8(a), (b), (c) and (d) are respectively series-inserted, voltage-derived; series-inserted, current-derived; parallel-inserted, voltage-derived and parallel-inserted, current-derived feedback. For easy reference these four forms of circuit may be referred to by the more economical descriptions *series-voltage, series-current, shunt-voltage* and *shunt-current* feedback respectively.

Analysing the series-voltage case first, the open-circuit and loaded voltage gains of the open-loop amplifier are defined as A_{v0} and A_v respectively. With reference to figure 10.8(a), application of Kirchhoff's voltage law to the input circuit yields

$$\mathbf{v}_i = \mathbf{v} + \beta_v \mathbf{v}_o$$

But

$$\mathbf{v}_o = A_v \mathbf{v}$$

10.4 Feedback

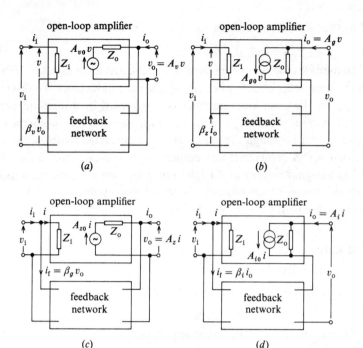

10.8 (a) Series-voltage, (b) series-current, (c) shunt-voltage and (d) shunt-current type of feedback network.

and so the closed-loop gain is

$$A_{vf} = v_o/v_i = A_v/(1 + \beta_v A_v) \tag{10.28}$$

in accordance with equation (10.26). The closed-loop input impedance is easily found as

$$Z_{if} = v_i/i_i = (v + \beta_v v_o)/(v/Z_i) = (1 + \beta_v A_v)Z_i \tag{10.29}$$

To find the closed-loop output impedance, a source $\mathscr{E}_S, Z_S$ is considered to be connected to the input in which case

$$\mathscr{E}_S = Z_S i_i + v + \beta_v v_o = [(Z_S + Z_i)/Z_i]v + \beta_v v_o$$

Assuming for simplicity that the feedback network negligibly loads the output

$$v_o = A_{v0} v + Z_o i_o$$

and so

$$\mathscr{E}_S = \{[(Z_S + Z_i)/A_{v0} Z_i] + \beta_v\} v_o - \left[\frac{(Z_S + Z_i)Z_o}{A_{v0} Z_i}\right] i_o$$

from which the closed-loop output impedance is

$$Z_{of} = \{1 + [\beta_v A_{v0} Z_i/(Z_S + Z_i)]\}^{-1} Z_o$$

238 Signal analysis

If further $Z_S \ll Z_i$, as is often the case,
$$Z_{of} = (1 + \beta_v A_{v0})^{-1} Z_o \tag{10.30}$$

In the case of series-current feedback, it is convenient to describe the gain of the open-loop amplifier in terms of mutual conductance. To this end the output of the open-loop amplifier is represented by a load-independent, signal, current source $A_{g0}\mathbf{v}$ in parallel with impedance Z_o while the output signal current through the load and series-connected feedback network is written $A_g \mathbf{v}$. A complementary nomenclature expresses the series-inserted, feedback, signal voltage as $\beta_z \mathbf{i}_o$, the subscript z on β indicating impedance dimensions. With reference to figure 10.8(b), application of Kirchhoff's voltage law to the input circuit yields

$$\mathbf{v}_i = \mathbf{v} + \beta_z \mathbf{i}_o$$

and since

$$\mathbf{i}_o = A_g \mathbf{v}$$

the closed-loop gain may be represented as

$$A_{gf} = \mathbf{i}_o / \mathbf{v}_i = A_g / (1 + \beta_z A_g) \tag{10.31}$$

This time the closed-loop input impedance is

$$Z_{if} = \mathbf{v}_i / \mathbf{i}_i = (\mathbf{v} + \beta_z \mathbf{i}_o)/\mathbf{v}/Z_i = (1 + \beta_z A_g) Z_i \tag{10.32}$$

With regard to the output impedance, if Z_β is the input impedance of the feedback network, Kirchhoff's laws applied to the output circuit give

$$\mathbf{v}_o = Z_\beta \mathbf{i}_o + Z_o (\mathbf{i}_o - A_{g0} \mathbf{v})$$

Assuming that the impedance of a source of e.m.f. $\mathscr{E}_S$ connected to the input is negligible compared with Z_i, Kirchhoff's voltage law applied to the input circuit yields

$$\mathscr{E}_S = \mathbf{v} + \beta_z \mathbf{i}_o$$

Hence

$$\mathbf{v}_o = [Z_\beta + (1 + \beta_z A_{g0}) Z_o] \mathbf{i}_o - A_{g0} Z_o \mathscr{E}_S$$

from which the closed-loop output impedance is

$$Z_{of} = Z_\beta + (1 + \beta_z A_{g0}) Z_o$$

Often the first term of this expression is negligible compared with the second in which case

$$Z_{of} = (1 + \beta_z A_{g0}) Z_o \tag{10.33}$$

In shunt-voltage feedback it is convenient to express the gain of the open-loop amplifier in terms of the gain between output signal voltage and input signal current. Writing this gain as A_{z0} off load and A_z on load and representing the fedback signal current i_f as $\beta_g v_o$ because the feedback fraction has conductance dimensions, the situation is as depicted in figure

10.4 Feedback

10.8(c). Application of Kirchhoff's current law at the input gives

$$i_i = i + i_f = i + \beta_g v_o$$

where

$$v_o = A_z i$$

Hence the closed-loop gain can be expressed as

$$A_{zf} = v_o/i_i = A_z/(1 + \beta_g A_z) \tag{10.34}$$

while the closed-loop input impedance is

$$Z_{if} = v_i/i_i = Z_i i/i_i = Z_i/(1 + \beta_g A_z) \tag{10.35}$$

With shunt-inserted feedback it is easier to find the output impedance when the input is connected to a source of infinite internal impedance, that is, to a constant-current source i_S so that $i_i = i_S$. Applying Kirchhoff's voltage law to the output circuit and assuming that the feedback network negligibly loads it

$$v_o = A_{zo} i + Z_o i_o$$

However, Kirchhoff's current law applied at the input gives

$$i_i = i_S = i + \beta_g v_o$$

and so

$$(1 + \beta_g A_{zo})v_o = A_{zo} i_S + Z_o i_o$$

from which the closed-loop output impedance is

$$Z_{of} = Z_o/(1 + \beta_g A_{zo}) \tag{10.36}$$

In the final case of shunt-current feedback it is helpful to adopt the Norton equivalent of the output circuit of the open-loop amplifier as shown in figure 10.8(d). The short-circuit and generally loaded current gains of the open-loop amplifier are written A_{i0} and A_i. It is left as an exercise to show that in terms of the overall notation of figure 10.8(d), the closed-loop current gain is

$$A_{if} = A_i/(1 + \beta_i A_i) \tag{10.37}$$

the closed-loop input impedance is

$$Z_{if} = Z_i/(1 + \beta_i A_i) \tag{10.38}$$

and that fed from a source of infinite internal impedance, the closed-loop output impedance is

$$Z_{of} = (1 + \beta_i A_{i0})Z_o \tag{10.39}$$

neglecting loading of the output by the feedback network.

The analysis of this section culminating in equations (10.28)–(10.39) establishes that the magnitude of the effect of feedback always depends on the appropriate loop gain βA. However, while the nature of the effect of feedback on gain is similar for each circuit arrangement, occurrence of a decreased or increased gain simply depending on whether the feedback is negative or positive, the character of the effect of feedback on the input and output impedance depends on whether the relevant circuit connection is series or parallel as well as on whether the feedback is negative or positive. Thus the effect on the *input* impedance depends on whether the feedback is series or parallel-*inserted* while the effect on the *output* impedance depends on whether the feedback is series or parallel-*derived*. Negative parallel-derived, that is, voltage, feedback always lowers the output impedance while negative series-derived, that is, current, feedback raises it. Negative series-inserted feedback always raises the input impedance while negative shunt-inserted feedback reduces it. In each case, positive feedback has the opposite tendency but important behaviour of a very different nature arises if βA becomes -1. Discussion of this aspect will be pursued in the last two sections of this chapter as already promised. For the moment, the enormous value of negative feedback will be illustrated by considering one fascinating area of application.

10.5 Operational amplifiers

Amplifiers that are well suited to performing mathematical operations on input signals when negative feedback is applied are termed *operational* amplifiers. An ideal operational amplifier would at all frequencies exhibit infinite gain, no electrical noise, zero output impedance and infinite input impedance so that the mathematical operation would be completely determined by the feedback network. Although real operational amplifiers can never provide such properties, a modern operational amplifier formed within a single chip of silicon comes impressively close to meeting the ideal specification. To maintain high gain down to *zero* frequency, the amplifier is directly coupled and to overcome the resulting problem of distinguishing between real and pseudo signals, due to, for example, bias or thermal variations, *differential* circuit techniques are adopted. The differential aspect means that an operational amplifier possesses a *noninverting* and an *inverting* input terminal, the sign of any signal applied between these terminals and the common rail being respectively preserved and reversed at the output. In principle only the difference signal between the noninverting and inverting inputs is amplified; like input signals at the two inputs create cancelling signals at the output. The degree of rejection of so-called *common-mode* signals is indicated by the *common-mode rejection ratio* which is just the ratio of the output signals obtained when a given input signal is applied first to one input only and then to both. Thermal and bias variations tend to create equal signals at the two inputs and therefore negligible output.

10.5 Operational amplifiers

High gain is achieved at low frequencies through incorporation of several differential stages of amplification. However, stability demands that the gain falls off at high frequency (see next section). Typically the gain becomes too small to be useful at frequencies above ~ 100 kHz. The input impedance also deteriorates at high frequencies. In the following analysis of the behaviour of operational amplifiers subjected to various forms of negative feedback, the open-loop amplifier will be assumed effectively ideal for simplicity. Clearly this analysis will only give the performance of corresponding practical circuits up to some critical frequency beyond which their response will be influenced by the significant departure of the operational amplifier from ideal behaviour.

Consider the basic so-called *inverting* and *noninverting* operational-amplifier configurations featuring negative feedback that are depicted in figures 10.9(a) and (b) respectively. Following standard practice the operational amplifier of gain A is denoted by a triangular symbol with the noninverting and inverting inputs marked by positive and negative signs respectively. Applying Kirchhoff's current law to node S of the inverting configuration and assuming that the input impedance between the inverting and noninverting terminals of the operational amplifier is high enough to neglect current through them compared with that through Z_1 and Z_2

$$(v_o + v_o/A)/Z_2 = -(v_o/A + v_i)/Z_1$$

Consequently provided that A is large enough

$$v_o/v_i = -Z_2/Z_1 \tag{10.40}$$

With respect to the noninverting configuration of figure 10.9(b), neglecting the current through the inverting terminal of the operational amplifier again compared with that through Z_1 and Z_2

$$v_o/(Z_1 + Z_2) = (v_i - v_o/A)/Z_1$$

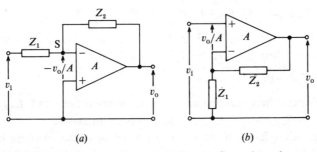

(a)   (b)

10.9 (a) Inverting and (b) noninverting configuration of an operational amplifier.

so that if A is large enough
$$\mathbf{v}_o/\mathbf{v}_i = (Z_1 + Z_2)/Z_1 \tag{10.41}$$

Notice that, while the theory of this section has been developed in terms of small signals $\mathbf{v}_i$ and $\mathbf{v}_o$, the fact that the input and output signals are related through the linear feedback components Z_1 and Z_2 means that the gains given by equations (10.40) and (10.41) apply to any magnitude of signal that does not swing beyond the fixed potentials of the bias supply. Moreover, because the amplification is maintained down to zero frequency and the output can be nulled for zero input, the right-hand sides of equations (10.40) and (10.41) give the ratios between the *total* output and input potential differences.

In terms of the general feedback models of the preceding section, the inverting configuration is an example of shunt-voltage feedback while the noninverting configuration is an example of series-voltage feedback. Applying equation (10.28) to the latter case with $\beta_v A_v \gg 1$ gives equation (10.41) again since $\beta_v = Z_1/(Z_1 + Z_2)$. According to equation (10.34), $\mathbf{v}_o/\mathbf{i}_i = 1/\beta_g$ for shunt-voltage feedback round an open-loop amplifier of high gain. But, for the inverting configuration, $\beta_g = -1/Z_2$ and $\mathbf{i}_i = \mathbf{v}_i/Z_1$ so that equation (10.40) is again obtained.

When Z_1 and Z_2 are resistances R_1 and R_2, both configurations execute a *scaling* operation between their input and output. However, although the output impedances of both configurations are very low, the input impedance of the noninverting configuration is very high while the input impedance of the inverting configuration is virtually Z_1 since the signal $\mathbf{v}_o/A$ at S is negligible compared with $\mathbf{v}_i$. The smallness of $\mathbf{v}_o/A$ causes the point S to be referred to as a *virtual earth*. In view of the virtual earth property, other signal currents can be introduced at the node S without disturbing the signal current through Z_1. In particular the various signal sources in the circuit of figure 10.10(a) act independently and for an effectively ideal operational amplifier

$$V_o/R_2 = -V_{i1}/R_{11} - V_{i2}/R_{21} - V_{i3}/R_{31} - \cdots - \mathbf{V}_{in}/R_{n1} \tag{10.42}$$

That is, the circuit of figure 10.10(a) fulfills the role of a *scaling adder*. In the special case when $R_{11} = R_{21} = R_{31} = \cdots = R_{n1} = R_2$

$$V_o = -(V_{i1} + V_{i2} + V_{i3} + \cdots + V_{in}) \tag{10.43}$$

and the circuit becomes just an *adder*. If, on the other hand, $R_{11} = R_{21} = R_{31} = \cdots = R_{n1} = nR_2$, the circuit performs as an *averager*.

The circuit of figure 10.10(b) is a version of the circuit of figure 10.9(b) in which $Z_1 = \infty$ and $Z_2 = 0$. Therefore, in accordance with equation (10.41), $\mathbf{v}_o = \mathbf{v}_i$ but, for the reasons already given above, the equality extends to

10.5 Operational amplifiers

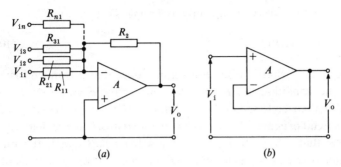

10.10 (a) Scaling adder and (b) voltage follower.

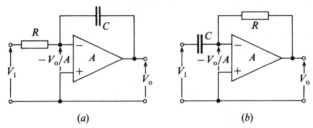

10.11 (a) Operational amplifier integrator and (b) operational amplifier differentiator.

much larger signals and, when properly nulled

$$V_o = V_i$$

Naturally this particular circuit is known as a *voltage follower*. The parallel-derived, series-inserted feedback with unity feedback fraction makes it a very valuable circuit that is much used as a *buffer* in electronics. Its high input impedance means that it does not load a signal source which it copies at its output. Because of the low output impedance, the output can deliver a substantial signal current without being altered.

While the sinusoidal responses of the circuits of figures 10.11(a) and (b) are represented by equation (10.40) with $Z_1 = R$, $Z_2 = 1/j\omega C$ and $Z_1 = 1/j\omega C$, $Z_2 = R$ respectively, their behaviour is better appreciated by returning to fundamentals. If in the circuit of figure 10.11(a) the input current of the operational amplifier is negligible compared with the current through resistance R, then

$$C\, d(V_o + V_o/A)/dt = -(V_o/A + V_i)/R \tag{10.45}$$

Now for a sinusoidal signal of pulsatance ω, the amplitude of dV_o/dt is ω times that of V_o. Therefore, as long as $\omega ARC \gg 1$ for the sinusoidal component of lowest frequency present in V_o, the first term on the right-hand side of equation (10.45) can be neglected compared with the first term on the left-hand side. Consequently assuming $A \gg 1$ as usual

$$V_o = -(1/RC) \int V_i \, dt \tag{10.46}$$

and the circuit behaves as an *integrator*. Notice that it is much easier to satisfy the condition, $\omega ARC \gg 1$, for this active circuit to act as an integrator, than it is to satisfy the condition, $\omega RC \gg 1$, for the corresponding passive circuit of figure 4.11(b) to act as an integrator. One problem with the circuit is that when the input is a low-frequency signal, the output amplitude becomes large and limiting may occur at the potentials of the bias supply. Another problem is drift of the output, in the absence of an input V_i, due to tiny bias currents at the input of the amplifier.

Neglecting the input current of the operational amplifier in the circuit of figure 10.11(b) compared with the current through resistance R

$$(V_o + V_o/A)/R = -C \, d(V_o/A + V_i)/dt \tag{10.47}$$

This time, if $A \gg 1$ and $\omega RC \ll A$ for the sinusoidal component of highest frequency present in V_o, the equation reduces to

$$V_o = -RC \, dV_i/dt \tag{10.48}$$

Clearly the circuit of figure 10.11(b) acts as a *differentiator* when $\omega RC \ll A$, which condition is much more easily satisfied than the condition $\omega RC \ll 1$ for the corresponding passive circuit of figure 4.11(a) to act as a differentiator. Differentiators based on operational amplifiers do not suffer from drift. However, because their gain is high at high frequencies they do tend to exhibit high-frequency instability. A small capacitance in parallel with the feedback resistance R and a small resistance in series with the capacitance C helps to overcome this problem.

In general the circuits of figures 10.11(a) and (b) behave as low and high-pass filters respectively, the responses being given by equation (10.40) with the appropriate values of Z_1 and Z_2 inserted in each case. Forming the impedances Z_1 and Z_2 of the inverting or noninverting configuration from more complicated combinations of reactances and resistances yields more intricate filtering. For example, bridging the capacitor of the integrator with a resistor restricts the gain to some reasonable maximum at low frequencies and inclusion of a resistor in series with the capacitor maintains a finite rather than negligible gain at high frequencies.

10.6 Nyquist's criterion and oscillators

Sinusoidal signals passing through an open-loop amplifier or feedback network generally experience a phase shift as well as a change in amplitude. Thus, as stressed near the beginning of section 10.4 when

10.6 Nyquist's criterion and oscillators

introducing the topic of feedback, the open-loop gain and feedback fraction are frequency-dependent complex quantities in general. Representing complex character by an asterisk, the closed-loop gain is expressible as

$$A_f^* = \frac{A^*}{1 + \beta^* A^*} \qquad (10.49)$$

where β^* is the complex feedback fraction and A^* the complex open-loop gain. In this notation positive feedback corresponds to $|1+\beta^*A^*|<1$ or $|\beta^*A^*|<0$ so that the gain is increased. Especially significant is the fact that the closed-loop gain becomes infinite if $(1+\beta^*A^*)=0$, that is, if the loop gain β^*A^* equals -1. Under this condition a spurious infinitesimal input signal quickly grows to create a substantial periodic output signal. Such electrical behaviour is described very appropriately as *oscillation*. Amplifier circuits that oscillate parasitically through *accidental* positive feedback are said to be *unstable* and, unless their design can be modified to render them stable, are useless for amplification purposes. Circuits which oscillate as a result of suitable positive feedback being applied deliberately round an open-loop amplifier are termed *oscillators*.

The critical condition regarding stability is embodied in *Nyquist's criterion*. This states that a closed-loop amplifier is stable if, in a plot of its loop gain in the complex plane, the locus of the tip of the β^*A^* vector does not enclose the coordinate point $(-1,0)$ as the frequency changes. Conversely, for oscillation to occur, the plot of the loop gain must enclose the point $(-1,0)$ as the frequency changes. The frequency scale is marked along the locus, of course. With respect to the examples of Nyquist plots shown in figure 10.12, locus L passes through the point $P=(-1,0)$ and the corresponding circuit therefore oscillates at the frequency corresponding to this point. Locus M corresponds to a circuit that exhibits real positive feedback at a frequency represented by the point Q but, as the length of OQ

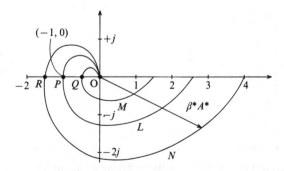

10.12 Nyquist plot of the loop gain β^*A^* in the complex plane.

is less than unity, oscillation cannot occur. The circuit corresponding to locus N exhibits real positive feedback at a frequency corresponding to the point R where $|\beta^* A^*| > 1$ and so oscillation can build up in amplitude in this circuit until some nonlinearity reduces the magnitude of $|\beta^* A^*|$ below unity.

If the condition for oscillation is met at only one frequency then essentially sinusoidal oscillation will occur at that frequency. Clearly, the nearer the magnitude of the loop gain is to unity for small signals that grow in amplitude, the less will be the distortion introduced by the amplitude limiting nonlinearity. A neat example of a sinusoidal oscillator is the *Wien bridge* oscillator shown in figure 10.13. It is convenient to regard the open-loop amplifier of this oscillator as the noninverting combination of the operational amplifier and negative feedback resistors R_1 and R_2, so that the open-loop gain is $(R_1 + R_2)/R_1$. Positive feedback between the open-loop output and noninverting input is provided by the Wien band-pass filter comprising components C and R. The transfer function of this filter has already been derived in section 8.3 and reference to equation (8.22) together with equation (8.23) or figure 8.8 establishes that the modulus of the feedback fraction reaches a maximum of $\frac{1}{3}$ and there is no phase shift when $\omega RC = 1$. Consequently, adjustment of the negative feedback loop R_1, R_2 such that the gain before positive feedback just exceeds three yields a close approximation to sinusoidal oscillation at frequency

$$f = 1/2\pi RC \qquad (10.50)$$

As an alternative to determining the frequency of oscillation through R–C filters as in the Wien bridge oscillator, sinusoidal oscillators can be constructed using series or parallel-resonant L–C filters in conjunction with positive feedback.

In a different class of important circuits, positive feedback is applied over a band of frequencies from zero frequency upwards. When such circuits

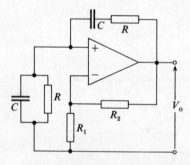

10.13 Circuit diagram of a basic Wien bridge oscillator involving feedback round an operational amplifier.

10.6 Nyquist's criterion and oscillators

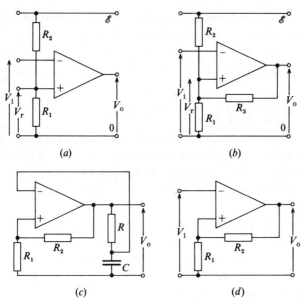

10.14 (a) Comparator, (b) Schmitt trigger, (c) astable multivibrator and (d) bistable multivibrator, each based on an operational amplifier.

oscillate, the waveform is far from sinusoidal, often being approximately rectangular or square. To understand the operation of these *multivibrators* or *relaxation oscillators*, consider first the *comparator* shown in figure 10.14(a) from which they can be derived. In the comparator if the input potential difference V_i is greater than the derived potential difference

$$V_r = [R_1/(R_1 + R_2)]\mathscr{E}$$

by any appreciable amount, the output is driven into negative saturation, that is, $V_o = -V_{sat}$. Complementarily, if V_i is slightly less than V_r, the output is driven into positive saturation, $V_o = V_{sat}$. Note that the saturation potential differences $\pm V_{sat}$ are close to the supply e.m.f.s $\pm \mathscr{E}$ biasing the operational amplifier. In general terms the circuit acts as a threshold detector but if $V_r = 0$ it acts as a zero-crossing detector.

Application of positive feedback to the comparator as shown in figure 10.14(b) creates a regenerative switch with hysteresis called the *Schmitt trigger*. Using Kirchhoff's current law at the noninverting input of this trigger gives

$$\frac{V_r}{R_1} = \frac{V_o - V_r}{R_3} + \frac{\mathscr{E} - V_r}{R_2}$$

so that

$$V_r = \left(\frac{\mathscr{E}}{R_2} + \frac{V_o}{R_3}\right) \bigg/ \left(\frac{1}{R_1} + \frac{1}{R_2} + \frac{1}{R_3}\right) \quad (10.51)$$

248 Signal analysis

Hence the output switches over to the opposite saturation state from positive and negative saturation when respectively

$$V_i > \left(\frac{\mathscr{E}}{R_2} + \frac{V_{sat}}{R_3}\right) \bigg/ \left(\frac{1}{R_1} + \frac{1}{R_2} + \frac{1}{R_3}\right)$$
$$V_i < \left(\frac{\mathscr{E}}{R_2} - \frac{V_{sat}}{R_3}\right) \bigg/ \left(\frac{1}{R_1} + \frac{1}{R_2} + \frac{1}{R_3}\right)$$
(10.52)

It is said that the circuit acts as a *discriminator*. It can, for example, generate a very rectangular wave from a noisy fluctuating input signal so that one use is at the inputs of logic circuitry. Notice that the presence of positive feedback makes the speed of switching independent of the rate of change of the input potential.

Another derivative of the comparator is the *astable* multivibrator shown in figure 10.14(c). Suppose that the capacitor C in this circuit is initially uncharged and, at switch on of the bias supplies to the operational amplifier, the positive feedback drives the output into positive saturation, V_{sat}. The positive feedback holds the output at V_{sat} while capacitor C charges through resistor R until the potential at the inverting input exceeds that, $R_1 V_{sat}/(R_1 + R_2)$, at the noninverting input when the positive feedback causes the output to be driven into negative saturation, $-V_{sat}$. Capacitor C now charges in the opposite sense while the positive feedback holds the output at $-V_{sat}$ until the potential at the inverting input falls below $-R_1 V_{sat}/(R_1 + R_2)$ causing the circuit output to switch back into positive saturation. This cycle of events continually repeats to create an essentially rectangular or square-wave output signal, the frequency of which is controlled by the time constant RC.

Connecting a diode in parallel with the capacitor C converts the astable multivibrator into a *monostable* multivibrator with one stable state. When the output is at that saturation level for which the diode is biased forward, the capacitor cannot charge and the output therefore remains at that saturation level. An input pulse at the noninverting input of such a sign and amplitude as to reverse the input signal to the open-loop operational amplifier, will cause the output to switch over to the opposite saturation level. Now the diode is reverse biased and the capacitor can charge until the feedback makes the output switch back to its original saturation level where it remains in the absence of a further trigger pulse. Evidently, in response to a suitable input triggering pulse, a monostable multivibrator creates a rectangular output pulse of amplitude $2V_{sat}$ and duration related to the time constant RC.

The *bistable* multivibrator or *flip-flop* shown in figure 10.14(d) has two stable states corresponding to the two possible saturated output levels

$\pm V_{sat}$. Normally the circuit exhibits a preference for one of them because of some asymmetry and is held in that state by the positive feedback network R_1, R_2 in the absence of an input triggering signal. Again an input trigger pulse of appropriate sign and amplitude will switch the circuit from one stable state to the other. Two suitable trigger pulses take the output through a complete cycle and the relevance to binary counting is abundantly clear.

10.7 Amplifier instability and Bode diagrams

Turning to the question of accidental positive feedback in amplifiers, it is important to appreciate the difficulties encountered in trying to form stable amplifiers through the incorporation of negative feedback. The network designed to provide negative feedback over the required operational frequency range can itself introduce positive feedback at other frequencies. Such positive feedback can occur because the original design was not sufficiently careful or because of additional phase shifts that could not reasonably be foreseen. Stray capacitance and inductance are especially troublesome with respect to stability at high frequencies where capacitive reactance is small and any mutual coupling induces relatively large e.m.f.s. Another possible origin of positive feedback is the finite internal impedance of the bias supply. How the negative-feedback loop can be designed to prevent its presence giving rise to instability will now be demonstrated in the context of operational amplifiers.

A convenient alternative presentation of the information inherent in a Nyquist plot is one in which the logarithm of the modulus of the loop gain, $\log |\beta^* A^*|$, and the phase shift, ϕ, round the loop are plotted separately as a function of the logarithm of the pulsatance, $\log \omega$. Figure 10.15 presents an example of such a *Bode* plot. Recall that parasitic oscillation takes place if the magnitude of the loop gain is greater than or equal to unity at the frequency at which the phase shift becomes $\pm \pi$. Thus, with reference to the figure, the degree of stability can be expressed as the *gain margin*, $\log |A_m|$, which is a measure of the reduction in loop gain below unity at the frequency at which the phase shift is $\pm \pi$. Alternatively, the degree of stability can be expressed by the *phase margin*, ϕ_m, which is the phase separation from $\pm \pi$ at the frequency at which the magnitude of the loop gain is unity.

While production of a Bode plot from separate knowledge of β^* and A^* can be awkward, linear approximations to $\log |A^*|$ and $\log |\beta^*|$ and to ϕ_A and ϕ_β are easily added to yield linear approximations to $\log |\beta^* A^*|$ and ϕ. For many circuits simplification of the Bode plot is feasible because the dependences of gain and phase on frequency are not independent. For

250 Signal analysis

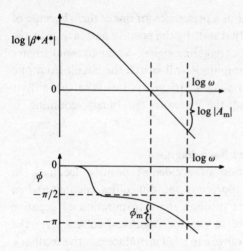

10.15 Illustrative example of a Bode plot.

instance, in the frequency range where a first-order network gives a gain variation of $\pm 6\,\mathrm{dB}$ per octave, the phase shift is $\pm \pi/2$. In such cases the phase information is superfluous because it is implied in the plot of the amplitude behaviour and only the latter is required. In particular, since oscillation is avoided if the phase shift is less than $\pm \pi$ at the frequency at which $|\beta^* A^*|$ falls to unity, it follows that a negative-feedback amplifier involving just first-order networks is stable if the slope of the plot of $\log|\beta^* A^*|$ versus $\log \omega$ is less than $12\,\mathrm{dB}$ per octave as the condition $|\beta^* A^*| = 1$ is approached. Now, if $\log|A^*|$ and $-\log|\beta^*|$ are presented on the same plot, their separation is the required $\log|\beta^* A^*|$ as illustrated in figure 10.16(a) and where they cross corresponds to the crucial condition $\log|\beta^* A^*| = 0$ or $|\beta^* A^*| = 1$. Hence amplifiers involving first-order networks are stable if the closing angle between plots of $\log|A^*|$ and $-\log|\beta^*|$ is less than $12\,\mathrm{dB}$ per octave as crossover is approached. When the negative feedback is large enough to satisfy $|\beta^* A^*| \gg 1$, as is usually the case at least over the intended, operational frequency range, the closed-loop gain is $A_f^* = A^*/(1 + \beta^* A^*) \approx 1/\beta^*$ and $\log|A_f^*| \approx -\log|\beta^*|$. In such cases stability just requires that the closed-loop gain approaches the open-loop gain at less than $12\,\mathrm{dB}$ per octave. Of course, stability is alternatively assured if the closed and open-loop plots intersect at a frequency at which the open-loop gain is less than unity.

As an example of the use of a Bode plot to ensure the stability of a negative-feedback amplifier, consider the operational amplifier differentiator of figure 10.11(b). If the plot of $\log|A^*|$ representing the magnitude of the open-loop gain is as shown in figure 10.16(b), the

10.7 Amplifier instability and Bode diagrams

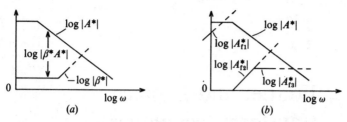

10.16 (a) Bode plots of open-loop gain and feedback fraction yielding Bode plot of loop gain. (b) Bode plots in connection with the stability of an operational amplifier differentiator.

differentiator can be made stable if R and C are chosen so that the plot representing the magnitude of the closed-loop gain is like $\log|A_{f1}^*|$ in the same figure. If R and C were to be chosen such that the closed-loop plot was like $\log|A_{f2}^*|$ approaching crossover with the falling portion of the $\log|A^*|$ plot, the differentiator would be unstable. As suggested in section 10.5, insertion of resistance R_s in series with the capacitance C limits the closed-loop gain at high frequencies that satisfy $\omega \gg 1/R_s C$ and if the closed-loop Bode plot is like that of $\log|A_{f3}^*|$ with an approach to the open-loop plot at 6 dB per octave, the amplifier is likely to be stable. For really reliable stability a design should have a rate of closure in the Bode plot of less than 6 dB per octave and all break points of the linear approximation should be at least a decade in frequency away from the intersection corresponding to $|\beta^* A^*| = 1$. Further addition of suitable capacitance in parallel with resistance R of figure 10.11(b), also as suggested in section 10.5, reduces the closure in the Bode plot to well below 6 dB per octave and thereby assures stability.

11

Fourier and Laplace transform techniques

11.1 Fourier analysis of periodic nonsinusoidal signals

Quite often the signal appearing in an electrical network is not sinusoidal. Such a signal can arise as a result of a nonsinusoidal input signal being applied or through the response of a nonlinear component to a sinusoidal signal. Examples of commonly occurring nonsinusoidal periodic waves in electrical networks include periodic square, ramp and exponentially decaying step functions and half and full-wave rectified sinewaves. Fortunately, all such waves can be expressed as the sum of a constant term and an infinite set of harmonic sinewaves, the sinewave of lowest frequency in the harmonic set being known as the *fundamental* and having the same period as the original nonsinusoidal wave. That is to say, any wave of period T can be represented by the *Fourier* series

$$F(t) = a_0 + \sum_{n=1}^{\infty} a_n \cos n\omega t + \sum_{n=1}^{\infty} b_n \sin n\omega t \tag{11.1}$$

where $\omega = 2\pi/T$, n is any integer and a_n and b_n are constants. Turning to the specific example of a continuous square wave of amplitude a and period T, it can be represented in terms of $\omega = 2\pi/T$ by

$$F(t) = (4a/\pi)(\sin \omega t + \tfrac{1}{3} \sin 3\omega t + \tfrac{1}{5} \sin 5\omega t + \cdots) \tag{11.2}$$

The validity of this particular harmonic representation is demonstrated in figure 11.1 where it is shown that summing the first three terms of expression (11.2) produces a waveform not far removed from square. Addition of successively higher odd harmonics steepens the wings and reduces the amplitude of the fluctuations in between. Notice that periodic waves that are *even* functions of time, that is, symmetric about the time origin, can be represented by a constant and just *cosine* waves, while waves that are *odd* functions of time, that is, reversed in sign about the time origin, can be represented by just *sine* waves. Accordingly, the time origin of the

11.1 Fourier analysis of periodic signals

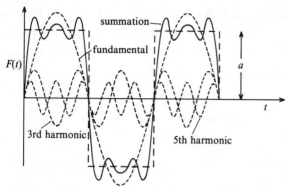

11.1 Addition of the first three terms in the Fourier harmonic series representation of a square wave.

square wave has been taken at one of its leading edges in arriving at the representation given by expression (11.2). A constant or direct component a_0 only appears in a representation of a periodic wave if the positive and negative excursions are unbalanced so that the mean is nonzero.

The usefulness of expressing a nonsinusoidal periodic voltage or current as an equivalent harmonic Fourier series has been alluded to on various occasions in the previous chapters. Once the Fourier spectrum of a signal is known, the overall response of any linear circuit to it can be found readily by superposing the responses to each harmonic component. Even when such a calculation is not followed through, the insight obtained from the Fourier series concept is often helpful.

Expressions for the amplitudes a_n and b_n of the harmonic components of any Fourier series are obtained by multiplying both sides of equation (11.1) by $\cos m\omega t$ or $\sin m\omega t$ and integrating over a complete period $T = 2\pi/\omega$. In the former case

$$\int_0^T F(t) \cos m\omega t \, dt$$

$$= \int_0^T a_0 \cos m\omega t \, dt + \int_0^T \cos m\omega t \sum_{n=1}^{\infty} a_n \cos n\omega t \, dt$$

$$+ \int_0^T \cos m\omega t \sum_{n=1}^{\infty} b_n \sin n\omega t \, dt \qquad (11.3)$$

When $m = 0$, all integrals on the right-hand side except the first are zero and therefore

$$a_0 = (1/T) \int_0^T F(t) \, dt \qquad (11.4)$$

This result just confirms, of course, that the steady component is the time average of the signal. When $m=n$, all integrals on the right-hand side of equation (11.3) are zero except that in which the integrand is $a_n \cos^2 n\omega t$, and since

$$\int_0^T a_n \cos^2 n\omega t \, dt = (a_n/n\omega) \int_0^{2n\pi} \cos^2 \theta \, d\theta$$

$$= (a_n/2n\omega) \int_0^{2n\pi} (1 + \cos 2\theta) \, d\theta$$

$$= a_n \pi/\omega = a_n T/2$$

it follows that for $n=1$ to $n=\infty$

$$a_n = (2/T) \int_0^T F(t) \cos n\omega t \, dt \tag{11.5}$$

Multiplication of equation (11.1) by $\sin m\omega t$ followed by integration over a period T similarly gives

$$\int_0^T F(t) \sin m\omega t \, dt = \int_0^T a_0 \sin m\omega t \, dt$$

$$+ \int_0^T \sin m\omega t \sum_{n=1}^\infty a_n \cos n\omega t \, dt$$

$$+ \int_0^T \sin m\omega t \sum_{n=1}^\infty b_n \sin n\omega t \, dt$$

When $m = n \neq 0$, all integrals on the right-hand side are zero except that in which the integrand is $b_n \sin^2 n\omega t$. Hence for $n=1$ to $n=\infty$

$$b_n = (2/T) \int_0^T F(t) \sin n\omega t \, dt \tag{11.6}$$

Calculation of the Fourier coefficients a_n and b_n from equations (11.4)–(11.6) will now be illustrated by finding their values for a half-wave rectified sinewave of pulsatance ω and amplitude a. It is convenient to make the wave an even function of time by choosing the time origin as shown in figure 11.2(a). Thus the wave is zero during the time interval $T/4$–$3T/4$ while it is $a \cos \omega t$ over the time intervals 0–$T/4$ and $3T/4$–T. From equation (11.4) it follows that

$$a_0 = (1/T) \left(\int_0^{T/4} a \cos \omega t \, dt + \int_{3T/4}^T a \cos \omega t \, dt \right)$$

$$= (1/T) \int_{-T/4}^{+T/4} a \cos \omega t \, dt$$

$$= (a/2\pi) \int_{-\pi/2}^{+\pi/2} \cos \omega t \, d(\omega t)$$

$$= (a/2\pi)[\sin \omega t]_{-\pi/2}^{+\pi/2}$$

$$= a/\pi$$

11.1 Fourier analysis of periodic signals

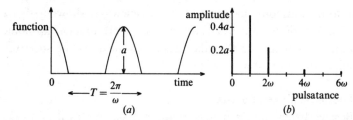

11.2 (a) Half-wave rectified sinewave and (b) its harmonic frequency spectrum.

Also for $n = 1$ to $n = \infty$, from equation (11.5)

$$a_n = (2/T) \int_{-T/4}^{+T/4} a \cos \omega t \cos n\omega t \, dt$$

$$= (a/\pi) \int_{-\pi/2}^{+\pi/2} \cos n\omega t \cos \omega t \, d(\omega t)$$

$$= (a/2\pi) \int_{-\pi/2}^{+\pi/2} [\cos(n+1)\omega t + \cos(n-1)\omega t] \, d(\omega t)$$

$$= \frac{a}{2\pi} \left[\frac{1}{(n+1)} \sin(n+1)\omega t + \frac{1}{(n-1)} \sin(n-1)\omega t \right]_{-\pi/2}^{+\pi/2}$$

When n is odd other than unity, $a_n = 0$, since the sine function of an even number of $\pm \pi/2$ is zero. The result for $n = 1$ is found by letting n tend to unity which shows that the second term in the expression is

$$\lim_{n \to 1} \left[\frac{a}{2\pi} \frac{\sin(n-1)\omega t}{(n-1)} \right]_{-\pi/2}^{+\pi/2} = \left[\frac{a(n-1)\omega t}{2\pi(n-1)} \right]_{-\pi/2}^{+\pi/2} = \frac{a}{2}$$

while the first term is zero. Hence

$$a_1 = a/2$$

When n is even, a_n is finite. In particular

$$a_2 = \frac{a}{2\pi} [\tfrac{1}{3}(-2) + \tfrac{1}{1}(2)] = \frac{2a}{3\pi}$$

and

$$a_4 = \frac{a}{2\pi} [\tfrac{1}{5}(2) + \tfrac{1}{3}(-2)] = \frac{-2a}{15\pi}$$

Since all the b_n coefficients are zero through having chosen the time origin so as to make the function even in time, the half-wave rectified sinewave of figure 11.2(a) is evidently equivalent to

$$F(t) = a/\pi + (a/2)\cos \omega t + (2a/3\pi)\cos 2\omega t$$
$$- (2a/15\pi)\cos 4\omega t + \cdots \quad (11.7)$$

In other words, the half-wave rectified sinewave shown in figure 11.2(a) is equivalent to the harmonic spectrum of pure cosinewaves shown in figure 11.2 (b). Every periodic wave is equivalent to some unique harmonic spectrum and it is suggested that the reader should draw the spectrum corresponding to a square wave as an exercise.

11.2 Fourier analysis of pulses

Besides being subjected to periodic signals, electrical networks often process nonperiodic signals or pulses. For the purpose of Fourier analysis, such pulses can be regarded conveniently as periodic nonsinusoidal signals of infinite period or zero frequency. As the frequency of a periodic nonsinusoidal signal becomes lower and lower, it can be seen with reference to figure 11.2(b) that its harmonic spectral components become bunched closer and closer together. Clearly in the limit of the frequency going to zero, the frequency separation of the harmonic components becomes infinitesimal. Thus it emerges that a pulse is equivalent to a *continuous* frequency spectrum of sinusoidal signals.

Before obtaining an expression for the continuous frequency spectrum corresponding to a nonperiodic signal, it is helpful to reformulate the theory of the previous section regarding the Fourier spectrum of a periodic signal in terms of complex exponential functions. To begin with, the harmonic Fourier series of equation (11.1) is equivalent to

$$F(t) = \sum_{n=-\infty}^{+\infty} c_n \exp jn\omega t \qquad (11.8)$$

where n takes all integral values between $-\infty$ and $+\infty$, $c_0 = a_0$, $c_m = (a_m - jb_m)/2$ and $c_{-m} = (a_m + jb_m)/2$. Multiplying both sides of equation (11.8) by $\exp -jn\omega t$ and integrating over a period of the fundamental, $T = 2\pi/\omega$, that is, from $t = -\pi/\omega = -T/2$ to $t = \pi/\omega = T/2$

$$\int_{-T/2}^{+T/2} F(t)(\exp -jn\omega t) \, dt$$
$$= \int_{-T/2}^{+T/2} \left(\sum_{n=-\infty}^{+\infty} c_n \exp jn\omega t \right)(\exp -jn\omega t) \, dt$$
$$= c_n T$$

Hence the equivalent expression to equations (11.4)–(11.6) giving the amplitudes of the harmonics of a Fourier series is

$$c_n = (1/T) \int_{-T/2}^{+T/2} F(t)(\exp -jn\omega t) \, dt \qquad (11.9)$$

Thinking, as already explained, of a nonperiodic signal as a periodic signal of zero frequency, it is apparent that its Fourier spectrum is given by the limit of equations (11.8) and (11.9) as the fundamental pulsatance ω goes

11.2 Fourier analysis of pulses

to zero, that is, as the period $T = 2\pi/\omega$ goes to infinity. Thus, with the help of a dummy variable t' for clarity, a nonperiodic signal may be represented by

$$F(t) = \lim_{\substack{\omega \to 0 \\ T \to \infty}} \sum_{n=-\infty}^{+\infty} \left[\frac{1}{T} \int_{-T/2}^{+T/2} F(t')(\exp -jn\omega t') \, dt' \right] \exp jn\omega t$$

Now the pulsatances of the nth and $(n+1)$th harmonics are

$$\omega_n = n\omega = n2\pi/T$$
$$\omega_{n+1} = (n+1)\omega = (n+1)2\pi/T$$

and so

$$\omega_{n+1} - \omega_n = \Delta\omega = 2\pi/T = \omega$$

Hence

$$F(t) = \lim_{\substack{\omega \to 0 \\ T \to \infty}} (1/2\pi) \sum_{n=-\infty}^{+\infty} \left[\int_{-T/2}^{+T/2} F(t')(\exp -j\omega_n t') \, dt' \right] (\exp j\omega_n t) \, \Delta\omega$$

$$= (1/2\pi) \int_{-\infty}^{+\infty} \left[\int_{-\infty}^{+\infty} F(t')(\exp -j\omega t') \, dt' \right] (\exp j\omega t) \, d\omega$$

or

$$F(t) = (1/2\pi) \int_{-\infty}^{+\infty} G(\omega)(\exp j\omega t) \, d\omega \qquad (11.10)$$

where, dropping the primes, since they are no longer necessary for clarity,

$$G(\omega) = \int_{-\infty}^{+\infty} F(t)(\exp -j\omega t) \, dt \qquad (11.11)$$

The function $G(\omega)$ giving the amplitude distribution in the equivalent continuous spectrum of the nonperiodic function $F(t)$ is said to be the *Fourier transform* of $F(t)$. In a similar way, $F(t)$ may be described as the *inverse Fourier transform* of $G(\omega)$.

Equation (11.11) will now be used to find the equivalent continuous spectrum of a rectangular pulse of height h and duration τ. In this case, with reference to the depiction of the pulse in figure 11.3(a),

$$G(\omega) = \int_{-\tau/2}^{+\tau/2} h(\exp -j\omega t) \, dt = (-h/j\omega)[\exp -j\omega t]_{-\tau/2}^{+\tau/2}$$

Thus

$$G(\omega) = \frac{h}{j\omega}\left[\exp\left(\frac{j\omega\tau}{2}\right) - \exp\left(-\frac{j\omega\tau}{2}\right)\right] = h\tau\left[\frac{\sin(\omega\tau/2)}{\omega\tau/2}\right] \qquad (11.12)$$

Figure 11.3(b) presents the frequency spectrum of the pulse, that is, $G(\omega)$ as a function of ω, according to equation (11.12). Negative pulsatances are to be regarded as arising from the mathematical processes and do not, of course, have physical significance. Most of the spectral distribution is confined to the central maximum which has a positive upper frequency boundary of $v =$

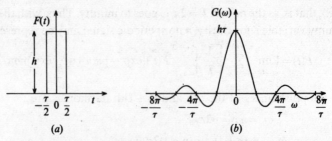

11.3 (a) A rectangular pulse of amplitude h and duration τ and (b) its continuous Fourier spectrum.

$\omega/2\pi = 1/\tau$. Thus an electronic system that passes signals at frequencies up to $1/\tau$ only modestly distorts a rectangular pulse of duration τ between its input and output. To render any distortion of such a pulse trivial, the bandwidth of the electronic system handling it will have to be greater than or equal to $\sim 10/\tau$. It is clear that the narrower a pulse becomes, the greater the bandwidth a system must possess in order to handle it properly. At the opposite extreme of a pulse of extremely long duration, the Fourier spectrum only contains extremely low frequencies. Indeed, in the limit, when the duration becomes infinite, only the zero frequency or direct component exists.

Determination of the response of a four-terminal network to an input pulse by means of the Fourier transform technique generally involves a mathematically difficult inverse Fourier transformation. However, the approach to such analysis will now be set out and the method illustrated by applying it to a trivial case for which the solution is already known from section 4.4. Suppose that an input signal $V_i(t)$ with Fourier transform $G_i(\omega)$ is applied to a four-terminal network for which the transfer function is $\mathcal{T}(\omega)$. The output response to the input spectrum $G_i(\omega)\exp j\omega t$ is $G_o(\omega)\exp j\omega t$ where

$$G_o(\omega) = \mathcal{T}(\omega)G_i(\omega) \tag{11.13}$$

Therefore the time dependence of the output signal is given by

$$V_o(t) = (1/2\pi)\int_{-\infty}^{+\infty} \mathcal{T}(\omega)G_i(\omega)(\exp j\omega t)\,d\omega \tag{11.14}$$

To see this method in action let the pulse of figure 11.3(a) be applied to the basic C–R low-pass filter of figure 4.11(b) with the capacitor initially uncharged. In this case

$$G_i(\omega) = h\tau\left[\frac{\sin(\omega\tau/2)}{\omega\tau/2}\right]$$

11.3 The Laplace transform

and if the time constant RC also happens to equal the pulse duration τ
$$\mathcal{T}(\omega) = (1+j\omega\tau)^{-1}$$
Thus
$$V_o(t) = \frac{h\tau}{2\pi} \int_{-\infty}^{+\infty} \left[\frac{\sin(\omega\tau/2)}{\omega\tau/2}\right]\left(\frac{\exp j\omega t}{1+j\omega\tau}\right) d\omega$$

As anticipated, this integral is difficult to evaluate but it does reduce to the solution expected from chapter 4 of

$$V_o(t) = 0 \qquad \text{when} \quad t < -\tau/2$$

$$V_o(t) = h\left\{1 - \exp\left[-\left(t+\frac{\tau}{2}\right)/\tau\right]\right\} \qquad \text{when} \quad -\tau/2 \leqslant t \leqslant \tau/2$$

$$V_o(t) = h[1 - \exp(-1)]\exp\left[-\left(t-\frac{\tau}{2}\right)/\tau\right] \qquad \text{when} \quad t \geqslant \tau/2$$

Help with difficult inverse Fourier transforms is often available from special tables.

11.3 The Laplace transform

A difficulty arises with the Fourier transform integral of equation (11.11) because it is indefinite for certain important time-dependent functions $F(t)$ such as the unit step function shown in figure 11.4(a). The problem can be overcome in the case of any straight step function by evaluating the Fourier transform of the corresponding exponentially decaying step function and then finding its limit as the rate of decay goes to zero. The exponentially decaying step function corresponding to unit step function is depicted in figure 11.4(b). It is defined by $F(t) = 0$ when $t < 0$ and by $F(t) = \exp-\sigma t$ when $t \geqslant 0$, where σ is real and positive. Thus its Fourier transform is just

$$G(\omega) = \int_0^\infty (\exp-\sigma t)(\exp-j\omega t)\,dt = 1/(\sigma+j\omega) \qquad (11.15)$$

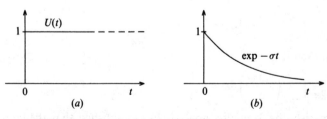

11.4 (a) The unit step function and (b) the exponentially decaying step function which becomes the unit step function in the limit as the positive real decay constant σ goes to zero.

and $F(t)$ is represented by

$$F(t) = \frac{1}{2\pi}\int_{-\infty}^{+\infty} \frac{1}{\sigma+j\omega}(\exp j\omega t)\,d\omega$$

$$= \frac{1}{2\pi}\int_{-\infty}^{+\infty}\left[\left(\frac{\sigma}{\sigma^2+\omega^2}\cos\omega t + \frac{\omega}{\sigma^2+\omega^2}\sin\omega t\right)\right.$$
$$\left. + j\left(\frac{\sigma}{\sigma^2+\omega^2}\sin\omega t - \frac{\omega}{\sigma^2+\omega^2}\cos\omega t\right)\right]d\omega$$

Both imaginary integrands in this expression are odd functions of ω and so the corresponding integrals between $-\infty$ and $+\infty$ are zero. Clearly, unit step function is equivalent to

$$F(t) = \lim_{\sigma\to 0}\frac{1}{2\pi}\int_{-\infty}^{+\infty}\left(\frac{\sigma}{\sigma^2+\omega^2}\cos\omega t + \frac{\omega}{\sigma^2+\omega^2}\sin\omega t\right)d\omega \quad (11.16)$$

and involves just real integrals as would be expected. Regarding the required limiting value of the first of these two integrals, when ω is sufficiently positive or negative to satisfy $\omega^2 \gg \sigma^2$, that is, at all finite ω, the integrand is zero. On the other hand, when ω is small enough to satisfy $\omega^2 \not\gg \sigma^2$, $\cos\omega t = 1$ and putting $\omega = \sigma\tan\theta$ the integral becomes

$$\lim_{\sigma\to 0}\frac{1}{2\pi}\int_{-\pi/2}^{+\pi/2}\frac{\sigma^2\sec^2\theta\,d\theta}{\sigma^2(1+\tan^2\theta)} = \frac{1}{2}$$

In spectral terms the first integral corresponds to a narrow line of infinite amplitude centred on zero frequency and it provides the mean level or direct component of the signal over all time, $-\infty < t < \infty$. The limiting value of the second integral is, from equation (11.16),

$$\frac{1}{2\pi}\int_{-\infty}^{+\infty}\left(\frac{\sin\omega t}{\omega}\right)d\omega$$

which represents the combined effect of a continuous spectrum of sinewaves with the amplitude inversely proportional to frequency. Introducing the variable $\phi = \omega t$, it is seen that when $t > 0$ this second contribution is

$$\frac{1}{2\pi}\int_{-\infty}^{+\infty}\left(\frac{\sin\phi}{\phi}\right)d\phi = \frac{1}{2}$$

but when $t < 0$ it is

$$\frac{1}{2\pi}\int_{+\infty}^{-\infty}\left(\frac{\sin\phi}{\phi}\right)d\phi = -\frac{1}{2}$$

To summarise, the foregoing analysis has established that the complete Fourier spectrum of a unit step function comprises a delta function centred on zero frequency that provides the mean direct level of one-half and a

11.4 Commonly required Laplace transforms

continuous spectrum of sinewaves with amplitude $1/2\pi\omega$ that provides the unit step about the mean.

Notice that effectively the spectrum of the unit step function has been derived by multiplying it by a factor $\exp-\sigma t$ which makes the Fourier integral converge. Fourier transform integrals of many other waveforms can also be rendered convergent by this means although it must be appreciated that the technique is not universally successful. Naturally, the technique fails whenever the function grows too rapidly with time compared with the decay of $\exp-\sigma t$. The technique also fails if the signal concerned extends through all time because $\exp-\sigma t$, where σ is positive, grows without limit as the time t becomes more negative. A corollary of this last point is that the technique is likely to succeed when the signal is switched on at some moment to which the time origin may be ascribed, the inference being that the signal is zero up to time $t=0$.

Introducing the convergence factor $\exp-\sigma t$, where $\sigma > 0$, in the general case of a signal $F(t)$ which is zero before time $t=0$, leads to a modified transform

$$G'(\omega) = \int_0^\infty F(t)[\exp-(\sigma+j\omega)t]\,dt \qquad (11.17)$$

which it is customary to express in terms of the complex variable

$$s = \sigma + j\omega \qquad (11.18)$$

as

$$G(s) = \int_0^\infty F(t)(\exp-st)\,dt \qquad (11.19)$$

This particular transform is known as the *Laplace transform*. Just as the Fourier transform defined by equation (11.11) involves analysing the signal $F(t)$ in terms of an infinite set of imaginary exponential terms $\exp j\omega t$, that is, in terms of infinite sets of sine and cosine waves, so the Laplace transform defined by equations (11.18) and (11.19) corresponds to analysing the signal in terms of an infinite set of functions of the form $\exp st$ where s is complex. These complex functions represent growing and decaying sine and cosine waves and growing and decaying exponentials as well as just sine and cosine waves of constant amplitude. Making the complex variable purely imaginary by putting $\sigma = 0$ in the Laplace transform means that the signal is again being analysed into just sine and cosine waves.

11.4 Commonly required Laplace transforms

Let the unit step function for which the step occurs at time $t=0$ as depicted in figure 11.4(*a*) be denoted by $U(t)$. Representing the operation of

Laplace transformation by $\mathscr{L}$, it immediately follows that

$$\mathscr{L}U(t) = \int_0^\infty 1 \times (\exp -st)\,\mathrm{d}t = [(\exp -st)/-s]_0^\infty = 1/s \quad (11.20)$$

Next consider a unit step function delayed by some time t_d as depicted in figure 11.5(a) and denote it by $U_d(t)$. Thus

$$U_d(t) = 1 \quad \text{for} \quad t - t_d \geq 0 \quad \text{or} \quad t \geq t_d$$
$$U_d(t) = 0 \quad \text{for} \quad t - t_d < 0 \quad \text{or} \quad t < t_d$$

If t' represents the time with respect to a new origin taken at time $t = t_d$ so that $t' = t - t_d$, then

$$U_d(t) = U(t - t_d) = U(t')$$

and of course

$$\mathscr{L}U(t') = \int_0^\infty U(t')(\exp -st')\,\mathrm{d}t' = 1/s = \mathscr{L}U(t)$$

But

$$\mathscr{L}U_d(t) = \int_0^\infty U_d(t)(\exp -st)\,\mathrm{d}t$$

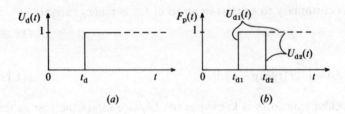

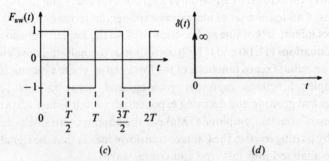

11.5 (a) Unit step function delayed by time t_d with respect to the origin of time, (b) unit pulse thought of as the difference between two unit step functions with differing delays t_{d1} and t_{d2}, (c) unit square wave thought of as a combination of unit step functions with delays that are multiples of half the period and (d) unit impulse function (see text).

11.4 Commonly required Laplace transforms

Hence

$$\mathscr{L}U_d(t) = \int_0^\infty U(t')(\exp-st')(\exp-st_d)\,dt'$$
$$= (\exp-t_d s)\mathscr{L}U(t') = (\exp-t_d s)/s \qquad (11.21)$$

From the point of view of finding its Laplace transform, a unit rectangular pulse may be conveniently regarded as the difference between two unit steps occurring at times t_{d1} and t_{d2} as shown in figure 11.5(b). Thus unit rectangular pulse is

$$F_p(t) = U_{d1}(t) - U_{d2}(t)$$

and its Laplace transform is

$$\mathscr{L}F_p(t) = \mathscr{L}U_{d1}(t) - \mathscr{L}U_{d2}(t)$$

or

$$\mathscr{L}F_p(t) = (1/s)(\exp-t_{d1}s - \exp-t_{d2}s) \qquad (11.22)$$

Treating the unit square-wave signal of figure 11.5(c) similarly, it is representable as

$$F_{sw}(t) = U(t) - 2U(t-T/2) + 2U(t-T) - 2U(t-3T/2) + \cdots$$

where T is the period. Consequently its Laplace transform is

$$\mathscr{L}F_{sw}(t) = (1/s)[1 - 2\exp(-Ts/2) + 2\exp(-Ts) - 2\exp(-3Ts/2) + \cdots]$$
$$= \frac{1}{s}\left[1 - \frac{2\exp(-Ts/2)}{1+\exp(-Ts/2)}\right]$$

or

$$\mathscr{L}F_{sw}(t) = (1/s)\tanh(Ts/4) \qquad (11.23)$$

The unit impulse function

$$\delta(t) = \lim_{\Delta t \to 0} \frac{U(t) - U(t-\Delta t)}{\Delta t} = \frac{d}{dt}U(t) \qquad (11.24)$$

is infinite for an infinitesimal time as indicated in figure 11.5(d) and its integral over all time is unity. Considering $U(t)$ to be

$$\lim_{\alpha \to \infty}[1 - \exp-\alpha t]$$

in the time interval $t \geq 0$, yields

$$\delta(t) = \lim_{\alpha \to \infty}[\alpha \exp-\alpha t]$$

in the same time interval. Hence the Laplace transform of unit impulse is

$$\mathscr{L}\delta(t) = \lim_{\alpha \to \infty}\int_0^\infty \alpha(\exp-\alpha t)(\exp-st)\,dt = \lim_{\alpha \to \infty}[\alpha/(s+\alpha)]$$

or

$$\mathscr{L}\delta(t) = 1 \qquad (11.25)$$

The Laplace transform of a delayed unit impulse is of course

$$\mathscr{L}\,\delta(t-t_\mathrm{d}) = \exp-t_\mathrm{d}s \tag{11.26}$$

For a signal of the exponential form $\exp \alpha t$ from time $t=0$ but zero before, the Laplace transform is

$$\mathscr{L}\exp\alpha t = \int_0^\infty \exp(\alpha-s)t\,\mathrm{d}t$$

Provided α is either negative, or is positive and less than s, the integral is convergent and

$$\mathscr{L}\exp\alpha t = 1/(s-\alpha) \tag{11.27}$$

Considering $\sin \omega t$ to be the imaginary part, $\mathscr{I}\exp\mathrm{j}\omega t$, of $\exp\mathrm{j}\omega t$, it immediately follows that

$$\mathscr{L}\sin\omega t = \mathscr{I}\int_0^\infty \exp(\mathrm{j}\omega-s)t\,\mathrm{d}t = \mathscr{I}(s-\mathrm{j}\omega)^{-1} = \omega/(s^2+\omega^2) \tag{11.28}$$

Similarly

$$\mathscr{L}\cos\omega t = \mathscr{R}\int_0^\infty \exp(\mathrm{j}\omega-s)t\,\mathrm{d}t = \mathscr{R}(s-\mathrm{j}\omega)^{-1} = s/(s^2+\omega^2) \tag{11.29}$$

Other useful transforms are those of differential and integral functions. In the former case

$$\mathscr{L}[(\mathrm{d}/\mathrm{d}t)F(t)] = \int_0^\infty (\exp-st)\,\mathrm{d}F = [(\exp-st)F]_0^\infty + s\int_0^\infty F(\exp-st)\,\mathrm{d}t$$

and assuming that $(\exp-st)F$ is zero when $t=\infty$

$$\mathscr{L}[(\mathrm{d}/\mathrm{d}t)F(t)] = [s\mathscr{L}F(t)] - F(0) \tag{11.30}$$

With regard to the Laplace transform of an integral function

$$\mathscr{L}\left[\int_0^t F(t)\,\mathrm{d}t\right] = \int_0^\infty \left[\int_0^t F(t)\,\mathrm{d}t\right](\exp-st)\,\mathrm{d}t$$

$$= \left[\left\{\int_0^t F(t)\,\mathrm{d}t\right\}\frac{\exp-st}{-s}\right]_0^\infty + (1/s)\int_0^\infty (\exp-st)F(t)\,\mathrm{d}t$$

and assuming that when $t=\infty$

$$\left[\int_0^t F(t)\,\mathrm{d}t\right]\exp-st = 0$$

the required transform is

$$\mathscr{L}\left[\int_0^t F(t)\,\mathrm{d}t\right] = \frac{1}{s}\mathscr{L}F(t) + \frac{1}{s}\left[\int_0^t F(t)\,\mathrm{d}t\right]_{t=0} \tag{11.31}$$

Note that $[\int_0^t F(t)\,dt]_{t=0}$ is to be interpreted as the initial value of the integral quantity.

11.5 Inverse Laplace transforms

One source of inverse Laplace transforms is of course direct comparison with known Laplace transforms. Expressing the operation of inverse Laplace transformation by $\mathscr{L}^{-1}$, particularly useful results are

$$\mathscr{L}^{-1}\frac{1}{s+\alpha} = \exp-\alpha t \tag{11.32}$$

$$\mathscr{L}^{-1}\frac{\omega}{s^2+\omega^2} = \sin \omega t \tag{11.33}$$

$$\mathscr{L}^{-1}\frac{s}{s^2+\omega^2} = \cos \omega t \tag{11.34}$$

$$\mathscr{L}^{-1}\frac{1}{(s+\alpha)^n} = \frac{1}{(n-1)!}\,t^{n-1}\exp-\alpha t \tag{11.35}$$

where the first three represent the inverse forms of equations (11.27), (11.28) and (11.29) obtained in the previous section. The last result follows from

$$\mathscr{L}\frac{1}{(n-1)!}\,t^{n-1}\exp-\alpha t = \frac{1}{(n-1)!}\int_0^\infty t^{n-1}[\exp-(s+\alpha)t]\,dt$$

which becomes on substituting x for $(s+\alpha)t$

$$\frac{1}{(n-1)!}\frac{1}{(s+\alpha)^n}\int_0^\infty x^{n-1}(\exp-x)\,dx = \frac{1}{(s+\alpha)^n}$$

Where it is required to find the inverse Laplace transform of a function $G(s)$ which may be expressed as the ratio of two polynomials, the problem is readily solved if the function can be split into partial fractions. Such splitting is feasible if the polynomial on the numerator is of lower degree than that on the denominator. Thus if

$$G(s) = \frac{a_0 + a_1 s + a_2 s^2 + \cdots + a_{n-1}s^{n-1}}{b_0 + b_1 s + b_2 s^2 + \cdots + s^n}$$

where the quantities a_i and b_i are constant coefficients then

$$G(s) = \frac{a_0 + a_1 s + a_2 s^2 + \cdots + a_{n-1}s^{n-1}}{(s-\alpha_1)(s-\alpha_2)(s-\alpha_3)\cdots(s-\alpha_n)}$$

or

$$G(s) = \frac{A_1}{(s-\alpha_1)} + \frac{A_2}{(s-\alpha_2)} + \cdots + \frac{A_n}{(s-\alpha_n)} \tag{11.36}$$

where the numerators A_i are constants. Once the function $G(s)$ is expressed in the form of equation (11.36), equation (11.32) can be used to find the inverse Laplace transform of each partial fraction. If the denominator of $G(s)$ contains repeated factors, the partial fraction expansion is slightly modified. Should, say, the factor $(s-\alpha_1)$ be repeated r times, then in terms of a set of constants B_i the expansion is expressible as

$$G(s) = \frac{B_1}{(s-\alpha_1)} + \frac{B_2}{(s-\alpha_1)^2} + \cdots + \frac{B_r}{(s-\alpha_1)^r} + \frac{B_{r+1}}{(s-\alpha_2)} + \cdots + \frac{B_{r+n-1}}{(s-\alpha_n)}$$
(11.37)

Restoration to a common denominator and comparison of the coefficients of powers of s in the numerator with those in the original numerator determines the constants B_i in equation (11.37) or A_i in equation (11.36).

To illustrate the partial fraction method, suppose that the inverse Laplace transform of the simple function

$$\frac{3s-1}{s^2+s-6}$$

is needed. Splitting this function into its partial fractions

$$\frac{3s-1}{s^2+s-6} = \frac{A_1}{s+3} + \frac{A_2}{s-2}$$

where

$$(s-2)A_1 + (s+3)A_2 = 3s-1$$

Now comparison of coefficients of s shows that

$$A_1 + A_2 = 3, \quad -2A_1 + 3A_2 = -1$$

or

$$A_1 = 2, \quad A_2 = 1$$

Hence

$$\frac{3s-1}{s^2+s-6} = \frac{2}{s+3} + \frac{1}{s-2}$$

and application of equation (11.32) reveals that

$$\mathscr{L}^{-1}\left(\frac{3s-1}{s^2+s-6}\right) = 2\exp{-3t} + \exp{2t}$$

In a similar way

$$\mathscr{L}^{-1}\left[\frac{3}{(s+2)(s+1)^2}\right] = 3\mathscr{L}^{-1}\left[\frac{1}{s+2} + \frac{1}{(s+1)^2} - \frac{1}{s+1}\right]$$
$$= 3(\exp{-2t} + t\exp{-t} - \exp{-t})$$

from equations (11.32) and (11.35).

In general in network analysis, the parameters α_i appearing in the partial fraction representation of equation (11.36) are either real, as in the examples just considered, or occur as complex conjugate pairs. Consider, for

11.5 Inverse Laplace transforms

example, the Laplace transform

$$G(s) = \frac{1}{s^2 + 2\alpha s + \beta}$$

This function factorises into

$$G(s) = \frac{1}{2j\omega}\left(\frac{1}{s+\alpha-j\omega} - \frac{1}{s+\alpha+j\omega}\right)$$

where $\alpha^2 + \omega^2 = \beta$. Thus, when $\beta > \alpha^2$, ω is real and complex conjugates appear in association with s in the denominators of the partial fractions. The inverse Laplace transform of this last expression is

$$\mathscr{L}^{-1}G(s) = (1/2j\omega)(\exp-\alpha t)(\exp j\omega t - \exp -j\omega t)$$

and so

$$\mathscr{L}^{-1}\left(\frac{1}{s^2+2\alpha s+\beta}\right) = \frac{1}{\omega}(\exp-\alpha t)\sin\omega t \tag{11.38}$$

if $\beta > \alpha^2$ where

$$\omega = +(\beta - \alpha^2)^{\frac{1}{2}} \tag{11.39}$$

An alternative means of finding the constant coefficients A_i in the partial fraction expansion of equation (11.36) emerges upon multiplying through that equation by the factor $(s - \alpha_m)$ to give

$$(s-\alpha_m)G(s) = \frac{s-\alpha_m}{s-\alpha_1}A_1 + \frac{s-\alpha_m}{s-\alpha_2}A_2 + \cdots + \frac{s-\alpha_m}{s-\alpha_m}A_m + \cdots + \frac{s-\alpha_m}{s-\alpha_n}A_n$$

which reveals that

$$A_m = [(s-\alpha_m)G(s)]_{s=\alpha_m} \tag{11.40}$$

Although this relation is not immediately helpful because the factor $(s - \alpha_m)$ is zero when $s = \alpha_m$, it must be appreciated that $G(s)$ here is the ratio, $N(s)/D(s)$, of a numerator to denominator polynomial in s in which

$$D(s) = (s - \alpha_m)Q(s)$$

where $Q(s)$ is another polynomial in s. Thus

$$A_m = [(s-\alpha_m)N(s)/D(s)]_{s=\alpha_m} = [N(s)/Q(s)]_{s=\alpha_m}$$

But

$$[(d/ds)D(s)]_{s=\alpha_m} = [(s-\alpha_m)(d/ds)Q(s)]_{s=\alpha_m} + [Q(s)]_{s=\alpha_m}$$
$$= [Q(s)]_{s=\alpha_m}$$

Hence a useful alternative expression for the partial fraction coefficients is

$$A_m = [N(s)/(d/ds)D(s)]_{s=\alpha_m} \tag{11.41}$$

in terms of which

$$G(s) = \sum_{m=1 \to n} [N(s)/(d/ds)D(s)]_{s=\alpha_m}[1/(s-\alpha_m)] \tag{11.42}$$

and

$$\mathcal{L}^{-1}G(s) = \sum_{m=1\to n} [N(s)(\exp st)/(d/ds)D(s)]_{s=\alpha_m} \quad (11.43)$$

The final result here is referred to as the *Heavyside expansion theorem*. Applying it to the particular example, $G(s) = (3s-1)/(s^2+s-6)$ already considered, $\alpha_1 = -3$ and $\alpha_2 = 2$ so that

$$\mathcal{L}^{-1}G(s) = \sum_{m=1,2} \left[\frac{3s-1}{2s+1} \exp st\right]_{s=\alpha_m} = \left[\frac{-10}{-5} \exp -3t\right] + \left[\frac{5}{5} \exp 2t\right]$$

$$= 2\exp -3t + \exp 2t$$

as before.

The Heavyside theorem also works when complex conjugate pairs of parameters α_i exist. Consider the transform

$$G(s) = \frac{A+jA'}{s+\alpha-j\omega} + \frac{A-jA'}{s+\alpha+j\omega} \quad (11.44)$$

Notice that the two numerators must also be complex conjugates in order for $G(s)$ to be real and its inverse Laplace transform correspondingly a real function of time. Both the real and imaginary coefficients A and A' in the partial fraction expansion are given by the Heavyside expansion theorem. Thus

$$A+jA' = [N(s)/(d/ds)D(s)]_{s=-\alpha+j\omega}$$

and

$$\mathcal{L}^{-1}G(s) = [(A+jA')\exp(-\alpha+j\omega)t]$$
$$+ [(A-jA')\exp(-\alpha-j\omega)t]$$
$$= 2(\exp-\alpha t)(A\cos\omega t - A'\sin\omega t)$$

or

$$\mathcal{L}^{-1}G(s) = 2(A^2+(A')^2)^{\frac{1}{2}}(\exp-\alpha t)[\cos(\omega t + \phi)] \quad (11.45)$$

where $\tan\phi = A'/A$.

When the denominator of $G(s)$ contains repeated factors, the Heavyside expansion theorem must be modified. Consider the transform

$$G(s) = \frac{B_1}{s-\alpha_1} + \frac{B_2}{(s-\alpha_1)^2} + \cdots + \frac{B_r}{(s-\alpha_1)^r}$$

Multiplying through by $(s-\alpha_1)^r$ it will be appreciated that

$$B_r = [(s-\alpha_1)^r G(s)]_{s=\alpha_1}$$

Furthermore

$$B_{r-1} = [(d/ds)\{(s-\alpha_1)^r G(s)\}]_{s=\alpha_1}$$
$$2B_{r-2} = [(d^2/ds^2)\{(s-\alpha_1)^r G(s)\}]_{s=\alpha_1}$$

and in general

$$B_{r-m} = \frac{1}{m!} \left[\frac{d^m}{ds^m} \{(s-\alpha_1)^r G(s)\} \right]_{s=\alpha_1} \tag{11.46}$$

11.6 Network analysis by Laplace transformation

In section 11.3 it was stressed that the Laplace transform of a signal is pertinent to practical situations in which the signal is switched on at some instant. Because of this the technique of Laplace transformation is relevant to deducing the transient responses of networks. In fact, Laplace transformation of an entire equation that has been obtained by invoking one of Kirchhoff's laws converts it from integro-differential form into algebraic form. Consequently the solution of awkward integro-differential equations, something of a stumbling block in the straightforward deduction of transient response, is avoided upon Laplace transformation, just as the solution of such equations is avoided in steady-state alternating current theory through the introduction of the phasor technique. In this section the facility of the Laplace transformation technique will be demonstrated by applying it to find the transient response in a few illustrative cases.

To begin with, consider just a simple series circuit embracing total inductance L and resistance R into which a steady e.m.f. $\mathscr{E}$ is suddenly introduced at time $t=0$. Application of Kirchhoff's voltage law gives

$$L\,dI/dt + RI = \mathscr{E}U(t)$$

for the current I. Taking the Laplace transformation with the help of relations (11.20) and (11.30)

$$L(s\mathscr{L}I - I_{t=0}) + R\mathscr{L}I = \mathscr{E}/s \tag{11.47}$$

Hence, assuming that the current is zero up to time $t=0$

$$\mathscr{L}I = \frac{\mathscr{E}}{s(R+Ls)} = \frac{\mathscr{E}}{R}\left[\frac{1}{s} - \frac{1}{s+R/L}\right]$$

and making the inverse Laplace transformation with the aid of relations (11.20) and (11.32)

$$I = \frac{\mathscr{E}}{R}\left\{1 - \exp\left[-\frac{t}{(L/R)}\right]\right\} \tag{11.48}$$

when $t \geqslant 0$, in accordance with equations (4.19) and (4.21) of section 4.3.

Next consider the situation when a switch is closed at time $t=0$ to connect a steady e.m.f. $\mathscr{E}$ to a series circuit comprising just capacitance C and resistance R. In these circumstances Kirchhoff's voltage law gives

$$(1/C)\int I\,dt + RI = \mathscr{E}U(t)$$

and taking the Laplace transform with the aid of relations (11.20) and (11.31), the corresponding equation

$$(1/sC)\left[\mathscr{L}I+\left(\int I\,dt\right)_{t=0}\right]+R\mathscr{L}I=\mathscr{E}/s \tag{11.49}$$

in $\mathscr{L}I$ is obtained. Therefore if there is no initial capacitive charging

$$\mathscr{L}I=\frac{\mathscr{E}}{R}\left(s+\frac{1}{RC}\right)^{-1}$$

and making the inverse transformation through relation (11.32)

$$I=\frac{\mathscr{E}}{R}\exp\left(-\frac{t}{RC}\right) \tag{11.50}$$

when $t\geqslant 0$, again in agreement with the theoretical result of section 4.3.

The two cases analysed so far have been chosen so as to clearly reveal, in a very simple context, the method of determining transient response through Laplace transformation. While little benefit is gained from the transformation in these trivial cases, great benefit accrues from transformation in more difficult cases where there is a more complicated network or input stimulus. Notice, too, that the Laplace transform of a relevant Kirchhoff equation can be written down immediately when there is no energy stored in the circuit initially. In this respect equations (11.47) and (11.49) reveal that inductance L and capacitance C respectively act like reactances of sL and $1/sC$ with respect to the transformed current $\mathscr{L}(I)$ compared with reactances of ωL and $1/\omega C$ with respect to the actual current I.

Turning to the circuit shown in figure 11.6(a), the Laplace transformation of Kirchhoff's voltage law in the two meshes gives

$$\left(R_1+\frac{1}{sC_1}\right)\mathscr{L}I_1-\frac{1}{sC_1}\mathscr{L}I_2=\frac{1}{s}$$

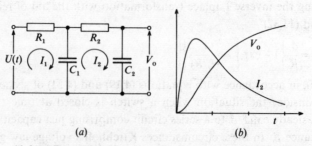

11.6 (a) Two-mesh circuit analysed with the aid of Laplace transformation in the text and (b) the solutions for I_2 and V_0 as a function of time for certain network parameters.

11.6 Network analysis by Laplace transformation

$$-\frac{1}{sC_1}\mathcal{L}I_1 + \left(R_2 + \frac{1}{sC_1} + \frac{1}{sC_2}\right)\mathcal{L}I_2 = 0$$

assuming that the capacitors are initially uncharged. Thus, writing C for the capacitance of C_1 in parallel with C_2 and eliminating $\mathcal{L}I_1$ between the two equations

$$1/s^2 C_1 = [(R_1 + 1/sC_1)(R_2 + 1/sC) - 1/s^2 C_1^2]\mathcal{L}I_2$$

or

$$\mathcal{L}I_2 = [(1/C)(R_1 C_1 s + 1)(R_2 C s + 1) - 1/C_1]^{-1}$$

$$= \{R_1 C_1 R_2 s^2 + [R_2 + (C_1/C)R_1]s + 1/C_2\}^{-1}$$

$$= \frac{C_2}{\tau_1 \tau_2}\left[s^2 + \frac{\tau_1 + \tau_2 + \tau_{12}}{\tau_1 \tau_2} s + \frac{1}{\tau_1 \tau_2}\right]^{-1}$$

where

$$R_1 C_1 = \tau_1, \quad R_2 C_2 = \tau_2 \quad \text{and} \quad R_1 C_2 = \tau_{12}$$

In terms of partial fractions it is convenient to express the Laplace transform of I_2 as

$$\mathcal{L}I_2 = \frac{C_2}{\tau_1 \tau_2 (s + \alpha_1)(s + \alpha_2)} = \frac{C_2}{(\alpha_2 - \alpha_1)\tau_1 \tau_2}\left(\frac{1}{s + \alpha_1} - \frac{1}{s + \alpha_2}\right)$$

so that taking the inverse Laplace transformation

$$I_2 = \frac{C_2}{(\alpha_2 - \alpha_1)\tau_1 \tau_2}(\exp-\alpha_1 t - \exp-\alpha_2 t) \tag{11.51}$$

where

$$\alpha_1, \alpha_2 = \frac{\tau_1 + \tau_2 + \tau_{12}}{2\tau_1 \tau_2} \mp \left[\left(\frac{\tau_1 + \tau_2 + \tau_{12}}{2\tau_1 \tau_2}\right)^2 - \frac{1}{\tau_1 \tau_2}\right]^{\frac{1}{2}} \tag{11.52}$$

Note that α_1 and α_2 are both positive and real since

$$\left[\frac{\tau_1 + \tau_2 + \tau_{12}}{2\tau_1 \tau_2}\right]^2 > \left[\frac{\tau_1 + \tau_2 + \tau_{12}}{2\tau_1 \tau_2}\right]^2 - \frac{1}{\tau_1 \tau_2}$$

and

$$\left[\frac{\tau_1 + \tau_2 + \tau_{12}}{2\tau_1 \tau_2}\right]^2 > \frac{1}{\tau_1 \tau_2}$$

The latter result follows because

$$(\tau_1 + \tau_2 + \tau_{12})^2 - 4\tau_1 \tau_2 = (\tau_1 - \tau_2)^2 + 2\tau_{12}(\tau_1 + \tau_2) + \tau_{12}^2 > 0$$

From equation (11.51) the time dependence of the output voltage is given by

$$V_o = \frac{1}{(\alpha_2 - \alpha_1)\tau_1 \tau_2}\left(\frac{\exp-\alpha_2 t}{\alpha_2} - \frac{\exp-\alpha_1 t}{\alpha_1}\right) + \text{constant}$$

But $V_o = 0$ when $t = 0$ assuming capacitor C_2 is uncharged initially. Hence

$$\text{constant} = 1/\alpha_1 \alpha_2 \tau_1 \tau_2$$

and

$$V_o = \frac{1}{\alpha_1\alpha_2\tau_1\tau_2}\left(1 + \frac{\alpha_1\exp-\alpha_2 t - \alpha_2\exp-\alpha_1 t}{\alpha_2-\alpha_1}\right) \quad (11.53)$$

Typical forms of response represented by equations (11.51) and (11.53) are shown in figure 11.6(b).

The Laplacian derivation of the response of a series resonant circuit comprising resistance R, inductance L and capacitance C, to an e.m.f. $\mathscr{E}$ suddenly applied at time $t=0$, is worthy of comparison with the direct derivation of the same response carried out in section 4.5 through the solution of appropriate differential equations. Kirchhoff's voltage law for the circuit means that

$$(R+sL+1/sC)\mathscr{L}I = \mathscr{L}\mathscr{E}U(t) = \mathscr{E}/s$$

assuming that there is no initial stored energy, that is, that there is no current in the circuit or charge stored on the capacitor at time $t=0$. Hence

$$\mathscr{L}I = (\mathscr{E}/L)(s^2+2\alpha s+\omega_r^2)^{-1} \quad (11.54)$$

where

$$\alpha = R/2L, \quad \omega_r^2 = 1/LC \quad (11.55)$$

and it should be recognised that ω_r represents the natural resonant pulsatance.

When $\omega_r^2 > \alpha^2$, that is, when $R^2/4L^2 < 1/LC$

$$\mathscr{L}I = \frac{\mathscr{E}}{2j\omega_0 L}\left(\frac{1}{s+\alpha-j\omega_0} - \frac{1}{s+\alpha+j\omega_0}\right)$$

where $\omega_0 = (\omega_r^2-\alpha^2)^{\frac{1}{2}}$ is real. Consequently, in accordance with equation (11.38),

$$I = (\mathscr{E}/\omega_0 L)\exp-\alpha t \sin\omega_0 t \quad (11.56)$$

which agrees with the expression for current obtained from the earlier equation (4.47). Equation (11.56) clearly represents damped oscillations at pulsatance $\omega_0 \neq \omega_r$. However, if the Q of the circuit, $(L/C)^{\frac{1}{2}}/R$, is large compared with $\frac{1}{2}$, that is, if $R^2/4L^2 \ll 1/LC$, then ω_0 is close to ω_r. In the limit when $R=0$, α is zero and undamped continuous oscillation takes place at the resonant pulsatance ω_r. While the potential difference across the resistance R is just $V_R = RI$, that across the capacitance C is $V_C = (\int I \, dt)/C$. It is left as an exercise for the reader to show that

$$V_C = \mathscr{E}\{1-[(\omega_r/\omega_0)\exp-\alpha t \sin(\omega_0 t+\phi)]\} \quad (11.57)$$

where $\tan\phi = \omega_0/\alpha$, again in agreement with equation (4.47) obtained before. Evidently, as pointed out in section 4.5, V_C approaches $\mathscr{E}$ via damped oscillations which are often referred to as ringing.

11.6 Network analysis by Laplace transformation

When $\omega_r^2 = \alpha^2$, that is, when $R^2/4L^2 = 1/LC$
$$\mathscr{L}I = (\mathscr{E}/L)(s+\alpha)^{-2}$$
so that from equation (11.35)
$$I = (\mathscr{E}t/L)\exp-\alpha t \qquad (11.58)$$
$$V_R = (R\mathscr{E}t/L)\exp-\alpha t \qquad (11.59)$$
and
$$V_C = \frac{1}{C}\int I\,dt = \frac{\mathscr{E}}{CL}\left[\left(-\frac{t}{\alpha}\exp-\alpha t\right)+\frac{1}{\alpha}\int(\exp-\alpha t)\,dt\right]$$
$$= \frac{\mathscr{E}}{CL}\left[\text{constant}-\left(\frac{1}{\alpha^2}+\frac{t}{\alpha}\right)\exp-\alpha t\right]$$

Assuming again that $V_C=0$ when $t=0$, the constant is $1/\alpha^2 = 1/\omega_r^2 = LC$. Consequently
$$V_C = \mathscr{E}[1-(1+\alpha t)\exp-\alpha t] \qquad (11.60)$$
in accordance with equation (4.49) obtained before. Oscillation is clearly just prevented and there is said to be critical damping (refer back to figure 4.15(b)).

When $\omega_r^2 < \alpha^2$ the quantity ω_0 becomes imaginary. Putting $\omega_0 = j\beta$
$$\mathscr{L}I = \frac{\mathscr{E}}{2\beta L}\left[\frac{1}{s+\alpha-\beta}-\frac{1}{s+\alpha+\beta}\right]$$
and making the inverse transformation via equation (11.32)
$$I = (\mathscr{E}/2\beta L)[\exp(-\alpha+\beta)t - \exp(-\alpha-\beta)t]$$
or
$$I = (\mathscr{E}/\beta L)\exp-\alpha t\,\sinh\beta t \qquad (11.61)$$

Again there is an absence of oscillation and the circuit in this condition is said to be overdamped (refer back to figure 4.15(b) again).

The final circuit example that will be analysed in this section is that of an e.m.f. $\mathscr{E}_0 \sin\omega t$ being suddenly applied to a series resonant circuit at time $t=0$. With the usual notation, and taking both the current in the circuit and the charge associated with the capacitance to be initially zero as before, Kirchhoff's voltage law gives
$$(R+sL+1/sC)\mathscr{L}I = \mathscr{L}\mathscr{E}_0 \sin\omega t$$

Thus, making use of the Laplace transform of $\sin\omega t$ given in equation (11.28), the Laplace transform of the current is
$$\mathscr{L}I = \frac{\omega\mathscr{E}_0 s}{L(s^2+2\alpha s+\omega_r^2)(s^2+\omega^2)} \qquad (11.62)$$

where again $\alpha = R/2L$ and $\omega_r^2 = 1/LC$. Expanding into partial fractions

$$\frac{\omega s}{(s^2+2\alpha s+\omega_r^2)(s^2+\omega^2)} = \frac{A_1 s + A_2}{s^2+2\alpha s+\omega_r^2} + \frac{A_3 s + A_4}{s^2+\omega^2}$$

and comparison of coefficients in the numerators establishes that

$$0 = A_1 + A_3$$
$$0 = A_2 + 2\alpha A_3 + A_4$$
$$\omega = \omega^2 A_1 + \omega_r^2 A_3 + 2\alpha A_4$$
$$0 = \omega^2 A_2 + \omega_r^2 A_4$$

or, following suitable algebraic manipulation, that

$$A_1 = -\frac{(\omega_r^2 - \omega^2)\omega}{(\omega_r^2 - \omega^2)^2 + 4\alpha^2 \omega^2}$$

$$A_2 = \left(\frac{2\alpha \omega_r^2}{\omega_r^2 - \omega^2}\right) A_1$$

$$A_3 = -A_1$$

$$A_4 = -\left(\frac{2\alpha \omega^2}{\omega_r^2 - \omega^2}\right) A_1$$

Consequently

$$\mathscr{L}I = \frac{A_1 \mathscr{E}_0}{L}\left[\frac{s+2\alpha\omega_r^2/(\omega_r^2-\omega^2)}{s^2+2\alpha s+\omega_r^2} - \frac{s+2\alpha\omega^2/(\omega_r^2-\omega^2)}{s^2+\omega^2}\right]$$

and taking the inverse transform with the help of equations (11.33), (11.34) and (11.38)

$$I = \frac{A_1 \mathscr{E}_0}{L}\left\{\left[\left(\frac{2\alpha\omega_r^2}{\omega_r^2-\omega^2}\right)\frac{1}{\omega_0}\exp-\alpha t \sin\omega_0 t\right]\right.$$
$$-\left[\cos\omega t\right] - \left[\frac{2\alpha\omega}{\omega_r^2-\omega^2}\sin\omega t\right]$$
$$\left.+\left[\mathscr{L}^{-1}\frac{1}{2j\omega_0}\left(\frac{\alpha+j\omega_0}{s+\alpha+j\omega_0} - \frac{\alpha-j\omega_0}{s+\alpha-j\omega_0}\right)\right]\right\}$$

where $\omega_0^2 = \omega_r^2 - \alpha^2$ as in the series, resonant, step response. Evaluating the remaining inverse transform

$$I = \frac{A_1 \mathscr{E}_0}{L}\left\{\left[\exp-\alpha t\right]\left[\frac{2\alpha\omega_r^2}{(\omega_r^2-\omega^2)\omega_0}\sin\omega_0 t\right.\right.$$
$$\left.+\frac{\alpha+j\omega_0}{2j\omega_0}\exp-j\omega_0 t - \frac{\alpha-j\omega_0}{2j\omega_0}\exp j\omega_0 t\right]$$
$$\left.-\left[\cos\omega t\right]-\left[\frac{2\alpha\omega}{\omega_r^2-\omega^2}\sin\omega t\right]\right\}$$

$$= \frac{A_1 \mathscr{E}_0}{L} \left\{ \left[\exp -\alpha t \right] \left[\frac{2\alpha \omega_r^2}{(\omega_r^2 - \omega^2)\omega_0} \sin \omega_0 t - \frac{\alpha}{\omega_0} \sin \omega_0 t + \cos \omega_0 t \right] \right.$$
$$\left. - \left[\cos \omega t \right] - \left[\frac{2\alpha \omega}{\omega_r^2 - \omega^2} \sin \omega t \right] \right\}$$

$$= \frac{A_1 \mathscr{E}_0}{L} \left\{ \left[\exp -\alpha t \right] \left[\frac{\alpha(\omega_r^2 + \omega^2)}{(\omega_r^2 - \omega^2)\omega_0} \sin \omega_0 t + \cos \omega_0 t \right] \right.$$
$$\left. - \left[\frac{2\alpha \omega}{\omega_r^2 - \omega^2} \sin \omega t + \cos \omega t \right] \right\}$$

$$= \frac{A_1 \mathscr{E}_0}{L} \left\{ \left[\frac{\alpha^2(\omega_r^2 + \omega^2)^2}{(\omega_r^2 - \omega^2)^2 \omega_0^2} + 1 \right]^{\frac{1}{2}} \exp -\alpha t \sin (\omega_0 t + \theta) \right.$$
$$\left. - \left[\frac{4\alpha^2 \omega^2}{(\omega_r^2 - \omega^2)^2} + 1 \right]^{\frac{1}{2}} \sin (\omega t + \phi) \right\}$$

where

$$\tan \theta = \frac{(\omega_r^2 - \omega^2)\omega_0}{(\omega_r^2 + \omega^2)\alpha}; \quad \tan \phi = \frac{\omega_r^2 - \omega^2}{2\alpha \omega} = Q\left(\frac{\omega_r}{\omega} - \frac{\omega}{\omega_r} \right)$$

Finally, substituting for A_1 leads to

$$I = \frac{\mathscr{E}_0}{L} \left\{ \frac{\omega}{[(\omega_r^2 - \omega^2)^2 + 4\alpha^2 \omega^2]^{\frac{1}{2}}} \sin (\omega t + \phi) \right.$$
$$\left. - \frac{\omega \omega_r}{\omega_0 [(\omega_r^2 - \omega^2)^2 + 4\alpha^2 \omega^2]^{\frac{1}{2}}} \exp -\alpha t \sin (\omega_0 t + \theta) \right\}$$

or

$$I = \frac{\mathscr{E}_0}{[R^2 + (\omega L - 1/\omega C)^2]^{\frac{1}{2}}} \left\{ \sin (\omega t + \phi) - \frac{\omega_r}{\omega_0} \exp -\alpha t \sin (\omega_0 t + \theta) \right\}$$

(11.63)

The second term here gives the decaying transient which is oscillatory if ω_0 is real, that is, if $R < 2(L/C)^{\frac{1}{2}}$, while the first term represents the steady-state response deduced way back in section 5.6.

11.7 Pole-zero plots in the complex s-plane

With regard to a Laplace transform function of the form

$$G(s) = \frac{A(s - z_1)(s - z_2) \cdots}{(s - p_1)(s - p_2) \cdots}$$

there are values of the complex frequency $s = \sigma + j\omega t = p_1, p_2, \ldots$, which make it infinite and other values $s = z_1, z_2, \ldots$, which make it zero. Such values are respectively known as the *poles* and *zeros* of the function and they obviously determine its essential form. In network analysis, because $G(s)$ relates to a signal that is a real function of time, the poles and zeros are

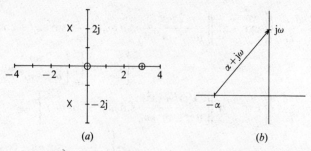

11.7 (a) The s-plane diagram of the function $s(s-3)/(s^2+2s+5)$ and (b) visualisation of the Fourier spectrum of an exponentially decaying unit step function through the behaviour of the vector between the pole $-\alpha$ and the point $j\omega$ in the s-plane.

either real or they occur as complex conjugate pairs. Their values may be depicted on an Argand diagram and it is normal practice to denote zeros by drawing circles and poles by marking crosses at relevant points. Such diagrams are graphically referred to as *s-plane* diagrams. Figure 11.7(a) shows the s-plane diagram of the function $s(s-3)/(s^2+2s+5)$, by way of an example.

To appreciate how an s-plane diagram can reveal the Fourier spectrum of a signal, first consider for simplicity the Laplace transform $G(s) = (s+\alpha)^{-1}$ corresponding to an exponentially decaying signal. The Fourier spectrum $G(\omega)$ follows from replacing s by $j\omega$ so that for the particular signal under consideration $G(\omega) = (\alpha + j\omega)^{-1}$. Now, as depicted in figure 11.7(b), $(\alpha + j\omega)$ is represented in the s-plane diagram by the vector which lies between the pole $s = -\alpha$ on the real axis and the point $j\omega$ on the imaginary axis. The length of this Argand vector representation of $(\alpha + j\omega)$ gives the magnitude of the denominator of $G(\omega)$ while its orientation gives the corresponding phase. Of course, the magnitude of $G(\omega)$ is just the reciprocal of the magnitude of $(\alpha + j\omega)$ and the phase of $G(\omega)$ is just that of $(\alpha + j\omega)$ but with opposite sign. Thus by considering how the vector representation of $(\alpha + j\omega)$ changes in the s-plane diagram as ω varies, the frequency dependences of both the magnitude and phase of $G(\omega)$ can be visualised. For the case under consideration, as ω goes from zero to infinity, the frequency spectrum features a fall in amplitude from $1/\alpha$ to zero and a phase shift that changes from zero to $-90°$. When there are several poles and zeros, the strength of the spectrum at a particular pulsatance ω is obtained by taking the product of the lengths of the various vectors drawn from the zeros to the point $j\omega$ on the imaginary axis and dividing this by the product of the lengths of the several vectors drawn from the poles to the same point on the imaginary axis. Similarly the overall phase is the sum of the phases of the

11.7 Pole-zero plots in the complex s-plane

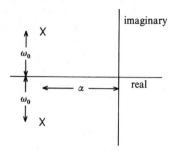

11.8 Plot in the s-plane of the poles of the Laplace transform of the current in a series resonant circuit subjected to a step e.m.f.

vectors drawn from the zeros minus the sum of the phases of the vectors drawn from the poles. Clearly a zero near the imaginary axis gives a minimum in the frequency spectrum at a pulsatance equal to the imaginary part of the zero point, while a pole close to the imaginary axis causes a similar maximum in the spectrum. Evidently the locations of the poles and zeros in the s-plane determine which frequency ranges are most significant in the spectrum of the signal. Notice that, since functions do not grow without limit in the real physical world, poles are restricted to the left-hand half of the s-plane in network analysis.

The poles, $s = -\alpha \pm j\omega_0$, of the Laplace transform of the current in a series resonant circuit subjected to a step e.m.f., which were deduced in the previous section, exhibit interesting behaviour. With reference to figure 11.8, their distance along the negative real axis of an s-plane plot gives the degree of damping while their separation from the real axis in the imaginary direction constitutes the ringing pulsatance. With the same notation as earlier, since $\omega_0^2 = (\omega_r^2 - \alpha^2)$, where $\omega_r^2 = 1/LC$ and $\alpha = R/2L$, as α becomes larger corresponding to more damping, ω_0 gets smaller and the poles converge on the real axis until when $\alpha^2 = \omega_r^2$, $\omega_0 = 0$ which is the critically damped condition and the poles coincide on the real axis. As α increases even further, the poles split again but now separate along the real axis which corresponds to the overdamped situation. With no damping, the poles simply lie on the imaginary axis.

Finally, note that arranging the poles and zeros to be coincident makes the Laplace transform independent of s so that the network behaves as an attenuator.

12

Filter synthesis

12.1 Introduction

An ideal filter would perfectly transmit signals at all desired frequencies and completely reject them at all other frequencies. In the particular case of an ideal low-pass filter, for example, the modulus of the transfer function, $|\mathcal{T}|$, would behave as shown in figure 12.1(a). Up to a certain critical pulsatance ω_c, $|\mathcal{T}|$ would be unity but above this pulsatance, $|\mathcal{T}|$ would be zero. Any practical filter can only approximate to such an ideal, of course.

In section 8.2 it was pointed out how $|\mathcal{T}|^2$ for a simple single-section L–R or C–R filter comprising just one reactive component only reaches a maximum rate of fall-off outside the pass band of 20 dB per decade of frequency compared with an infinite rate of fall-off for an ideal filter. Remember that the significance of $|\mathcal{T}|^2$ is that it indicates the power in the load for a fixed amplitude of input signal. Increasing the number of reactive components in the filter stage to two, as in the simple low-pass L–C filter of figure 8.7(a), causes $|\mathcal{T}|^2$ to reach a maximum rate of fall-off outside the pass band of 40 dB per decade of frequency. With n reactive components in the filter stage, the maximum rate of fall-off of $|\mathcal{T}|^2$ outside the pass band becomes $20n$ dB per decade of frequency and the filter is accordingly said to be of *nth order*.

Further improvement in the sharpness of the cut-off response of practical filters can be achieved by cascading sections and the design of filters comprising multiple identical sections was considered at some length in chapter 9. In particular, simplification of the design of such ladder filters through the technique of loading the last section with the characteristic impedance was discussed. As pointed out before, this procedure ensures that every section is so loaded. Consequently each identical section

12.2 Butterworth filters

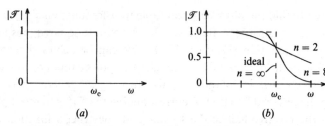

12.1 (a) Amplitude response of an ideal low-pass filter and (b) comparison of the amplitude responses of differing orders of Butterworth filter with the ideal.

responds in the same way and, if $\mathcal{T}$ is the transfer function of one section, the transfer function of the complete ladder of m sections is $\mathcal{T}^m$.

Especially note that the forms of frequency dependence of the transfer functions of filters considered in previous sections were simply accepted for what they were. In a radically different approach to filter design, a filter is *synthesised* so as to provide some preconceived functional form of frequency response that exhibits certain desirable features.

12.2 Butterworth filters

One way of describing the ideal low-pass response depicted in figure 12.1(a) is through the relation

$$|\mathcal{T}| = [1 + (\omega/\omega_c)^\infty]^{-1} \tag{12.1}$$

This suggested to *Butterworth* that

$$|\mathcal{T}|^2 = [1 + (\omega/\omega_c)^{2n}]^{-1} \tag{12.2}$$

where n is a large integer, ought to constitute a good response to synthesise from the point of view of creating high-performance low-pass filters. Figure 12.1(b) displays the frequency response of $|\mathcal{T}|$ represented by equation (12.2). A very valuable feature of this Butterworth response is its maximal initial flatness. Notice that when $\omega \gg \omega_c$, $|\mathcal{T}|^2$ falls off as $(1/\omega)^{2n}$, that is, at a rate of $20n$ dB per decade of frequency. Consequently the integer n is just the order of filter needed to synthesise the response of equation (12.2).

In order to gain an understanding of the synthesis procedure, consider first the elementary problem of designing a first-order Butterworth filter for which, from equation (12.2)

$$|\mathcal{T}|^2 = [1 + (\omega/\omega_c)^2]^{-1} \tag{12.3}$$

or in terms of the parameter $s = j\omega$

$$|\mathcal{T}|^2 = [1 - (s/\omega_c)^2]^{-1} \tag{12.4}$$

280   Filter synthesis

It is required to find the physically realisable transfer function $\mathcal{T}(s)$ which will yield the first-order, Butterworth, amplitude-squared response represented by equation (12.3) or (12.4). This can be achieved through consideration of the poles of $|\mathcal{T}|^2$ although these poles cannot of course be reached through variation of the real pulsatance ω. The general procedure for finding the physical function $\mathcal{T}$ corresponding to $|\mathcal{T}|^2$ is to reject poles of $|\mathcal{T}|^2$ in the positive half of the s-plane and construct a function that possesses just those poles of $|\mathcal{T}|^2$ that lie in the negative half of the s-plane. In the present case the poles of $|\mathcal{T}|^2$ are from equation (12.4) simply

$$s = \pm \omega_c \tag{12.5}$$

and rejecting the pole $s = +\omega_c$, the procedure for finding $\mathcal{T}(s)$ yields

$$\mathcal{T}(s) = [1 + (s/\omega_c)]^{-1} \tag{12.6}$$

This being the physically realisable transfer function, $s = j\omega$ in it is restricted to being imaginary and, checking back, when expression (12.6) is indeed the transfer function

$$|\mathcal{T}(s)|^2 = |[1+(s/\omega_c)]^{-1}|^2 = |[1+j(\omega/\omega_c)]^{-1}|^2 = [1+(\omega/\omega_c)^2]^{-1}$$

in agreement with equation (12.3). Having obtained the appropriate transfer function, the next step is to appreciate that first-order response is provided by a filter incorporating just one reactive component. Noting that the transfer function relevant to some filtering inductance L in series with load resistance R is

$$\mathcal{T}(s) = [1 + (L/R)s]^{-1} \tag{12.7}$$

it is seen that all that is needed to synthesise the transfer function of equation (12.6) is to introduce series inductance related to the load resistance R and desired critical pulsatance ω_c by

$$L = R/\omega_c \tag{12.8}$$

Turning to the design of a second-order Butterworth filter, according to equation (12.2)

$$|\mathcal{T}|^2 = [1 + (\omega/\omega_c)^4]^{-1} = [1 + (s/\omega_c)^4]^{-1} \tag{12.9}$$

is required. Since the poles of this function are given by

$$(s/\omega_c)^4 = -1 = \exp j(2p+1)\pi$$

where p is any integer including zero, which is equivalent to

$$s = \pm \omega_c \exp(\pm j\pi/4) \tag{12.10}$$

rejecting poles in the positive half of the s-plane reveals that the transfer function to be provided is

$$\mathcal{T}(s) = \left(\frac{s}{\omega_c} + \frac{1}{\sqrt{2}} + j\frac{1}{\sqrt{2}}\right)^{-1} \left(\frac{s}{\omega_c} + \frac{1}{\sqrt{2}} - j\frac{1}{\sqrt{2}}\right)^{-1}$$

12.2 Butterworth filters

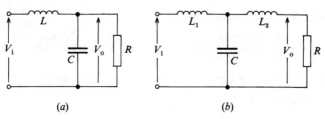

12.2 (a) Second-order and (b) third-order low-pass filter.

or
$$\mathcal{T}(s) = [(s/\omega_c)^2 + \sqrt{2}(s/\omega_c) + 1]^{-1} \quad (12.11)$$

Two reactive components are needed to achieve second-order response and so the relevant low-pass filter in conjunction with the load resistance R is as shown in figure 12.2(a). The transfer function between the input and output of this network is

$$\mathcal{T} = \frac{V_o}{V_i} = \left(\frac{R}{1+RCs}\right) \Big/ \left(Ls + \frac{R}{1+RCs}\right)$$

$$= [LCs^2 + (L/R)s + 1]^{-1} \quad (12.12)$$

Consequently, second-order Butterworth response is achieved with it provided that

$$L/R = \sqrt{2}/\omega_c \quad (12.13)$$
$$LC = 1/\omega_c^2 \quad (12.14)$$

For a third-order Butterworth filter
$$|\mathcal{T}|^2 = [1 + (\omega/\omega_c)^6]^{-1} = [1 - (s/\omega_c)^6]^{-1} \quad (12.15)$$

the poles of which are given by
$$(s/\omega_c)^6 = +1 = \exp j2p\pi$$

where p is any integer including zero, or by
$$s = \pm \omega_c \exp(jp\pi/3) \quad (12.16)$$

where $p = 0, 1, 2$. Forming the transfer function incorporating just those poles in the negative half of the s-plane.

$$\mathcal{T} = \left(\frac{s}{\omega_c} + 1\right)^{-1} \left(\frac{s}{\omega_c} + \frac{1}{2} + j\frac{\sqrt{3}}{2}\right)^{-1} \left(\frac{s}{\omega_c} + \frac{1}{2} - j\frac{\sqrt{3}}{2}\right)^{-1}$$

or
$$\mathcal{T} = [(s/\omega_c)^3 + 2(s/\omega_c)^2 + 2(s/\omega_c) + 1]^{-1} \quad (12.17)$$

An appropriate form of third-order low-pass filter is displayed in figure 12.2(b). With a little effort, the transfer function between its input and load resistance R may be shown to be

$$\mathcal{T} = \frac{V_o}{V_i} = \left(\frac{L_1 L_2 C}{R} s^3 + L_1 C s^2 + \frac{L_1 + L_2}{R} s + 1 \right)^{-1} \quad (12.18)$$

Thus it performs as a third-order Butterworth filter when

$$(L_1 + L_2)/R = 2/\omega_c \quad (12.19)$$
$$L_1 C = 2/\omega_c^2 \quad (12.20)$$
$$L_1 L_2 C/R = 1/\omega_c^3 \quad (12.21)$$

Often expressions for the values of the reactive components of Butterworth filters are quoted corresponding to unit load resistance and unit cut-off pulsatance. Inspection of equations (12.8), (12.13), (12.14), (12.19), (12.20) and (12.21) reveals that inductances must be scaled by R/ω_c and capacitances by $1/\omega_c R$ when the load resistance is R rather than unity and the cut-off pulsatance ω_c rather than unity.

12.3 Chebyshev filters

A very useful alternative approximation to the ideal low-pass filter response has been devised by *Chebyshev*. It is

$$|\mathcal{T}| = [1 + \xi^2 C_n^2(\omega)]^{-\frac{1}{2}} \quad (12.22)$$

where ξ^2 is a small constant and $C_n(\omega)$ is a Chebyshev polynomial of order n defined by

$$\left. \begin{array}{l} C_n(\omega) = \cos\,[n \cos^{-1}(\omega/\omega_c)] \quad \text{for } 0 \leqslant \omega \leqslant \omega_c \\ C_n(\omega) = \cosh\,[n \cosh^{-1}(\omega/\omega_c)] \quad \text{for } \omega \geqslant \omega_c \end{array} \right\} \quad (12.23)$$

In this definition of $C_n(\omega)$, ω_c is the cut-off pulsatance as before. Particularly note that because $\cos n\theta$ is expressible as a polynomial of order n in $\cos \theta$, $C_n(\omega)$ is a polynomial of order n in ω/ω_c. Consequently, when $\omega \gg \omega_c$, $|\mathcal{T}|^2 \propto \omega^{-2n}$, which means that the fall-off in response is once again $20n$ dB per decade of frequency and the order n corresponds to the order of practical filter needed to synthesise the Chebyshev response.

Now consider the nature of the Chebyshev response defined by equations (12.22) and (12.23) in more detail. If $n = 0$, $|\mathcal{T}|$ is simply $(1 + \xi^2)^{-\frac{1}{2}}$, which is independent of frequency and therefore not of interest. When $n = 1$

$$|\mathcal{T}| = [1 + (\xi\omega/\omega_c)^2]^{-\frac{1}{2}} \quad (12.24)$$

so that if the parameter ξ^2 were to be unity, the response would be identical to first-order Butterworth. With ξ small, $|\mathcal{T}|$ decreases monotonically with ω, from unity when $\omega = 0$, passing through the value $(1 + \xi^2)^{-\frac{1}{2}}$ when $\omega = \omega_c$, as depicted in figure 12.3.

12.3 Chebyshev filters

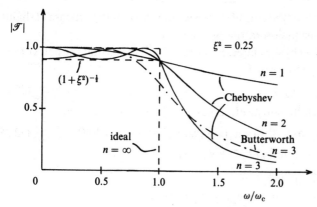

12.3 First, second and third-order Chebyshev responses compared with the third-order Butterworth and ideal low-pass responses.

When $n=2$, if $0 \leqslant \omega \leqslant \omega_c$

$$C_n(\omega) = \cos 2[\cos^{-1}(\omega/\omega_c)] = \{2\cos^2[\cos^{-1}(\omega/\omega_c)]\} - 1$$
$$= 2(\omega/\omega_c)^2 - 1$$

and similarly if $\omega \geqslant \omega_c$

$$C_n(\omega) = \cosh 2[\cosh^{-1}(\omega/\omega_c)] = \{2\cosh^2[\cosh^{-1}(\omega/\omega_c)]\} - 1$$
$$= 2(\omega/\omega_c)^2 - 1$$

Hence

$$|\mathscr{T}| = [1 + \xi^2(2\omega^2/\omega_c^2 - 1)^2]^{-\frac{1}{2}} \tag{12.25}$$

which is $(1+\xi^2)^{-\frac{1}{2}}$ at $\omega = 0$ and at $\omega = \omega_c$ and reaches a maximum of unity in between at $\omega = \omega_c/\sqrt{2}$, as shown in figure 12.3. For $\omega > \omega_c$ the fall-off in $|\mathscr{T}|$ with ω is more rapid than that for first order.

Similar analysis establishes that when $n=3$

$$|\mathscr{T}| = [1 + \xi^2(4\omega^3/\omega_c^3 - 3\omega/\omega_c)^2]^{-\frac{1}{2}} \tag{12.26}$$

and since

$$(d/d\omega)(4\omega^3 - 3\omega_c^2\omega)^2 = 2(4\omega^3 - 3\omega_c^2\omega)(12\omega^2 - 3\omega_c^2)$$

it follows that the third-order response exhibits maxima of unity at $\omega = 0$ and at $\omega = \sqrt{3}\omega_c/2$ and a minimum of $(1+\xi^2)^{-\frac{1}{2}}$ at $\omega = \omega_c/2$. Once again the third-order Chebyshev response is presented in figure 12.3.

From the responses just deduced for the first three Chebyshev orders it may be appreciated that, for all orders, the transfer function ripples between unity and $(1+\xi^2)^{-\frac{1}{2}}$ in the pass band. Clearly higher-order filters have a steeper cut-off. As already stated, well beyond cut-off their attenuation increases at a rate of $20n$ dB per decade of frequency. The smaller ξ, the smaller the ripple in the pass band but the less the attenuation in the stop band. Compared with a Butterworth filter of the same order, the cut-off

284   Filter synthesis

may be steeper near the cut-off frequency but this is at the expense of slightly oscillating transmission in the pass band.

To create a Chebyshev filter, equation (12.22) shows that the poles of $|\mathcal{T}|^2$ must be arranged to satisfy

$$C_n(s/j) = \pm j/\xi \tag{12.27}$$

Making the helpful substitution $\cos^{-1}(s/j\omega_c) = \gamma + j\beta$ so that

$$s = j\omega_c \cos(\gamma + j\beta) = \omega_c[\sin\gamma \sinh\beta + j\cos\gamma \cosh\beta] \tag{12.28}$$

the required condition (12.27) becomes

$$\cos(n\gamma + jn\beta) = \pm j/\xi$$

or on equating real and imaginary parts

$$\cos n\gamma \cosh n\beta = 0$$

$$-\sin n\gamma \sinh n\beta = \pm 1/\xi$$

Since $\cosh n\beta$ cannot be zero, $\cos n\gamma = 0$ and $\sin n\gamma = \pm 1$ and it follows that the poles of $|\mathcal{T}|^2$ are given by equation (12.28) subject to the two conditions

$$\gamma = (2p+1)\pi/2n \tag{12.29}$$

where p is any integer (positive or negative) including zero, and

$$\sinh n\beta = \pm 1/\xi \tag{12.30}$$

Of course, the poles can also be obtained through similar substitution for $\cosh^{-1}(s/j\omega_c)$ rather than $\cos^{-1}(s/j\omega_c)$. Having obtained the poles of $|\mathcal{T}|^2$, the relevant physical transfer function $\mathcal{T}$ is deduced by rejecting poles in the positive half of the s-plane and a network is synthesised so as to generate that transfer function.

To illustrate the synthesis procedure, consider the synthesis of a second-order Chebyshev network for which $\xi^2 = 0.25$ as in the plots of figure 12.3. In this case $n = 2$ and equation (12.29) gives $\gamma = (2p+1)\pi/4$ so that $\sin\gamma = \pm 1/\sqrt{2}$ and $\cos\gamma = \pm 1/\sqrt{2}$. Also from equation (12.30), $\beta = \pm 0.722$. The poles of $|\mathcal{T}|^2$ are therefore given by

$$s = (\pm 0.556 \pm 0.900j)\omega_c \tag{12.31}$$

and the appropriate transfer function is consequently

$$\mathcal{T} = (s/\omega_c + 0.556 + 0.900j)^{-1}(s/\omega_c + 0.556 - 0.900j)^{-1}$$

or

$$\mathcal{T} = [(s/\omega_c)^2 + 1.11(s/\omega_c) + 1.12]^{-1} \tag{12.32}$$

Comparison of this response with that of equation (12.12) establishes that the network of figure 12.2(a) achieves second-order Chebyshev response corresponding to $\xi^2 = 0.25$ provided that $L = 0.99R/\omega_c$ and $LC = 1/1.12\omega_c^2$. As an alternative to finding the transfer function (12.32) from the general results of equations (12.28), (12.29) and (12.30), it may be found from the particular second-order form of $|\mathcal{T}|^2$ given in equation (12.25). According to this particular result, the required poles of $|\mathcal{T}|^2$ are given by

$$\xi^2(2s^2/\omega_c^2 + 1)^2 = -1$$

or since $\xi^2 = 0.25$

$$(s/\omega_c)^2 = -0.5 \pm j = 1.25^{\frac{1}{2}} \exp j(\theta + 2p\pi)$$

where $\theta = 180° \pm \tan^{-1} 2 = 116.56°$ or $243.43°$. Hence

$$s/\omega_c = \pm 1.057 \exp j(58.28° \text{ or } 121.72°)$$
$$= \pm(\pm 0.556 + 0.899j)$$

in agreement with equation (12.31) obtained before.

12.4 Synthesis of high-pass filters

The low-pass Butterworth and Chebyshev filter designs of the previous two sections can readily be adapted to create corresponding high-pass filters. Firstly notice that replacing ω/ω_c by its inverse ω_c/ω in equation (12.2) or (12.23) converts the modelled pass approximation from low pass to high pass while maintaining the cut-off pulsatance at ω_c and the dependence of transmission on $1/\omega$ just what it was on ω. A filter to provide such high-pass transmission can again be synthesised. Compared with a low-pass filter of given order and type, the corresponding high-pass version will feature a capacitor in place of each inductor and an inductor in place of each capacitor.

To discover how to find the component values of a synthesised high-pass filter, consider the particular case of a second-order type. Let the prototype low-pass transfer function be

$$\mathcal{T}_{lp} = [a(s/\omega_c)^2 + b(s/\omega_c) + 1]^{-1} \tag{12.33}$$

so that the planned high-pass transfer function is

$$\mathcal{T}_{hp} = [a(\omega_c/s)^2 + b(\omega_c/s) + 1]^{-1} \tag{12.34}$$

where a and b are certain constants, for example for the Butterworth type, $a = 1$ and $b = \sqrt{2}$ (see equation (12.11)). The transfer function of the second-order low-pass filter shown in figure 12.2(a) is (see equation (12.12))

$$\mathcal{T}_{lp} = [LCs^2 + (L/R)s + 1]^{-1} \tag{12.35}$$

while that of the corresponding high-pass filter with capacitance C' in place of inductance L and inductance L' in place of capacitance C is

$$\mathcal{T}_{hp} = \frac{sL'R/(R+sL')}{1/sC' + sL'R/(R+sL')}$$

$$= \left[\frac{1}{L'C'}\left(\frac{1}{s}\right)^2 + \frac{1}{RC'}\left(\frac{1}{s}\right) + 1\right]^{-1} \tag{12.36}$$

From equations (12.33) and (12.35) it is apparent that to synthesise the prototype low-pass response with unit cut-off pulsatance, the inductance L and capacitance C must satisfy

$$L/R = b \tag{12.37}$$

$$LC = a \tag{12.38}$$

Equations (12.34) and (12.36) similarly show that to synthesise the planned high-pass response with cut-off pulsatance ω_c, the capacitance C' and inductance L' must satisfy

$$1/RC' = b\omega_c \tag{12.39}$$

$$1/L'C' = a\omega_c^2 \tag{12.40}$$

Combining equation (12.37) with equation (12.39) and equation (12.38) with equation (12.40) reveals that

$$C' = 1/\omega_c L \tag{12.41}$$

$$L' = 1/\omega_c^2 LCC' = 1/\omega_c C \tag{12.42}$$

Thus high-pass filters are easily derived from low-pass designs. The results of equations (12.41) and (12.42) are particularly neatly expressed and generalised by saying that the reactances of a synthesised high-pass filter at the cut-off pulsatance must equal the reactances of their counterparts in the prototype low-pass filter at unit pulsatance. In the case of a second-order filter, $1/\omega_c C' = L$ and $\omega_c L' = 1/C$.

12.5 Band filter synthesis

Consider Butterworth and Chebyshev low-pass filters designed to cut off at unit pulsatance, that is, at $\omega = \pm 1$ or $s = \pm j$, theoretically, where the negative value does not have physical significance. Suppose now that s is replaced by

$$(s + \omega_1 \omega_2 / s)/(\omega_2 - \omega_1) \tag{12.43}$$

in the transfer function. This will cause the cut-off to be shifted to pulsatances that satisfy

$$(s + \omega_1 \omega_2 / s)/(\omega_2 - \omega_1) = \pm j$$

or

$$\omega^2 \pm (\omega_2 - \omega_1)\omega - \omega_1 \omega_2 = 0$$

which has solutions

$$\omega = \pm \omega_1, \quad \pm \omega_2 \tag{12.44}$$

Again the negative solutions are not physically meaningful of course. The positive solutions represent the two cut-off pulsatances of a band-pass response. This may be appreciated from the fact that when ω is the

12.5 Band filter synthesis

geometric mean pulsatance $(\omega_1\omega_2)^{\frac{1}{2}}$, expression (12.43) is zero and therefore $|\mathcal{T}|^2$ is unity or near unity. To synthesise such band-pass response, notice that if an inductance of the prototype low-pass filter with unit cut-off pulsatance was L_n then its reactance sL_n must become

$$(s+\omega_1\omega_2/s)L_n/(\omega_2-\omega_1) = sL_s + 1/sC_s$$

where

$$L_s = L_n/(\omega_2-\omega_1) \tag{12.45}$$
$$C_s = (\omega_2-\omega_1)/\omega_1\omega_2 L_n \tag{12.46}$$

Thus it is clear that an inductance L_n of the low-pass prototype must be replaced by a series combination of an inductance L_s and capacitance C_s, the values of which are given by equations (12.45) and (12.46). In a similar way, the reactance $1/sC_n$ of a capacitance C_n of the prototype low-pass filter must be replaced by

$$(\omega_2-\omega_1)/C_n(s+\omega_1\omega_2/s)$$

which is of the form

$$(sL_p \times 1/sC_p)/(sL_p + 1/sC_p)$$

where

$$L_p = (\omega_2-\omega_1)/\omega_1\omega_2 C_n \tag{12.47}$$
$$C_p = C_n/(\omega_2-\omega_1) \tag{12.48}$$

Apparently, to synthesise the band-pass response, any capacitance C_n of the prototype must be replaced by capacitance C_p in parallel with inductance L_p, the values of which are given by equations (12.47) and (12.48). Note that the combinations L_s, C_s and L_p, C_p have the same resonant pulsatance $(\omega_1\omega_2)^{\frac{1}{2}}$ which is the geometric mean of the pass-band limits ω_1 and ω_2.

To further illustrate synthesis of a band-pass filter, consider development of the second-order low-pass filter of figure 12.2(a) into the band-pass form of figure 12.4. It is abundantly clear that the network of figure 12.4 passes signals in the vicinity of the resonant frequencies of the series and parallel LC combinations but rejects at both low and high frequencies. The transfer function of the low-pass prototype is given by equation (12.12) and substituting expression (12.43) for s generates the modified transfer function

$$\mathcal{T} = \left[\frac{LC}{(\omega_2-\omega_1)^2}\left(s^2+2\omega_1\omega_2+\frac{\omega_1^2\omega_2^2}{s^2}\right)+\frac{L}{(\omega_2-\omega_1)R}\left(s+\frac{\omega_1\omega_2}{s}\right)+1\right]^{-1}$$

On the other hand, direct analysis of the network of figure (12.4) yields

$$\mathcal{T} = \frac{Z_p}{Z_s+Z_p} = \left[1+\frac{Z_s}{Z_p}\right]^{-1} = \left[1+\left(sL_s+\frac{1}{sC_s}\right)\left(\frac{1}{R}+\frac{1}{sL_p}+sC_p\right)\right]^{-1}$$

or

$$\mathcal{T} = \left[1+\frac{L_s}{R}s+\frac{1}{RC_s}\frac{1}{s}+L_sC_ps^2+\frac{C_p}{C_s}+\frac{L_s}{L_p}+\frac{1}{L_pC_s}\frac{1}{s^2}\right]^{-1}$$

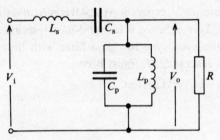

12.4 Band-pass filter.

which is of course of the same form. Comparison of coefficients of s^2, s, $1/s$ and $1/s^2$ in the denominators of the two expressions for $\mathcal{T}$ confirms equations (12.45)–(12.48).

Replacing s by

$$(\omega_2 - \omega_1)/(s + \omega_1\omega_2/s) \tag{12.49}$$

in the low-pass transfer function creates a band-stop form of response with cut-off pulsatances satisfying

$$(\omega_2 - \omega_1) = \pm j(s + \omega_1\omega_2/s)$$

when the low-pass cut-off occurs at unit pulsatance. The physical cut-off pulsatances are once again ω_1 and ω_2 but, to synthesise the band-stop response, inductive reactance sL_n in the low-pass prototype must become reactance

$$(\omega_2 - \omega_1)L_n/(s + \omega_1\omega_2/s)$$

This represents the reactance of inductance L_p in parallel with capacitance C_p where

$$L_p = (\omega_2 - \omega_1)L_n/\omega_1\omega_2 \tag{12.50}$$
$$C_p = 1/(\omega_2 - \omega_1)L_n \tag{12.51}$$

Similarly, capacitive reactance $1/sC_n$ in the prototype becomes reactance

$$(s + \omega_1\omega_2/s)/(\omega_2 - \omega_1)C_n$$

which represents the reactance of inductance L_s in series with capacitance C_s where

$$L_s = 1/(\omega_2 - \omega_1)C_n \tag{12.52}$$
$$C_s = (\omega_2 - \omega_1)C_n/\omega_1\omega_2 \tag{12.53}$$

MATHEMATICAL BACKGROUND APPENDICES

1 Harmonic functions

With reference to the right-angled triangle depicted in figure A1.1(a), sine and cosine functions are defined by

$$\sin \theta = a/h \tag{A1}$$

$$\cos \theta = b/h \tag{A2}$$

Thus additionally,

$$\sin \theta = \cos \phi = \cos (90 - \theta) \tag{A3}$$

$$\cos \theta = \sin \phi = \sin (90 - \theta) \tag{A4}$$

Also from figure A1.1(b),

$$\sin (\theta_1 + \theta_2) = \frac{a_1 + a_2 \cos \theta_1}{h_2} = \frac{h_1 \sin \theta_1 + h_2 \sin \theta_2 \cos \theta_1}{h_2}$$

$$= \sin \theta_1 \cos \theta_2 + \cos \theta_1 \sin \theta_2 \tag{A5}$$

$$\cos (\theta_1 + \theta_2) = \frac{b_1 - a_2 \sin \theta_1}{h_2} = \frac{h_1 \cos \theta_1 - h_2 \sin \theta_2 \sin \theta_1}{h_2}$$

$$= \cos \theta_1 \cos \theta_2 - \sin \theta_1 \sin \theta_2 \tag{A6}$$

Turning to the differential behaviour of harmonic functions, it is convenient to let $y = \sin \theta$ and $z = \cos \theta$. Small changes Δy in y and Δz in z corresponding to a small change $\Delta \theta$ in θ are given by

$$y + \Delta y = \sin (\theta + \Delta \theta) = \sin \theta \cos \Delta \theta + \cos \theta \sin \Delta \theta$$

$$z + \Delta z = \cos (\theta + \Delta \theta) = \cos \theta \cos \Delta \theta - \sin \theta \sin \Delta \theta$$

where use has been made of relations (A5) and (A6). Thus

$$\frac{dy}{d\theta} = \lim_{\Delta \theta \to 0} \frac{\Delta y}{\Delta \theta} = \lim_{\Delta \theta \to 0} \left(\frac{\sin \theta \cos \Delta \theta - \sin \theta + \cos \theta \sin \Delta \theta}{\Delta \theta} \right)$$

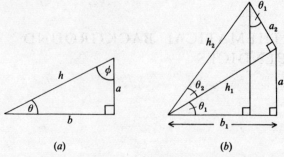

A1.1 (a) Right-angled triangle with respect to which sine and cosine functions are defined and (b) figure used to deduce expansions of sine and cosine functions of $(\theta_1 + \theta_2)$ in terms of $\sin\theta_1$, $\cos\theta_1$, $\sin\theta_2$ and $\cos\theta_2$.

$$\frac{dz}{d\theta} = \operatorname*{Lim}_{\Delta\theta\to 0}\frac{\Delta z}{\Delta\theta} = \operatorname*{Lim}_{\Delta\theta\to 0}\left(\frac{\cos\theta\cos\Delta\theta - \cos\theta - \sin\theta\sin\Delta\theta}{\Delta\theta}\right)$$

But

$$\operatorname*{Lim}_{\Delta\theta\to 0}\cos\Delta\theta = 1, \quad \operatorname*{Lim}_{\Delta\theta\to 0}(\sin\Delta\theta)/\Delta\theta = 1$$

Hence

$$dy/d\theta = \cos\theta \tag{A7}$$

$$dz/d\theta = -\sin\theta \tag{A8}$$

Differentiating a second time

$$\frac{d^2y}{d\theta^2} = \frac{d}{d\theta}(\cos\theta) = \frac{dz}{d\theta} = -\sin\theta = -y$$

$$\frac{d^2z}{d\theta^2} = -\frac{d}{d\theta}(\sin\theta) = -\frac{dy}{d\theta} = -\cos\theta = -z$$

Evidently sine and cosine functions of θ satisfy the simple differential equation

$$\frac{d^2f}{d\theta^2} = -f \tag{A9}$$

Now if f is expressed as a polynomial in θ

$$f = A_0 + A_1\theta + A_2\theta^2 + \cdots + A_n\theta^n + \cdots$$

such that the various coefficients A_i are all independent of θ, then

$$df/d\theta = A_1 + 2A_2\theta + 3A_3\theta^2 + \cdots + nA_n\theta^{n-1} + \cdots$$

and

$$d^2f/d\theta^2 = 2\times 1 A_2 + 3\times 2 A_3\theta + \cdots + n(n-1)A_n\theta^{n-2} + \cdots$$

Consequently to satisfy equation (A9)

$$A_2 = -\frac{A_0}{2\times 1}, \quad A_3 = -\frac{A_1}{2\times 3}, \quad A_4 = -\frac{A_2}{3\times 4}, \quad \ldots, \quad A_n = -\frac{A_{n-2}}{(n-1)n}$$

and

$$f = A_0\left(1 - \frac{\theta^2}{2\times 1} + \frac{\theta^4}{4\times 3\times 2\times 1} - \cdots\right)$$

$$+ A_1\left(\theta - \frac{\theta^3}{3\times 2\times 1} + \frac{\theta^5}{5\times 4\times 3\times 2\times 1} - \cdots\right)$$

But if $f = \sin\theta$, then $f = 0$ when $\theta = 0$ so that $A_0 = 0$, while $f = \theta$ when θ is small so that $A_1 = 1$. Thus

$$\sin\theta = \theta - \frac{\theta^3}{3!} + \frac{\theta^5}{5!} - \cdots \tag{A10}$$

On the other hand, if $f = \cos\theta$, then $f = 1$ when $\theta = 0$ so that $A_0 = 1$, while $df/d\theta = 0$ when $\theta = 0$ so that $A_1 = 0$. Thus

$$\cos\theta = 1 - \frac{\theta^2}{2!} + \frac{\theta^4}{4!} - \cdots \tag{A11}$$

In conclusion, figure A1.2 presents plots of $\sin\theta$ and $\cos\theta$ as a function of θ.

2   Exponential functions

The exponential function of the variable x is defined as that function f which satisfies the simple differential equation

$$df/dx = f \tag{A12}$$

and has unit value when $x = 0$. One accepted means of denoting the exponential function of a variable x is to write $\exp x$. If $\exp x$ is expressed as a polynomial in x

$$\exp x = A_0 + A_1 x + A_2 x^2 + \cdots + A_n x^n + \cdots \tag{A13}$$

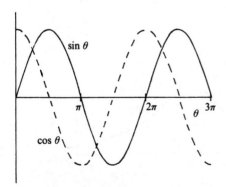

A1.2 Plots of $\sin\theta$ and $\cos\theta$ as a function of θ.

in which the various coefficients A_i are all independent of x then

$$\frac{d(\exp x)}{dx} = A_1 + 2A_2 x + 3A_3 x^2 + \cdots + nA_n x^{n-1} + \cdots$$

Consequently to satisfy equation (A12)

$$A_1 = A_0, \quad A_2 = A_1/2 = A_0/2, \quad A_3 = A_2/3 = A_0/2 \times 3$$

and

$$\exp x = A_0 \left(1 + x + \frac{x^2}{2!} + \cdots + \frac{x^n}{n!} + \cdots \right)$$

But $\exp x = 1$ when $x = 0$ so that $A_0 = 1$ and

$$\exp x = 1 + x + \frac{x^2}{2!} + \cdots + \frac{x^n}{n!} + \cdots \tag{A14}$$

Clearly from equation (A14), the function $\exp -x$ exists where

$$\exp -x = 1 - x + \frac{x^2}{2!} - \cdots + (-1)^n \frac{x^n}{n!} + \cdots \tag{A15}$$

and it will be appreciated that this function satisfies the differential equation

$$df/dx = -f \tag{A16}$$

Figure A2.1 presents plots of $\exp x$ and $\exp -x$ as a function of x.

Now consider the function

$$f = (\exp u)(\exp v)$$

Differentiating with respect to $(u+v)$

$$\frac{df}{d(u+v)} = (\exp u) \frac{d}{d(u+v)} (\exp v) + (\exp v) \frac{d}{d(u+v)} (\exp u)$$

$$= (\exp u)(\exp v) \frac{dv}{d(u+v)} + (\exp v)(\exp u) \frac{du}{d(u+v)} = f$$

But, when u and v are both zero, both $\exp u$ and $\exp v$ are unity so that f is

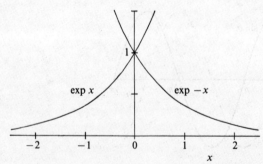

A2.1 Plots of $\exp x$ and $\exp -x$ as a function of x.

3 *Phasors and complex representation*

unity. Evidently, f is alternatively expressible as $\exp(u+v)$, that is,

$$\exp(u+v) = (\exp u)(\exp v) \qquad (A17)$$

revealing that it is permissible to express the exponential function $\exp x$ as some base e raised to the power x, that is

$$\exp x = e^x \qquad (A18)$$

Putting $x = 1$ in equation (A14) establishes that the base e is 2.718 correct to three decimal places.

From the foregoing, the complex functions $e^{\pm jx}$ where $j = \sqrt{-1}$ are equivalent to the polynomials

$$e^{\pm jx} = 1 \pm jx - \frac{x^2}{2!} \mp \frac{jx^3}{3!} + \frac{x^4}{4!} \cdots$$

$$= \left[1 - \frac{x^2}{2!} + \frac{x^4}{4!} - \frac{x^6}{6!} + \cdots\right] \pm j\left[x - \frac{x^3}{3!} + \frac{x^5}{5!} - \frac{x^7}{7!} + \cdots\right]$$

Comparison with equations (A10) and (A11) therefore shows that

$$e^{\pm j\theta} = \cos\theta \pm j\sin\theta \qquad (A19)$$

which is known as Euler's identity. Useful corollaries are that

$$\cos\theta = \frac{1}{2}(e^{j\theta} + e^{-j\theta}) \qquad (A20)$$

$$\sin\theta = \frac{1}{2j}(e^{j\theta} - e^{-j\theta}) \qquad (A21)$$

Analogous expressions to the right-hand sides of these last two equations in terms of real exponential functions lead to the convenient definitions

$$\cosh\theta = \frac{1}{2}(e^{\theta} + e^{-\theta}) = 1 + \frac{\theta^2}{2!} + \frac{\theta^4}{4!} + \cdots \qquad (A22)$$

$$\sinh\theta = \frac{1}{2}(e^{\theta} - e^{-\theta}) = \theta + \frac{\theta^3}{3!} + \frac{\theta^5}{5!} + \cdots \qquad (A23)$$

Furthermore

$$\tan\theta = \sin\theta/\cos\theta \qquad (A24)$$

$$\tanh\theta = \sinh\theta/\cosh\theta \qquad (A25)$$

3 Phasors and complex representation

In the analysis of the steady-state behaviour of any linear electrical network energised by a single sinusoidal source of e.m.f. or current, it is required to add and/or subtract scalar quantities, each of which varies sinusoidally with time at the frequency of the source but in general has a different amplitude and phase. This requirement stems from the application of Kirchhoff's current and voltage laws and from the linear nature of the network. Consideration of how such addition and/or subtraction can be accomplished reveals the relevance of complex representation to network analysis.

For simplicity of explanation, consider the addition of just two sinusoidal quantities of the same frequency, say, $A_1 \sin(\omega t + \phi_1)$ and $A_2 \sin(\omega t + \phi_2)$. Making use of the trigonometrical relation (A5), the sum is seen to be

$$S = A_1 \sin(\omega t + \phi_1) + A_2 \sin(\omega t + \phi_2)$$
$$= (A_1 \cos \phi_1 + A_2 \cos \phi_2) \sin \omega t + (A_1 \sin \phi_1 + A_2 \sin \phi_2) \cos \omega t$$

or

$$S = A \sin(\omega t + \phi) \tag{A26}$$

where

$$A = [(A_1 \cos \phi_1 + A_2 \cos \phi_2)^2 + (A_1 \sin \phi_1 + A_2 \sin \phi_2)^2]^{\frac{1}{2}} \tag{A27}$$

and

$$\tan \phi = (A_1 \sin \phi_1 + A_2 \sin \phi_2)/(A_1 \cos \phi_1 + A_2 \cos \phi_2) \tag{A28}$$

The addition of the two sinusoidal quantities can also be implemented with the help of a suitable diagram. With reference to figure A3.1(a), the required sum is the sum of the projections of the lines of length A_1 and A_2 onto the y axis which is just the projection of OP onto the y axis. Consideration of the figure at the convenient time $t=0$ shows that the length of OP is precisely the quantity A given by equation (A27) while the relative phase of OP is ϕ given by equation (A28). Thus the sum of the two sinusoidal functions is $A \sin(\omega t + \phi)$ as before.

By now it should be abundantly clear that to find the required sum, which inevitably varies sinusoidally at pulsatance ω, all that is needed is to evaluate its amplitude A and relative phase ϕ. This can be done very simply by drawing the diagram of figure A3.1(a) at time $t=0$ as is done in figure A3.1(b). Lines of length A_1 and A_2 are drawn at angles ϕ_1 and ϕ_2

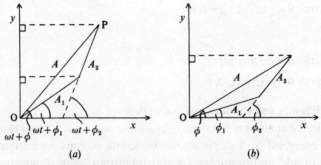

A3.1 (a) Finding the sum of two sinusoidal quantities of the same frequency from projections in a diagram and (b) the corresponding diagram at time $t=0$ giving the amplitude A and phase ϕ of the resultant.

3 Phasors and complex representation

respectively with respect to a reference direction (conveniently the x axis) and the length of the resultant line or vector sum constitutes the amplitude of the sum and the angle it makes with the reference direction its relative phase. The complete solution is of course the projection of the resultant line onto the y axis as the resultant rotates at angular frequency ω. Particularly note that the lines in these diagrams do not represent vector quantities. Each line denotes a magnitude and phase rather than a magnitude and direction. Very appropriately, lines representing amplitude or r.m.s. value and phase are known as *phasors*. Scalar and vector products of phasors are of course meaningless.

Through complex algebra the features of a phasor diagram can be expressed without the need to draw it. The essential point is that the operation of multiplying by $j = \sqrt{-1}$ is equivalent to rotation through 90° in a graphical plot because the operation j repeated, that is, $j^2 = -1$, is equivalent to rotation through 180°. Evidently the *complex* quantity $(a+jb)$ where a and b are real can be thought of as a displacement 'a' along an axis known as *real* followed by a displacement 'b' along an orthogonal axis described as *imaginary*. That is, $(a+jb)$ can be thought of as the combined effect of displacements 'a' and 'jb'. This is illustrated in figure A3.2. Such representations of the real and imaginary aspects of complex quantities are known as *Argand* diagrams. By Pythagoras' theorem, the length of $(a+jb)$, that is, its magnitude, is $(a^2+b^2)^{\frac{1}{2}}$, while its orientation ϕ relative to the orientation of 'a' is given by $\tan \phi = b/a$.

Clearly the phasor (A_1, ϕ_1) of figure A3.1(b) can be represented by $(a+jb)$ where $a = A_1 \cos \phi_1$ and $b = A_1 \sin \phi_1$ so that its magnitude is $(a^2+b^2)^{\frac{1}{2}} = (A_1^2 \cos^2 \phi_1 + A_1^2 \sin^2 \phi_1)^{\frac{1}{2}} = A_1$ and its phase is $\tan^{-1}(b/a) = \tan^{-1}(A_1 \sin \phi_1 / A_1 \cos \phi_1) = \phi_1$. If the phasor (A_2, ϕ_2) is represented similarly by $(c+jd)$ where $c = A_2 \cos \phi_2$ and $d = A_2 \sin \phi_2$ then the sum of

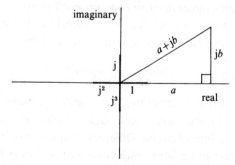

A3.2 Argand diagram of the complex quantity $(a+jb)$.

the two phasors is
$$(a+jb)+(c+jd)=(a+c)+j(b+d)$$
and has magnitude $[(a+c)^2+(b+d)^2]^{\frac{1}{2}}$ by Pythagoras' theorem and phase ϕ where $\tan\phi=(b+d)/(a+c)$. These results are again in accordance with those given in equations (A27) and (A28).

Making use of equation (A19), a neat alternative representation of the information inherent in figure A3.1(a) is
$$A\exp j(\omega t+\phi)=A_1\exp j(\omega t+\phi_1)+A_2\exp j(\omega t+\phi_2)$$
Dividing through by $\exp j\omega t$ yields
$$A\exp j\phi=A_1\exp j\phi_1+A_2\exp j\phi_2$$
which is a neat representation of the information contained in figure A3.1(b). Consideration of the real and imaginary parts of either of these two equations shows that
$$A\cos\phi=A_1\cos\phi_1+A_2\cos\phi_2$$
$$A\sin\phi=A_1\sin\phi_1+A_2\sin\phi_2$$
once more in agreement with equations (A27) and (A28) obtained originally.

The complex algebraic representation is especially helpful in network analysis in connection with the potential difference that arises from a sinusoidal current flowing through a complex impedance. If a current represented by $I=I_0\exp j(\omega t+\alpha)$ flows through a complex impedance $Z=Z_0\exp j\theta$, it creates a potential difference across it represented by
$$V=ZI=Z_0I_0\exp j(\omega t+\alpha+\theta) \quad (A29)$$
This result clearly demonstrates that the amplitude of the potential difference is $Z_0\,(=|Z|)$ times the amplitude I_0 of the current and that the potential difference is advanced in phase by an angle θ relative to the phase of the current.

4 Linear differential equations with constant coefficients

Complete responses of linear electrical networks to input signals may be deduced by solving differential equations of the form
$$c_n\frac{d^n x}{dt^n}+c_{n-1}\frac{d^{n-1}x}{dt^{n-1}}+\cdots+c_1\frac{dx}{dt}+c_0 x=f(t) \quad (A30)$$
where all the coefficients c_i are constants. When such equations refer to electrical networks, x denotes some associated current or potential difference, while t denotes time, of course. Differential equations of the form of equation (A30) are said to be *linear* with *constant coefficients* and it is customary in any particular case to refer to the *order*, meaning the highest

4 Linear differential equations with constant coefficients

order of differentiation involved. Note that a differential equation is linear if all the derivatives in it including that of zero order are raised to unit power and never occur as products.

In solving equations of the form of equation (A30), it is helpful to consider first the solution of

$$c_n \frac{d^n x}{dt^n} + c_{n-1} \frac{d^{n-1} x}{dt^{n-1}} + \cdots + c_1 \frac{dx}{dt} + c_0 x = 0 \tag{A31}$$

Substitution of the simple function

$$x = A \exp mt \tag{A32}$$

where m is independent of t into equation (A31) readily establishes that it is a solution provided that A is an arbitrary constant and m satisfies the *auxiliary* equation

$$c_n m^n + c_{n-1} m^{n-1} + \cdots + c_1 m + c_0 = 0 \tag{A33}$$

Since from this auxiliary equation m can take n values $m_1, m_2, m_3, \ldots, m_n$ in the solution and since each value may be associated with a different arbitrary constant, the complete solution is

$$x = A_1 \exp m_1 t + A_2 \exp m_2 t + \cdots + A_n \exp m_n t \tag{A34}$$

Illustrating the foregoing by treating the case of a second-order, linear, differential equation

$$c_2 \frac{d^2 x}{dt^2} + c_1 \frac{dx}{dt} + c_0 x = 0 \tag{A35}$$

with constant coefficients c_0, c_1 and c_2, its solution is

$$x = A_1 \exp m_1 t + A_2 \exp m_2 t \tag{A36}$$

where m_1 and m_2 are the roots of

$$c_2 m^2 + c_1 m + c_0 = 0 \tag{A37}$$

Substitution of equation (A36) back into the left-hand side of equation (A35) gives

$$c_2(m_1^2 A_1 \exp m_1 t + m_2^2 A_2 \exp m_2 t)$$
$$+ c_1(m_1 A_1 \exp m_1 t + m_2 A_2 \exp m_2 t)$$
$$+ c_0(A_1 \exp m_1 t + A_2 \exp m_2 t)$$
$$= (c_2 m_1^2 + c_1 m_1 + c_0) A_1 \exp m_1 t + (c_2 m_2^2 + c_1 m_2 + c_0) A_2 \exp m_2 t$$

which is zero from equation (A37), confirming that equation (A36) is indeed the solution of equation (A35). Particularly notice that when the auxiliary roots m_1 and m_2 are complex conjugates, say, $p \pm jq$

$$x = [\exp pt][A_1 \exp jqt + A_2 \exp -jqt]$$

Using equation (A19) this solution can be expressed alternatively as

$$x = [\exp pt][(A_1 + A_2)\cos qt + j(A_1 - A_2)\sin qt]$$
$$= B \exp pt \cos(qt - \phi) \tag{A38}$$

where B and ϕ are new arbitrary constants such that $B = 2(A_1 A_2)^{\frac{1}{2}}$ and $\tan \phi = j(A_1 - A_2)/(A_1 + A_2)$. In cases where the roots of the auxiliary equation are real and distinct, the solution is simply the sum of two real exponential functions of time. A problem arises if the roots m_1 and m_2 of the auxiliary equation are identical; how to overcome it is expounded later.

The solution of equation (A30) when $f(t) \neq 0$ is very simply related to the solution when $f(t)$ is zero. Suppose that when $f(t) \neq 0$ a solution x_p, called the *particular integral*, can be found that does not involve arbitrary constants. In this case

$$c_n \frac{d^n x_p}{dt^n} + c_{n-1} \frac{d^{n-1} x_p}{dt^{n-1}} + \cdots + c_1 \frac{dx_p}{dt} + c_0 x_p = f(t) \tag{A39}$$

Putting $x = x_p + x_q$ in equation (A30) and subtracting equation (A39) yields

$$c_n \frac{d^n x_q}{dt^n} + c_{n-1} \frac{d^{n-1} x_q}{dt^{n-1}} + \cdots + c_1 \frac{dx_q}{dt} + c_0 x_q = 0 \tag{A40}$$

The solution to equation (A40) in x_q is the same as that to equation (A31) in x which has already been found as equation (A34) involving n arbitrary constants. It is customary to refer to x_q as the *complementary function* and the complete solution to equation (A30) is the sum of the particular integral and complementary function. Clearly the number of arbitrary constants in the complementary function is equal to the order of the differential equation. Values of the arbitrary constants are determined by *boundary conditions* such as the value of x at the origin of time.

In finding particular integrals it is convenient to work in terms of the operator $\mathscr{D} = d/dt$. Because

$$\mathscr{D}(u + v) = \mathscr{D}u + \mathscr{D}v$$
$$\mathscr{D}^m \mathscr{D}^n u = \mathscr{D}^{m+n} u$$

and

$$\mathscr{D}(cu) = c\mathscr{D}(u)$$

provided that c is constant (although not otherwise), the operator $\mathscr{D}$ can be treated according to the fundamental laws of algebra when dealing with linear differential equations with constant coefficients. Expressing equation (A30) in terms of $\mathscr{D}$ as

$$[F(\mathscr{D})]x = f(t) \tag{A41}$$

certain useful results follow with respect to the operation $F(\mathscr{D})$. To begin

4 Linear differential equations with constant coefficients

with

$$[F(\mathscr{D})]\exp at = \left(c_n\frac{d^n}{dt^n}+c_{n-1}\frac{d^{n-1}}{dt^{n-1}}+\cdots+c_1\frac{d}{dt}+c_0\right)\exp at$$
$$= (c_n a^n + c_{n-1}a^{n-1}+\cdots+c_1 a+c_0)\exp at$$

or

$$[F(\mathscr{D})]\exp at = F(a)\exp at \tag{A42}$$

Similarly it is readily shown that

$$[F(\mathscr{D}^2)]\cos at = F(-a^2)\cos at \tag{A43}$$
$$[F(\mathscr{D}^2)]\sin at = F(-a^2)\sin at \tag{A44}$$
$$[F(\mathscr{D})][(\exp at)U] = (\exp at)\{[F(\mathscr{D}+a)]U\} \tag{A45}$$

where U is any function of t in the last equation.

Digressing for a moment to deal with the situation when the second-order auxiliary equation (A37) has equal roots, say α, the original differential equation is

$$(\mathscr{D}^2 - 2\alpha\mathscr{D} + \alpha^2)x = (\mathscr{D}-\alpha)^2 x = 0 \tag{A46}$$

Since $x = A\exp \alpha t$ is one solution, try as a more general solution $x = (\exp \alpha t)U$, where U is some function of t. Now from equation (A45)

$$(\mathscr{D}-\alpha)^2[(\exp \alpha t)U] = (\exp \alpha t)[(\mathscr{D}+\alpha-\alpha)^2 U] = (\exp \alpha t)\mathscr{D}^2 U$$

Thus equation (A46) becomes

$$\mathscr{D}^2 U = 0$$

so that

$$U = A + Bt$$

where A and B are arbitrary constants and

$$x = (A+Bt)\exp \alpha t \tag{A47}$$

In terms of the notation of equation (A41), a particular integral of equation (A30) is given by

$$x_p = [1/F(\mathscr{D})]f(t) \tag{A48}$$

But equation (A42) suggests that, as long as $F(a) \neq 0$,

$$[1/F(\mathscr{D})]\exp at = [1/F(a)]\exp at$$

Consequently, when $f(t)$ is $\exp at$, the particular integral can be obtained as

$$x_p = [1/F(a)]\exp at \tag{A49}$$

The correctness of this particular integral is easily checked by insertion in equation (A41) followed by application of equation (A42).

When $f(t)$ is $\cos at$ or $\sin at$, equations (A43) and (A44) indicate that the particular integral is found by replacing $\mathscr{D}^2$ by $-a^2$ everywhere it occurs on the right-hand side of equation (A48). When $f(t)$ is t^n, where n is a positive integer, $[1/F(\mathscr{D})]$ is expanded as a series in ascending powers of $\mathscr{D}$. Operating on t^n, a finite series in powers of t is then obtained for the particular integral.

In conclusion, three particular linear differential equations with constant coefficients will be solved by way of example.

(1) $\quad \dfrac{d^2x}{dt^2} + 3\dfrac{dx}{dt} + 2x = 2\exp{-3t}$

In terms of the operator $\mathscr{D} = d/dt$, the complementary function is the general solution of

$$(\mathscr{D}+2)(\mathscr{D}+1)x = 0$$

Thus the auxiliary equation is

$$(m+2)(m+1) = 0$$

and the complementary function is

$$x_q = A_1 \exp{-t} + A_2 \exp{-2t}$$

where A_1 and A_2 are arbitrary constants. The particular integral is

$$\dfrac{1}{\mathscr{D}^2 + 3\mathscr{D} + 2}(2\exp{-3t}) = \dfrac{1}{9 - 9 + 2}(2\exp{-3t}) = \exp{-3t}$$

Combining the complementary function with the particular integral, the complete solution is

$$x = A_1 \exp{-t} + A_2 \exp{-2t} + \exp{-3t}$$

(2) $\quad \dfrac{d^2x}{dt^2} + \dfrac{dx}{dt} + 1 = \cos 2t$

The auxiliary equation is

$$m^2 + m + 1 = \left(m + \dfrac{1}{2} + j\dfrac{\sqrt{3}}{2}\right)\left(m + \dfrac{1}{2} - j\dfrac{\sqrt{3}}{2}\right) = 0$$

so that the complementary function is

$$x_q = A_1 \exp\left(-\dfrac{1}{2} - j\dfrac{\sqrt{3}}{2}\right)t + A_2 \exp\left(-\dfrac{1}{2} + j\dfrac{\sqrt{3}}{2}\right)t$$

where A_1 and A_2 are arbitrary constants or

$$x_q = B\exp\left(-\dfrac{1}{2}t\right)\cos\left(\dfrac{\sqrt{3}}{2}t - \phi\right)$$

where B and ϕ are arbitrary constants. Since the particular integral is

$$[\mathscr{D}^2 + \mathscr{D} + 1]^{-1}\cos 2t = (\mathscr{D} - 3)^{-1}\cos 2t = [(\mathscr{D}+3)/(\mathscr{D}^2 - 9)]\cos 2t$$
$$= (2\sin 2t - 3\cos 2t)/13$$

the complete solution is
$$x = B\exp\left(-\frac{1}{2}t\right)\cos\left(\frac{\sqrt{3}}{2}t - \phi\right) + \frac{2\sin 2t - 3\cos 2t}{13}$$

(3) $\quad \dfrac{d^2x}{dt^2} + 4\dfrac{dx}{dt} + 4x = t^2$

This time the auxiliary equation is
$$(m+2)^2 = 0$$
and has equal roots of -2. Thus the complementary function is
$$x_q = (A + Bt)\exp-2t$$
where A and B are arbitrary constants. Since the particular integral is
$$x_p = (\mathscr{D}+2)^{-2}t^2 = \tfrac{1}{4}(1+\mathscr{D}/2)^{-2}t^2$$
$$= \tfrac{1}{4}(1 - \mathscr{D} + 3\mathscr{D}^2/4 - \cdots)t^2$$
$$= \tfrac{1}{4}(t^2 - 2t + \tfrac{3}{2})$$
the complete solution is
$$x = x_q + x_p = (A+Bt)\exp-2t + \tfrac{1}{4}(t^2 - 2t + \tfrac{3}{2})$$

PROBLEMS

The following collection of problems has been included to enable readers to discover whether they can apply the theory of the text in a fairly direct manner. Over the years the author has found that all students gain greatly in confidence through solving straightforward problems whereas the morale of some is adversely affected as a result of failing to cope with tricky questions. Questions that test general intelligence more than expertise in the subject being studied have been deliberately avoided in composing the present set. Answers to all the problems are listed between pages 313 and 317 while outline solutions to those marked with an asterisk are presented between pages 318 and 330.

*1.1 Positive point charges of 1.414×10^{-9} C each are situated at the corners of a square of side 10 cm in a medium of dielectric constant 3.6. Calculate the electric field and potential at the centre of the square. What is the effect of changing the sign of the charges at the ends of one side?

2.1 A certain semiconducting specimen, 2.4 cm long and of uniform cross-section 6 mm^2, is provided with ohmic end contacts. The resistance between the contacts is 50 Ω. Find the mobility of holes in the specimen given that they are the predominant current carriers and their density is 10^{16} cm^{-3}. If the effective mass of the holes is 3.2×10^{-31} kg, what is the mean free time?

2.2 The current I in nanoamps through a particular diode is related to the potential difference V in volts across it by
$$I = 10[(\exp 40V) - 1]$$
Calculate the small-signal resistance under (a) forward bias of 0.25 V, (b) zero bias and (c) reverse bias of 0.25 V.

*2.3 A direct electrical source connected in series with a variable resistance R delivers a current of 1 A when $R = 1\,\Omega$ and a current of 0.3 A when $R = 4.5\,\Omega$. Deduce the nature of the source and the maximum power that can be developed in R by it.

3.1 Use the formulae for the resistances of series and parallel combinations to find the resistance between the terminals of the network shown.

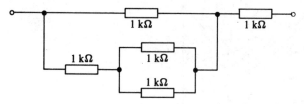

*3.2 Twelve resistors, each of 12 kΩ resistance, are connected together to form the edges of a cube. What is the resistance between diagonally opposite corners?

*3.3 By means of mesh current analysis, deduce the current flowing in branch XA of the network shown. Check your result by an alternative method that makes use of symmetry and the principle of superposition.

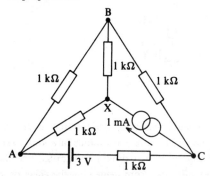

3.4 Find the potentials of nodes A and B with respect to the common line C in the circuit shown by the method of node-pair analysis.

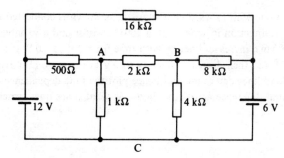

3.5 Apply Thévenin's theorem to find the current through the detector of resistance 4 kΩ in the unbalanced Wheatstone bridge network depicted.

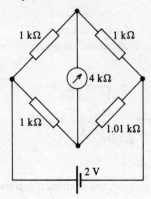

3.6 Derive the Norton equivalent circuit between terminals A and B of the network given.

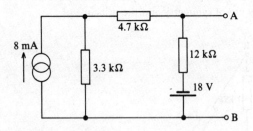

3.7 A voltmeter and an ammeter, respectively of internal resistance 50 kΩ and 50 Ω, are to be used to measure a resistance of about 1 kΩ. Deduce the circuit arrangement that should be adopted and the approximate error that will arise from neglecting the effect of meter internal resistance.

*3.8 A tunnel diode with forward characteristic as shown is connected in the forward direction in series with a 100 Ω resistor and a variable e.m.f. The e.m.f. increases linearly with time from 0 V to 1.0 V at a rate of 10 V s^{-1} and then decreases linearly to 0 V at the same rate. By drawing load lines determine and hence plot the time dependence of the potential difference across the diode. At what times does switching occur?

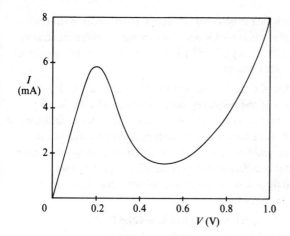

4.1 Find the inductance and capacitance between the terminals of the networks shown. (Note that solution of the capacitive circuit is greatly eased if it is appreciated from the outset that the charges carried by C_1 and C_5, by C_2 and C_4 and by C_6 and C_8 are equal on account of the symmetry.)

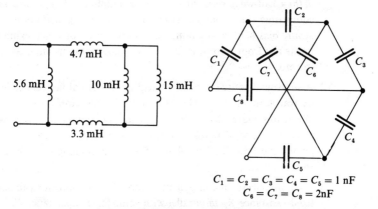

$C_1 = C_2 = C_3 = C_4 = C_5 = 1$ nF
$C_6 = C_7 = C_8 = 2$ nF

4.2 A $1\,\mu F$ and a $2\,\mu F$ capacitor are separately charged to potential differences of 10 V and 4 V respectively. How much electrical energy is stored in the pair of capacitors? If the charged capacitors are subsequently connected in parallel with their positive plates common, what is the potential difference across them and by how much has the stored electrical energy fallen? What becomes of the electrical energy loss? Throughout this question assume negligible discharge associated with leakage resistance.

*4.3 An e.m.f. of 24 V charges a 10 nF capacitor through a $10\,k\Omega$ resistor. Connected in parallel with the capacitor is a device that remains effectively open circuit until the potential difference across it reaches

17 V following which it becomes effectively short circuit until the potential difference falls to 3 V and the original open-circuit state is restored. What frequency of relaxation oscillation does this particular circuit arrangement generate?

*4.4 A certain source of e.m.f. has an internal resistance of 600 Ω and generates signals in the frequency range 40 Hz to 15 kHz. What minimum capacitance will satisfactorily couple this source to a load of 400 Ω? Also estimate how much capacitance may be inserted between the same source and load if the potential difference across the load is to be a reasonably differentiated version with respect to time of the source e.m.f. Why is it not a good idea to make the capacitance smaller than necessary for differentiation?

4.5 A charged 0.1 μF capacitor of negligible loss and inductance is suddenly connected across an inductor of 1 mH inductance and 40 Ω resistance. Calculate the frequency and decrement of the ensuing electrical oscillation. How would the oscillatory discharge be modified if a 120 Ω resistor were inserted in series with the circuit and what minimum series resistance would just prevent oscillation?

5.1 Find the r.m.s. value of (a) a full-wave rectified sinewave of amplitude a, (b) a half-wave rectified sinewave of amplitude a, (c) a function comprising sinusoidal oscillation of amplitude a superimposed on steady bias b, (d) a sawtooth wave, each period of which exhibits a linear rise from zero to magnitude a followed by instantaneous return to zero.

5.2 A sinusoidal source of e.m.f. of negligible internal impedance is derived by transforming the 50 Hz, 240 V r.m.s. mains supply down by 20:1. When this source is connected to a series circuit comprising 150 mH inductance and 47 Ω resistance, what amplitude of current flows in the circuit and how does the phase of the current compare with that of the source e.m.f?

*5.3 A certain electrical component is known to be equivalent to some fixed resistance R_C in parallel with some fixed capacitance C. Observations are made with the component connected through a 100 kΩ resistor of negligible reactance to a sinusoidal source of e.m.f. of adjustable frequency and negligible internal impedance. At sufficiently low frequencies the amplitude of the potential difference across the component is found to be independent of frequency and equal to four-fifths of the amplitude of the source e.m.f. On increasing the frequency to 100 kHz, the amplitude of the potential difference across the component falls to 1% of the amplitude of the source e.m.f. Deduce the values of the resistance R_C and capacitance C.

5.4 A circuit consists of 2.7 mH inductance, 3.3 nF capacitance and 22 Ω resistance in series. Calculate the resonant frequency and Q-factor. If the frequency of some sinusoidal e.m.f. introduced into the circuit is detuned by 1% from resonance, what is the phase difference between the current and e.m.f?

*5.5 On connection to the output of a certain tunable sinusoidal oscillator, a series combination of an inductor, capacitor and resistor is found to resonate at 33.9 kHz and to dissipate half of maximum power at frequencies of 32.9 kHz and 34.9 kHz. Given that the internal resistance of the oscillatory source is 11.8 Ω and that the resonant current is 17 mA r.m.s. when the e.m.f. is 1 V r.m.s., deduce as far as possible the magnitudes of the circuit components.

5.6 Design a parallel resonant circuit, based on a coil of 560 μH inductance and 15 Ω resistance, that will resonate at 200 kHz and have a Q-factor of 20. What load resistance connected across the circuit will halve the Q-factor?

5.7 A 20 W, 100 V lamp is to be lit by suitable connection to the 240 V r.m.s., 50 Hz mains supply. What capacitance inserted in series with the lamp and mains supply will cause the lamp to run at its designed rating? Why is capacitance preferable to resistance for this purpose?

*5.8 What average electrical power is delivered by a 50 Hz, 3 V r.m.s. source of negligible internal impedance to

(a) a 100 μF capacitor having 100 kΩ leakage resistance
(b) an inductor of 127 mH inductance and 30 Ω resistance?

Also evaluate the power factor in each case.

6.1 A primary coil of inductance 10 mH is magnetically coupled to a secondary coil of inductance 160 mH such that there is mutual inductance of 30 mH between them. Assuming that losses can be neglected, find the voltage and current transformation ratio between the secondary and primary when the secondary is loaded with a magnetically screened inductance of 80 mH.

*6.2 A closely coupled transformer of negligible loss is to be used to match a 12 Ω resistive load to a 1 kHz source of 10 V r.m.s. e.m.f. and 300 Ω internal resistance. If the inductance of the primary winding to which the source is to be connected is chosen to be 500 mH, what must be the inductance of the secondary? Also deduce the current and average power that the primary will draw from the source (a) in the matched condition and (b) with the secondary open-circuit.

6.3 A pair of identical circuits that separately resonate at a frequency of 100 kHz comprise coils of 1 mH inductance and 6.3 Ω resistance connected in series with appropriate capacitors of negligible loss. Estimate the degree of magnetic coupling that needs to be introduced between the coils of these two circuits in order to obtain a secondary response with a bandwidth of 6 kHz. When coupled to this extent, what is the ratio between the maximum amplitude of secondary current and that at 100 kHz?

7.1 A certain inductor is connected in series with a 100 Ω standard resistor and the output of a tunable sinusoidal oscillator. Measurements show that, when the frequency is adjusted to 400 Hz, the r.m.s. potential differences across the inductor, standard resistor and terminals of the oscillator are 1.26 V, 1.73 V and 2.45 V respectively. Deduce the circuit parameters of the inductor from this data.

7.2 Deduce the balance conditions for the Hay bridge shown in the figure. Hence show how it may be arranged to operate as a linear frequency bridge.

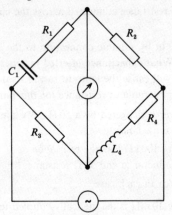

*7.3 The Anderson bridge shown in the figure constitutes a modification of the Wheatstone form that may be used to determine the inductance L and resistance R_L of an inductor. Find the balance conditions and indicate which components are most suitable for varying to achieve balance in such inductor determinations.

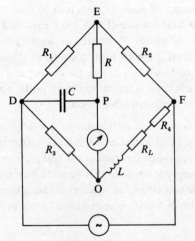

Problems

8.1 Design a simple single-stage R–C attenuator that will reduce the amplitude of an input signal by a factor of 100. The attenuation must be maintained constant to within about 1% irrespective of signal frequency as the load changes exhibiting resistance down to 1 kΩ and capacitance up to 100 pF.

8.2 Calculate the component values of a purely resistive bridged-T attenuator that, when inserted between a 600 Ω load resistance and any signal source, will introduce attenuation of 12 dB.

*8.3 A simple, first-order, low-pass, R–C filter is required to filter signals from a load resistance that may become as small as 1 kΩ. If the specification demands that the 3 dB point in the frequency response occurs at 6 kHz, that the attenuation never exceeds 0.5 dB in the low-frequency limit and that the highest possible input impedance is exhibited at all times, what values must the components of the filter take?

8.4 Design a twin-T, R–C, rejection filter for rejecting 1 kHz signals. The output of the filter is to feed into a 10 kΩ load and 80% of the input signal voltage must be transmitted to the output in the low-frequency limit.

9.1 What must the components of a constant-k, low-pass, ladder filter be in order that it will cut off at a frequency of 15 kHz and be suitable for operation into a terminating load resistance of 75 Ω? If such a filter comprises three sections, what attenuation will it introduce at 30 kHz and by how long will it delay a 10 ms pulse?

*9.2 Draw the circuit diagram of a T-section of a constant-k, band-pass filter that passes signals between 22.5 and 27.5 kHz and is properly terminated by 50 Ω resistance.

*9.3 Design and draw the circuit diagram of a Π-type, m-derived, high-pass filter section that is properly terminated by 600 Ω resistance, has a cut-off frequency of 10 kHz and resonates 2% below this frequency.

9.4 A certain low-pass, constant-k, T-type, filter section has a cut-off frequency of 800 Hz and is designed to operate into a fixed load resistance of 75 Ω. Deduce the circuit diagram of an m-derived, T-type, half-section that will significantly improve the termination when interposed between the output of the constant-k section and 75 Ω load resistance.

9.5 Find the image impedances of the network shown in terms of the inductance L and capacitance C at pulsatances $(1/LC)^{\frac{1}{2}}$, $(2/LC)^{\frac{1}{2}}$, $(2.5/LC)^{\frac{1}{2}}$ and $(4/LC)^{\frac{1}{2}}$.

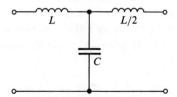

*9.6 Measurements on a 1.39 m sample length of low-loss, radio-frequency, transmission cable show that, at a frequency of 100 MHz, its input impedance under short-circuit termination corresponds to reactance of 137.4 Ω while its input impedance under open-circuit termination corresponds to reactance of 18.2 Ω. In the former case the potential difference leads the current by 90° in phase whereas in the latter case it lags behind the current by 90°. As the frequency is increased, these particular imaginary impedances first recur when the frequency reaches 172 MHz. Deduce the inductance and capacitance parameters of the cable from these results. What delay would a short pulse of around 100 ns duration experience in passing along a 15 m length of such cable?

*9.7 A voltage standing wave ratio of 4 exists on a very low-loss transmission line. Measuring from the terminated end of the line, the first and second nodes occur at distances of 16 cm and 96 cm. Where should an appropriate shunt stub be connected in order to remove the standing wave on the source side of the connection and thereby correctly terminate up to it?

*10.1 Below is shown the circuit diagram of a simple common-emitter amplifier together with typical output characteristics for the bipolar junction transistor involved. Apply the technique of load-line analysis to find the output operating point and the amplitude of the output signal potential difference developed across the load. It may be assumed that $V_{be} = 0.6$ V, that the reactances of the coupling capacitors are negligible and that the input resistance of the transistor is very much less than 118 kΩ.

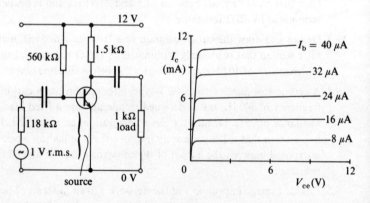

10.2 The small-signal behaviour of a certain four-terminal network is described in terms of h-parameters by $h_i = 3$ kΩ, $h_f = 200$ and $h_o = 2 \times 10^{-5}$ Ω^{-1}, h_r being negligible. What r.m.s. signal current flows through a resistance of 100 kΩ connected across the output terminals when a signal source of e.m.f. 1.5 mV r.m.s. and internal resistance 2 kΩ is connected between the input terminals. How much gain is there in signal voltage and signal power between the input and output?

10.3 At low frequencies an amplifier exhibits voltage gain of 150 and signal inversion between its input terminal pair I and C and output terminal pair O and C. Unfortunately, although the output resistance is very low and the input resistance quite high at 1 MΩ, there is capacitance of 10 pF between the output terminal O and input terminal I. At what frequency will the output of the amplifier be 3 dB down from its low-frequency level given that (a) the input is connected to a source of variable frequency but constant amplitude and 1 kΩ internal resistance and (b) the fall in gain at high frequency is solely due to the capacitance of 10 pF.

10.4 Series-voltage feedback is applied round an amplifier of nominal open-loop gain 2000 such that the feedback fraction is $\frac{1}{40}$. What are the magnitudes of the loop gain and closed-loop gain and how are the input and output impedances affected by the feedback? If the feedback fraction is held constant to better than 0.1% in a batch of such amplifiers but the open-loop gain shows $\pm 50\%$ variation, how much variation does the closed-loop gain exhibit?

*10.5 Design an operational amplifier circuit that will integrate a 100 Hz square-wave signal corresponding to instantaneous switching between potential differences of $+0.1$ V and -0.1 V. Make the input resistance of the circuit equal to 10 kΩ, allow for the saturation levels of the operational amplifier being as limited as ± 5 V and take the open-loop gain of the operational amplifier to be 10^5 at 100 Hz.

10.6 Select appropriate feedback components with which to convert an operational amplifier into

(a) a 2 kHz Wien bridge oscillator

(b) a Schmitt trigger with switching levels of 3 V and 6 V.

In the latter case assume that the bias supplies and saturation levels of the operational amplifier are ± 15 V and ± 14 V respectively.

*11.1 Obtain the Fourier spectrum of the time-base of a cathode-ray oscilloscope. Assume that the deflecting potential rises linearly with time and that the fly-back is instantaneous.

11.2 A certain signal is described by the function $h \cos(2\pi t/T)$ in the time interval $t = -T/4$ to $t = +T/4$ but is zero at all other times. Find the equivalent Fourier spectrum of this pulse and hence estimate the bandwidth that an amplifier would need to possess in order to amplify it without significant distortion.

11.3 Obtain the Laplace transforms of the following functions of time t:
(a) $3t$, (b) $\sin(\omega t + \phi)$, (c) $\sinh \alpha t$, (d) $(\exp -\alpha t) \sin \omega t$.

11.4 Find the inverse Laplace transforms of the following 's' functions:
(a) $2/s$, (b) $5/(s+2)$, (c) $6/(s^2+2s+10)$, (d) $1/s^2(s+1)$.

*11.5 A resistor and capacitor are connected in series with a source of negligible internal impedance that delivers e.m.f. $h(1-t/T)$ at times t in the interval $0-T$ but zero e.m.f. at all other times. Apply the Laplace transformation technique to determine the time dependence of the potential difference across the capacitor. It may be assumed that the capacitor and resistor are ideal components and that the capacitor is uncharged prior to time $t=0$. If the time constant of the circuit happens to be precisely T, what is the potential difference across the capacitor at time $t=2T$?

*12.1 Design a low-pass Butterworth filter that will cut off at 3 kHz with 60 dB per decade roll-off when inserted between a constant-current source and a 500 Ω resistive load.

*12.2 A filter is required to synthesise second-order, Chebyshev, high-pass response with 2 dB ripple in the pass band and a cut-off frequency of 300 Hz when it is connected between a constant-voltage source and a 1 kΩ load resistance. Design a network to meet this specification.

12.3 It is required to implement band-pass filtering between a source of extremely low impedance and a 100 Ω resistive load. Given that the desired cut-off frequencies are 8 kHz and 10 kHz, develop a suitable filter from a second-order, Butterworth, low-pass filter.

ANSWERS

1.1 At the centre of the square when all the point charges are positive, the field is zero and the potential is 200 V with respect to infinity. Changing the sign of the charges at the ends of one side makes the potential zero and the field 2000 V m^{-1} directed towards that side.

2.1 The mobility is 500 cm^2 V^{-1} s^{-1} and the mean free time is 10^{-13} s.

2.2 (a) 113.5 Ω, (b) 2.5 MΩ and (c) 55 GΩ.

2.3 The source is an e.m.f. of 1.5 V in series with 0.5 Ω resistance *or* a constant current of 3 A in parallel with 0.5 Ω resistance. Maximum power of 1.125 W is developed in R when $R = 0.5$ Ω.

3.1 1.6 kΩ.

3.2 10 kΩ.

3.3 $\frac{1}{8}$ mA from X to A.

3.4 $V_{AC} = 7.6$ V; $V_{BC} = 5.2$ V.

3.5 0.995 μA from the lower to upper terminal.

3.6 The Norton equivalent circuit comprises a constant current source, delivering 1.8 mA from B towards A, in parallel with 4.8 kΩ resistance.

3.7 The unknown resistance is connected in series with a suitable direct supply and the ammeter. Less error arises if the voltmeter is connected directly across the unknown resistance than if the voltmeter is connected across the series combination of ammeter and unknown resistance. The apparent resistance indicated by dividing the voltmeter reading by the ammeter reading in the better arrangement will be about 2% low on account of the small current that flows through the voltmeter.

3.8 Switching occurs 80 ms and 140 ms after the e.m.f. commences increasing. The time dependence of the potential difference across the diode is shown in the solution.

4.1 4 mH; $1\frac{2}{29}$ nF.

4.2 The initial electrical energy equals 66 μJ. When connected in parallel, the common potential difference becomes 6 V and the stored electrical energy 54 μJ. The lost electrical energy has been transformed into heat energy in the connecting leads.

4.3 9.09 kHz.

4.4 Minimum coupling capacitance is around 40 μF. Maximum capacitance for reasonable differentiation is about 1 nF. The signal in the load becomes too small if the capacitance becomes too small.

4.5 With 40 Ω series resistance, the oscillatory frequency f_0 is 15.6 kHz while the decrement δ is 3.6. With total series resistance of 160 Ω, $f_0 = 9.55$ kHz, $\delta = 4346$. Total series resistance of 200 Ω just prevents oscillation.

5.1 (a) $a/\sqrt{2}$, (b) $a/2$, (c) $[(a^2/2) + b^2]^{\frac{1}{2}}$, (d) $a/\sqrt{3}$.

5.2 The amplitude of the current is 0.255 A and the current lags the e.m.f. by 45° 4′.

5.3 $R_C = 400$ kΩ, $C = 1.59$ nF.

5.4 Resonant frequency = 53.4 kHz, $Q = 41.2$. The current lags or leads the e.m.f. by 39.5° depending on whether the frequency is 1% greater or less than the resonant frequency.

5.5 Inductance of inductor is 4.7 mH, capacitance of capacitor is 4.7 nF and net series resistance of inductor, capacitor and resistor is 47 Ω.

5.6 Capacitance of 1.13 nF should be connected across a series combination of the coil and 20.2 Ω resistance. The load resistance that halves the Q-factor is the parallel resistance of the completely parallel equivalent at resonance, which is 14.115 kΩ.

5.7 Required capacitance is 2.92 μF. Capacitance is preferable for controlling the power because no power is dissipated in it. A series resistance would waste electrical power.

5.8 (a) Average power = 90 μW; power factor = 0.00032
    (b) Average power = 108 mW; power factor = 0.6.

6.1 The secondary to primary current ratio is $-\frac{1}{8}$ while the secondary to primary voltage ratio is 1.6.

6.2 The secondary inductance must be 20 mH. When matched, the primary current is about 16.7 mA r.m.s. while the power delivered to the primary terminals is 83 mW. With the secondary open circuit, no average power is delivered to the primary terminals but a primary current of about 3.2 mA r.m.s. flows.

6.3 It is necessary for the coefficient of coupling to be about 0.06. With this coupling, the ratio between the maximum amplitude of secondary current and that at 100 kHz is about 3.1.

7.1 The parameters of the inductor are $R=23.8\,\Omega$, $L=27.4\,\text{mH}$.

7.2 The balance conditions are
$$R_1 R_4 + L_4/C_1 = R_2 R_3$$
$$\omega^2 L_4 C_1 R_1 = R_4$$
Fixing the magnitudes of L_4, C_1, R_2 and R_3 renders the balance conditions expressible as
$$R_1 = K_1/R_4$$
$$f = \omega/2\pi = K_2 R_4$$
where K_1 and K_2 are constants. If balance is achieved through adjustment of R_1 and R_4 then the frequency is obtained as $K_2 R_4$.

7.3 The balance conditions are
$$R_1/R_2 = R_3/(R_4 + R_L)$$
$$L = CR_3[R_2 + (1 + R_2/R_1)R]$$
To achieve balance it is most convenient to vary R_4 and R.

8.1 The required circuit is that of figure 8.1(d) with $R_1 = 10\,\Omega$, $R_2 = 1\,\text{k}\Omega$, $C_1 = 10\,\text{nF}$ and $C_2 = 100\,\text{pF}$.

8.2 With reference to figure 8.2(e), the component values of the attenuator are $R_L = 600\,\Omega$, $R_1 = 201.3\,\Omega$ and $R_2 = 1789\,\Omega$.

8.3 The required filter is that of figure 8.3(a) with $R = 59\,\Omega$, $C = 0.45\,\mu\text{F}$.

8.4 The required filter is that of figure 8.10(a) with $R = 1.25\,\text{k}\Omega$, $C = 0.127\,\mu\text{F}$.

9.1 The form of the constant-k, low-pass, ladder filter is shown in figures 9.3(a), (c) and (d) and its components must be $L = 1.59\,\text{mH}$, $C = 0.283\,\mu\text{F}$. At 30 kHz the modulus of the transfer function of three sections is 0.00037 which corresponds to attenuation of signal power by 68.6 dB. The duration of a 10 ms input pulse is long enough for all appreciable Fourier components to be of much lower frequency than the cut-off frequency. Consequently the entire pulse is transmitted essentially undistorted and with a delay of $6/2\pi f_c = 63.7\,\mu\text{s}$.

9.2 The required circuit diagram is that of figure 9.6(a) with $L_1 = 3.18\,\text{mH}$, $L_2 = 32.1\,\mu\text{H}$, $C_1 = 12.8\,\text{nF}$ and $C_2 = 1.27\,\mu\text{F}$.

9.3 The parallel arms must each comprise 47.9 mH inductance. The series arm comprises 3.95 mH inductance in parallel with 2.64 nF capacitance.

9.4 With reference to figure 9.11(b), the series arm contains 8.9 mH inductance while the parallel arm has 15.9 mH inductance in series with 1.59 μF capacitance.

9.5 The image impedances for the left and right-hand terminal pairs at pulsatances $(1/LC)^{\frac{1}{2}}$, $(2/LC)^{\frac{1}{2}}$, $(2.5/LC)^{\frac{1}{2}}$ and $(4/LC)^{\frac{1}{2}}$ are respectively zero impedance and infinite resistance, infinite reactance and zero impedance, $(3L/2C)^{\frac{1}{2}}$ resistance and $(L/24C)^{\frac{1}{2}}$ resistance, $(3L/2C)^{\frac{1}{2}}$ reactance and $(L/6C)^{\frac{1}{2}}$ reactance.

9.6 The inductance parameter is $0.25\,\mu\text{H m}^{-1}$, the capacitance parameter is $100\,\text{pF m}^{-1}$ and the delay along 15 m of cable is 75 ns.

9.7 Possible positions to connect the appropriate shunt stub are 44.2, 67.8, 124.2, 147.8 cm etc. from the terminated end. Clearly the greatest amount of line is correctly terminated by connecting the appropriate shunt stub 44.2 cm from the terminated end of the line.

10.1 The operating point is approximately $V_{ce} = 4.3$ V, $I_c = 5.0$ mA and the amplitude of the output signal potential difference developed across the load is about 1.9 V.

10.2 There is output signal current amounting to $20\,\mu\text{A}$ r.m.s. The signal voltage gain is 2.22×10^3 while the signal power gain is 1.48×10^5.

10.3 Three decibels fall occurs in the output at a frequency of 105 kHz.

10.4 The nominal loop gain is 50 and the nominal closed-loop gain is 39.22. The series-voltage feedback respectively increases the input impedance and decreases the output impedance by the factor 51. As the open-loop gain varies by $\pm 50\%$, the closed-loop gain ranges from 38.46 to 39.47, a total variation of about 2.5%. By comparison 0.1% change in the feedback fraction only makes about 0.1% difference to the closed-loop gain which can be neglected.

10.5 The required circuit is that of figure 10.11(a) with $R = 10\,\text{k}\Omega$, $C > 5\,\text{nF}$. Drift is reduced by connecting, say, $4.7\,\text{M}\Omega$ in parallel with C.

10.6 (a) With reference to figure 10.13 showing the required feedback arrangement, appropriate component values might be $C = 36\,\text{nF}$, $R = R_1 = 2.2\,\text{k}\Omega$ and $R_2 = 3.9\,\text{k}\Omega$ in series with $1\,\text{k}\Omega$ variable. The variable resistor would need adjusting for the nearest approach to sinusoidal oscillation. To achieve $C = 36\,\text{nF}$, capacitor tolerance would have to be borne in mind and combinations of capacitors, possibly including trimmers, would be needed. Also note that for a stable amplitude of oscillation and really low distortion, the negative feedback path R_1, R_2 would have to be modified to incorporate some form of automatic gain control.

(b) With reference to figure 10.14(b) showing the required feedback arrangement, appropriate resistor values might be $R_1 = 5.1\,\text{k}\Omega$, $R_2 = 10\,\text{k}\Omega$ and $R_3 = 28\,\text{k}\Omega$. In any event, the resistor values should be in this ratio.

11.1 If T and V_0 are respectively the period and peak-to-peak variation of the time-base, it may be represented by

$$V_0\left[\frac{1}{2} - \frac{1}{n\pi}\sin(2\pi nt/T)\right]$$

11.2 In the equivalent Fourier spectrum, the continuous amplitude distribution as a function of frequency f is

$$\frac{hT}{4}\left[\frac{\sin(Tf-1)\pi/2}{(Tf-1)\pi/2} + \frac{\sin(Tf+1)\pi/2}{(Tf+1)\pi/2}\right]$$

For an amplifier to amplify the pulse without significant distortion, its bandwidth would need to extend from zero frequency to at least $\sim 3/T$.

11.3 (a) $3/s^2$, (b) $(s\sin\phi + \omega\cos\phi)/(s^2+\omega^2)$, (c) $\alpha/(s^2-\alpha^2)$, (d) $\omega/[(s+\alpha)^2+\omega^2]$.

11.4 (a) 2, (b) $5\exp-2t$, (c) $2\exp-t\sin 3t$, (d) $t-1+\exp-t$.

11.5 In terms of the resistance R and capacitance C of the two series components, the potential difference across the capacitor is given by

$$h\left\{\left[\left(1+\frac{RC}{T}\right)\left(1-\exp-\frac{t}{RC}\right)-\frac{t}{T}\right]U(t) \right.$$
$$\left. +\left[\left(\frac{t-T}{T}\right)-\frac{RC}{T}\left(1-\exp-\frac{(t-T)}{RC}\right)\right]U(t-T)\right\}$$

or

$$h\left[\left(1+\frac{RC}{T}\right)\left(1-\exp-\frac{t}{RC}\right)-\frac{t}{T}\right] \quad \text{when } 0 \leqslant t \leqslant T$$

$$h\left[\left(1+\frac{RC}{T}\right)\left(1-\exp-\frac{T}{RC}\right)-1\right]\left[\exp-\frac{(t-T)}{RC}\right] \quad \text{when } t \geqslant T$$

If $RC = T$, then when $t = 2T$

$$V_C = h[(\exp-1)-(2\exp-2)] \approx 0.097h$$

12.1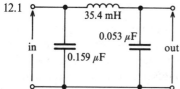

12.2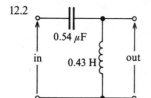

12.3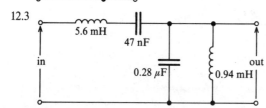

SOLUTIONS

1.1 The vector fields $\mathbf{E}_1$, $\mathbf{E}_2$, $\mathbf{E}_3$ and $\mathbf{E}_4$ at the centre of the square, due to the individual positive point charges q_1, q_2, q_3 and q_4, each have magnitude

$$\left[\frac{1.414 \times 10^{-9}}{4\pi \times 3.6 \times (10^{-9}/36\pi) \times 0.1^2 \times 0.5}\right] \text{V m}^{-1} = 707 \text{ V m}^{-1}$$

but, as illustrated in the figure, are oppositely directed in pairs so that the resultant field is zero. The scalar potential is simply the sum of that due to each charge and is therefore

$$\left[4 \times \frac{1.414 \times 10^{-9}}{4\pi \times 3.6 \times (10^{-9}/36\pi) \times (0.1/1.414)}\right] \text{V} = 200 \text{ V}$$

with respect to that at infinity.

Changing the sign of charges q_3 and q_4, say, makes their contribution to the potential opposite in sign so that the net potential at the centre of the square becomes zero. It also reverses the directions of $\mathbf{E}_3$ and $\mathbf{E}_4$ so that the resultant field is directed towards and perpendicular to the side on which the reversed charges are located. The magnitude of the field is now

$$4 \times 707 \cos 45 = 2000 \text{ V m}^{-1}.$$

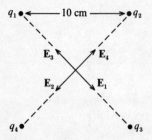

Solutions

2.3 Let the source be an e.m.f. $\mathscr{E}$ in series with internal resistance r. Then

$$\mathscr{E} = (r+1) \times 1$$
$$\mathscr{E} = (r+4.5) \times 0.3$$

simultaneous solution of which gives $0.7r = (4.5 \times 0.3) - 1$ or $r = 0.5\,\Omega$ and $\mathscr{E} = 1.5\,\text{V}$. This is equivalent to a source of $(1.5/0.5)\,\text{A} = 3\,\text{A}$ in parallel with $0.5\,\Omega$. Maximum power develops in the load in the matched condition, that is, when $R = r = 0.5\,\Omega$. In this condition the power is $[(1.5/2)^2/0.5]\,\text{W} = 1.125\,\text{W}$.

3.2 With an e.m.f. imposed or some current delivered between a pair of diagonally opposite corners, the network features seven independent node pairs or seven independent meshes. Although either mesh or node-pair analysis yields the required resistance, it is cumbersome and tedious unless advantage is taken of the symmetry associated with the disposition of the equal resistances. If current I is delivered between opposite corners, it divides equally into $I/3$ along each cube edge adjacent to the input and output corner. Each of these current components then divides equally at the next corner encountered, as shown in the figure. Thus the required resistance is

$$[(I/3) + (I/6) + (I/3)](12/I)\,\text{k}\Omega = (5/6)12\,\text{k}\Omega = 10\,\text{k}\Omega.$$

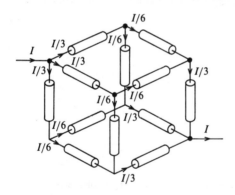

3.3 Let the clockwise mesh currents in milliamps in meshes ABX, BCX and CAX be denoted by I_1, I_2 and I_3 respectively. Kirchhoff's current law applied to meshes ABX and ABC gives

$$3I_1 - I_2 - I_3 = 0$$

and

$$I_1 + I_2 + I_3 = 3$$

respectively while the constant current source in branch CX forces

$$I_2 - I_3 = 1$$

Simultaneous solution of these three equations yields

$$I_1 = \tfrac{3}{4}, \quad I_2 = 1\tfrac{5}{8}, \quad I_3 = \tfrac{5}{8}$$

so that the current in branch XA is

$$(I_1 - I_3) = \tfrac{1}{8}$$

that is $\tfrac{1}{8}$ mA from X towards A.

The current delivered by the 3 V e.m.f. acting alone is found by open-circuiting branch XC so that it amounts to

$$[3/(1+1+\tfrac{2}{3})] \text{ mA} = \tfrac{9}{8} \text{ mA}$$

Of this, $\tfrac{1}{3}$ flows along AX. The current delivered along XA by the 1 mA source acting alone is found by shorting out the 3 V e.m.f. From symmetry it is $\tfrac{1}{2}$ mA. Hence the total current along XA with both sources acting is, by the principle of superposition,

$$(\tfrac{1}{2} - \tfrac{3}{8}) \text{ mA} = \tfrac{1}{8} \text{ mA}$$

from X towards A.

3.8 The e.m.f. increases by 0.2 V every 20 ms until it reaches 1 V following which it falls to zero at the same steady rate. Since the slope of the load line is $-(1/100)\,\Omega^{-1} = -10$ mA V^{-1}, the load line at 20 ms intervals is as shown in the first graph. From the intersections of the load line with the characteristic, the time dependence of the potential difference across the diode is as shown in the second graph. Switching clearly occurs at 80 ms and 140 ms after the e.m.f. commences increasing.

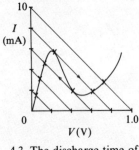
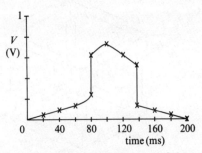

4.3 The discharge time of the capacitor is negligible because of the effective short circuit. During charging, the potential difference V across the capacitor is related to time t by

$$V = \mathscr{E}[1 - \exp(-t/RC)]$$

where $\mathscr{E}$ is the e.m.f., R the resistance and C the capacitance. It follows that the time $t_2 - t_1$ to charge from potential difference V_1 to V_2 is given by

$$t_2 - t_1 = RC \ln \frac{\mathscr{E}}{\mathscr{E} - V_2} - RC \ln \frac{\mathscr{E}}{\mathscr{E} - V_1} = RC \ln \frac{\mathscr{E} - V_1}{\mathscr{E} - V_2}$$

Now $RC = 10 \times 10^3 \times 10 \times 10^{-9}$ s $= 100\,\mu$s and $(\mathscr{E} - V_1)/(\mathscr{E} - V_2) = 21/7 = 3$. Hence $t_2 - t_1 = 110\,\mu$s and the frequency of oscillation is 9.09 kHz.

4.4 For the amplitude of the coupled signal to be essentially independent of frequency, the capacitive reactance has to be negligible compared with the total series resistance. Thus it is required that $1/\omega C \ll 1\,\text{k}\Omega$. This is most demanding at the lowest frequency of 40 Hz and the capacitance C must accordingly satisfy

$$C \gg (2\pi \times 40 \times 10^3)^{-1}\,\text{F} \approx 3.98\,\mu\text{F}$$

Applying a 'rule of thumb' of ten times being sufficiently greater than, C should be about $40\,\mu\text{F}$. The next, higher, readily available, preferred value of $47\,\mu\text{F}$ would normally be used.

For reasonable differentiation $1/\omega C \gg 1\,\text{k}\Omega$ is needed which is most demanding at the highest frequency. Accordingly the capacitance C should satisfy

$$C \ll (2\pi \times 15 \times 10^3 \times 10^3)^{-1}\,\text{F} \approx 10.6\,\text{nF}$$

and a 1 nF capacitor would be appropriate. Making the capacitance even smaller would render the output signal unnecessarily small in amplitude.

5.3 The impedance of the component is

$$Z = (R_C/\text{j}\omega C)/(R_C + 1/\text{j}\omega C) = R_C/(1 + \text{j}\omega R_C C)$$

Thus with reference to the diagram

$$\mathbf{V} = Z\mathscr{E}/(10^5 + Z)$$

or

$$\mathbf{V}/\mathscr{E} = R_C/(R_C + 10^5 + \text{j}\omega 10^5 R_C C)$$

When $\omega \to 0$, $\mathbf{V}/\mathscr{E} \to 4/5$. Therefore

$$R_C/(R_C + 10^5) = 4/5$$

or

$$R_C = 400\,\text{k}\Omega.$$

In order for $|\mathbf{V}/\mathscr{E}|$ to be 0.01, $1/\omega C \ll 100\,\text{k}\Omega$ which means that $1/\omega C \ll R_C$ also. Thus to a good-enough approximation

$$(2\pi \times 10^5 \times 10^5 C)^{-1} = 0.01$$

and C is $(2\pi \times 10^8)^{-1}\,\text{F}$ or 1.59 nF.

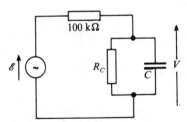

5.5 Let the inductance, capacitance and resistance of the series combination of inductor, capacitor and resistor be L, C and R as usual. The magnitude of the resonant current shows that

$$R + 11.8 = (17 \times 10^{-3})^{-1} \, \Omega$$

or $R = 47 \, \Omega$. The frequencies at which resonance occurs and half of maximum power is dissipated establishes that

$$Q = f_r/(f_2 - f_1) = 33.9/2 = 16.95 = (1/58.8)(L/C)^{\frac{1}{2}}$$

or

$$(L/C)^{\frac{1}{2}} = 997 \, \Omega$$

and that

$$1/(LC)^{\frac{1}{2}} = 2\pi \times 33.9 \times 10^3 \text{ rad s}^{-1}$$

Consequently

$$L = 997(2\pi \times 33.9 \times 10^3)^{-1} \text{ H} = 4.7 \text{ mH}$$
$$C = (997 \times 2\pi \times 33.9 \times 10^3)^{-1} \text{ F} = 4.7 \text{ nF}$$

From the information given it cannot be determined just how L, C and R are distributed between the inductor, capacitor and resistor.

5.8 (a) Because, on average, power is not developed in the capacitance but only in the resistance, the average power is $(3^2/10^5)\text{W} = 90 \, \mu\text{W}$.

Note that it is possible to convert the capacitance and parallel resistance to an equivalent capacitance and series resistance and apply the formula, average power $= \mathscr{E}_{\text{rms}} I_{\text{rms}} \cos \phi$. However, such conversion is tedious and unnecessary.

The complex impedance of parallel capacitance C_p and resistance R_p is $R_p/(1 + j\omega C_p R_p)$ and the r.m.s. current I_{rms} driven through it by a source e.m.f. $\mathscr{E}_{\text{rms}}$ is $(1 + \omega^2 C_p^2 R_p^2)^{\frac{1}{2}} \mathscr{E}_{\text{rms}}/R_p$. Now

$$\text{Average power} = \mathscr{E}_{\text{rms}} I_{\text{rms}} \cos \phi = \mathscr{E}_{\text{rms}}^2 / R_p$$

Hence the power factor $\cos \phi$ is given by

$$\cos \phi = \mathscr{E}_{\text{rms}}/R_p I_{\text{rms}} = 1/(1 + \omega^2 C_p^2 R_p^2)^{\frac{1}{2}}$$

where

$$\omega C_p R_p = 2\pi \times 50 \times 100 \times 10^{-6} \times 10^5 = 1000\pi$$

To a good approximation, then

$$\cos \phi = 1/1000\pi = 0.00032$$

(b) Average power is

$$P_{\text{av}} = R I_{\text{rms}}^2 = 30 \times \left[\frac{3^2}{30^2 + (2\pi \times 50 \times 0.127)^2} \right] \text{mW} = 108 \text{ mW}$$

and so

$$\cos \phi = P_{\text{av}}/\mathscr{E}_{\text{rms}} I_{\text{rms}} = 108 \times 10^{-3}/(3 \times 3 \div 49.9) = 0.6$$

Alternatively, from the formula for $\cos \phi$ in terms of R and L

$$\cos \phi = R/(R^2 + \omega^2 L^2)^{\frac{1}{2}} = \left[1 + \left(\frac{2\pi \times 50 \times 0.127}{30} \right)^2 \right]^{-\frac{1}{2}} = 0.6$$

so that the average power is

$$\mathscr{E}_{\text{rms}} I_{\text{rms}} \cos \phi = 108 \text{ mW}$$

6.2 The square of the turns ratio between the secondary and primary must equal $12/300 = 1/25$ and the secondary inductance must be $(500/25)\,\text{mH} = 20\,\text{mH}$.

(a) In the matched condition, the reflected resistance of $300\,\Omega$ in the primary is in parallel with the primary inductive reactance of $2\pi \times 10^3 \times 0.5\,\Omega = \pi\,\text{k}\Omega$. Neglecting current through the inductive reactance compared with that through the reflected resistance, the r.m.s. current drawn from the source is $(10/600)\,\text{A} \approx 16.7\,\text{mA}$ and the power delivered to the primary terminals is $(5^2/300)\,\text{W} \approx 83\,\text{mW}$. All of this power is dissipated in the reflected load which is equivalent to saying that it is dissipated in the secondary load.

(b) With the secondary open-circuit, no power is delivered to the primary terminals because there is only inductive reactance between them. Some power is, however, developed in the internal resistance of the source by virtue of the primary current that flows through it. The complex primary circuit impedance amounts to $(300 + \text{j}2\pi \times 10^3 \times 0.5)\Omega = (300 + \text{j}\pi \times 10^3)\,\Omega$. It is a good approximation here to neglect the resistive component of the impedance compared with the reactive component and so the r.m.s. primary current is $[10/(\pi \times 10^3)]\,\text{A} \approx 3.2\,\text{mA}$.

7.3 Let the potentials of nodes D, E and F with respect to node O be denoted by phasors $\mathbf{V}_D$, $\mathbf{V}_E$ and $\mathbf{V}_F$ respectively. At balance, because there is neither current nor potential difference between nodes O and P, application of Kirchhoff's current law to nodes E, P and O respectively yields

$$\frac{\mathbf{V}_E - \mathbf{V}_D}{R_1} + \frac{\mathbf{V}_E}{R} + \frac{\mathbf{V}_E - \mathbf{V}_F}{R_2} = 0$$

$$\frac{\mathbf{V}_E}{R} + \text{j}\omega C \mathbf{V}_D = 0$$

$$\frac{\mathbf{V}_D}{R_3} + \frac{\mathbf{V}_F}{R_4 + R_L + \text{j}\omega L} = 0$$

Eliminating any two of $\mathbf{V}_D$, $\mathbf{V}_E$ and $\mathbf{V}_F$ between these equations leads to

$$\left(\frac{1}{R_1} + \frac{1}{R_2} + \frac{1}{R}\right) + \frac{1}{\text{j}\omega C R_1 R} - \frac{R_4 + R_L + \text{j}\omega L}{\text{j}\omega C R_2 R_3 R} = 0$$

or upon equating the real and imaginary parts, to balance conditions

$$R_1/R_2 = R_3/(R_4 + R_L)$$
$$L = CR_3[R_2 + (1 + R_2/R_1)R]$$

Since resistances R_1, R_2 and R_3 feature in both balance conditions, to independently satisfy them when determining R_L and L, two of R_4, R

and C must be varied. The greater ease of varying resistance means that it is most sensible to balance through adjustment of R_4 and R.

8.3 The required filter circuit is shown in figure 8.3(a). In the low-frequency limit its transfer function is

$$\mathcal{T} = [R_L/(R_L+R)]$$

The low-frequency specification demands that when R_L falls to 1 kΩ

$$-0.5 \leqslant 10\log_{10}[R_L/(R_L+R)]^2 \leqslant 0$$

that is

$$0 \leqslant \log_{10}(R_L+R)/R_L \leqslant 0.025$$

or

$$1 \leqslant (R_L+R)/R_L \leqslant 1.059$$

For maximum input impedance at all frequencies, the largest value of R that satisfies this condition must be chosen. Thus $R = 59\,\Omega$.

The 3 dB point in the frequency response occurs when

$$10\log_{10}[1/(1+\omega^2C^2R'^2)] = -3$$

where R' represents the resistance of R in parallel with R_L. This is satisfied when $\omega CR' = 1$ and, since R' is close to 59 Ω as R_L varies, the required capacitance is

$$C = (2\pi \times 6 \times 10^3 \times 59)^{-1}\,\text{F} = 0.45\,\mu\text{F}$$

9.2 The required circuit diagram is that of figure 9.6(a), L_1, C_1, L_2 and C_2 taking appropriate values as follows. With reference to equations (9.48) and (9.56)

$$L_2/C_1 = L_1/C_2 = k^2 = 2500\,\Omega^2$$

while from equation (9.52)

$$\omega_1, \omega_2 = \frac{k}{L_1}\left[\left(1+\frac{L_1}{L_2}\right)^{\frac{1}{2}} \mp 1\right]$$

which yields

$$\omega_2 - \omega_1 = 2k/L_1$$

and

$$\omega_2 + \omega_1 = \frac{2k}{L_1}\left(1+\frac{L_1}{L_2}\right)^{\frac{1}{2}}$$

From the former

$$L_1 = \frac{k}{\pi(f_2-f_1)} = \frac{50}{\pi \times 5 \times 10^3}\,\text{H} = 3.18\,\text{mH}$$

and from the two combined

$$\left(1+\frac{L_1}{L_2}\right)^{\frac{1}{2}} = \frac{\omega_2+\omega_1}{\omega_2-\omega_1} = \frac{f_2+f_1}{f_2-f_1} = 10$$

so that

$$L_2 = L_1/99 = 32.1\,\mu\text{H}$$

Solutions

$$C_2 = L_1/k^2 = 1.27 \, \mu F$$
$$C_1 = L_2/k^2 = 12.8 \, nF$$

9.3 With the usual notation, the prototype must have $Z_1 = 1/j\omega C$, $Z_2 = j\omega L$ where L and C satisfy $\omega_c = 1/2(LC)^{\frac{1}{2}}$ and $Z_{k\Pi} = [L/C(1 - \omega_c^2/\omega^2)]^{\frac{1}{2}}$. Thus

$$1/(LC)^{\frac{1}{2}} = 4\pi \times 10^4 \, \text{rad s}^{-1}, \quad (L/C)^{\frac{1}{2}} = 600 \, \Omega$$

from which

$$L = 4.77 \, mH, \quad C = 13.25 \, nF$$

For the corresponding m-derived section, $1 - m^2 = (\omega_\infty/\omega_c)^2$ from which $m = 0.199$ and the required Π-section is as shown.

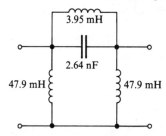

9.6 The impedance measurements reveal that in the notation of section 9.5

$$(j \tan \beta l)\left(\frac{L}{C}\right)^{\frac{1}{2}} = j137.4$$

$$(-j \cot \beta l)\left(\frac{L}{C}\right)^{\frac{1}{2}} = -j18.2$$

Hence
$$(L/C)^{\frac{1}{2}} = (137.4 \times 18.2)^{\frac{1}{2}} = 50 \, \Omega$$

and
$$\tan \beta l = (137.4/18.2)^{\frac{1}{2}} = 2.748 \quad \text{or} \quad \beta l = n\pi + 70°$$

But $\beta = \omega(LC)^{\frac{1}{2}}$, so that in terms of an integer m

$$2\pi \times 100 \times 10^6 \times (LC)^{\frac{1}{2}} \times l = m\pi + 70°$$
$$2\pi \times 172 \times 10^6 \times (LC)^{\frac{1}{2}} \times l = (m+1)\pi + 70°$$

from which

$$(LC)^{\frac{1}{2}} = (2 \times 72 \times 10^6 \times 1.39)^{-1} \, \text{s m}^{-1} \approx 5 \, \text{ns m}^{-1}$$

The delay through 15 m of cable is therefore $15 \times 5 \, \text{ns} = 75 \, \text{ns}$ and

$$L = 50 \times 5 \, \text{nH m}^{-1} = 0.25 \, \mu H \, m^{-1}$$
$$C = (5/50) \, nF \, m^{-1} = 100 \, pF \, m^{-1}$$

9.7 In the notation of section 9.5, nodes occur when

$$l - x = [\theta/\pi - (2n+1)]\lambda/4$$

Thus
$$\lambda/2 = 96 - 16 = 80 \, cm$$

and
$$16 = [\theta/\pi - 1]40 \quad \text{or} \quad \theta = 1.4\pi$$

But from equation (9.104), the impedance at any point of the line is

$$Z = \left[\frac{1 + \Gamma_0 \cos\phi + j\Gamma_0 \sin\phi}{1 - \Gamma_0 \cos\phi - j\Gamma_0 \sin\phi}\right](L/C)^{\frac{1}{2}}$$

where
$$\phi = 2\beta(x-l) + \theta$$
Hence

$$\mathcal{R}(Z) = \left[\frac{1 - \Gamma_0^2 \cos^2\phi - \Gamma_0^2 \sin^2\phi}{(1 - \Gamma_0 \cos\phi)^2 + \Gamma_0^2 \sin^2\phi}\right](L/C)^{\frac{1}{2}}$$

$$= \left[\frac{1 - \Gamma_0^2}{1 + \Gamma_0^2 - 2\Gamma_0 \cos\phi}\right](L/C)^{\frac{1}{2}}$$

from which it is seen that the real part of the impedance is equal to the characteristic impedance $(L/C)^{\frac{1}{2}}$ when

$$(1 - \Gamma_0^2)/(1 + \Gamma_0^2 - 2\Gamma_0 \cos\phi) = 1$$

or
$$\cos\phi = \Gamma_0$$

In this particular case, the voltage standing wave ratio is 4 and so, according to equation (9.100)

$$(1 + \Gamma_0)/(1 - \Gamma_0) = 4$$

and the real part of the impedance is equal to $(L/C)^{\frac{1}{2}}$ when

$$\cos\phi = \Gamma_0 = 0.6$$

or
$$\phi = \pm 0.927, \quad \pm 5.356 \text{ rad} \quad \text{etc.}$$

Because $l - x = (\theta - \phi)/2\beta = (\theta - \phi)\lambda/4\pi$, this means that the appropriate shunt stubs for tuning out the reactive component of the impedance must be connected 44.2 cm, 67.8 cm, 124.2 cm, 147.8 cm etc. from the terminated end of the line. Whichever of these points such stub tuning is carried out at, normally that closest to the terminated end of the line, the line will be correctly terminated up to it.

10.1 The direct, output, load line has intercept $V_{ce} = 12$ V and slope $-(1.5 \text{ k}\Omega)^{-1}$ so that its intercept on the I_c axis is 8 mA. Since $V_{be} = 0.6$ V, the direct base current is $I_b = [(12 - 0.6)/(560 \times 10^3)]$ A $\approx 20\,\mu$A. Estimating the characteristic for $I_b = 20\,\mu$A to be half-way between those for $I_b = 16\,\mu$A and $I_b = 24\,\mu$A, the intercept between this characteristic and the direct load line occurs at $V_{ce} = 4.3$ V, $I_c = 5$ mA, which is the operating point. Because coupling reactance can be neglected and the input resistance of the transistor is small compared with 118 kΩ, the input signal base current is virtually $(1/0.118)\,\mu$A r.m.s., that is, close to 12 μA in amplitude. Thus the base current swings between 8 μA and 32 μA while the slope of the signal load line is $-(1.5\,\text{k}\Omega\,\|\,1\,\text{k}\Omega)^{-1}=-(600\,\Omega)^{-1}$. Finding the intersections of this signal load line with the characteristics corresponding to $I_\text{b}=8\,\mu\text{A}$ and $I_\text{b}=32\,\mu\text{A}$, the output signal voltage is determined as swinging between about 2.4 V and 6.2 V, that is, the amplitude of the output signal potential difference is about 1.9 V.

10.5 The required form of circuit with the operational amplifier in the inverting configuration is that of figure 10.11(a). Making $R=10\,\text{k}\Omega$ achieves input resistance of 10 kΩ. Assuming that the input impedance of the operational amplifier is adequate, the fundamental requirement for integration is $ARC \gg 1/2\pi f$ or $C \gg (2\pi \times 10^2 \times 10^5 \times 10^4)^{-1}\,\text{F} \sim 1\,\text{pF}$. However, the output voltage swing is $(1/RC)\int_0^{5\times 10^{-3}} 0.1\,dt\,\text{V} = (5/10^4 RC)\,\text{V}$ and this has to be less than 10 V. Thus $C > [5/(10^4 \times 10^4 \times 10)]\,\text{F} = 5\,\text{nF}$ which is a more stringent condition on C. Hence make $C=10\,\text{nF}$, say; a larger capacitance also eases the drift problem. The reactance of C at 100 Hz is $[1/(2\pi \times 100 \times 10^{-8})]\,\Omega = (1/2\pi)\,\text{M}\Omega$. Connecting a resistor of, say, 4.7 MΩ in parallel with the capacitor does not interfere significantly with integration of the square wave but limits the direct gain and further eases the drift problem.

11.1 Let the period of the repetitive time base be $T=2\pi/\omega$ and choose the origin of time such that the time-base is represented by $V_0 t/T$ in the time interval $t=0$ to $t=T$, where V_0 is independent of t. In the notation of section 11.1

$$a_0 = \frac{1}{T}\int_0^T \frac{V_0 t}{T}\,dt = \frac{V_0}{2}$$

$$a_{n\neq 0} = 0$$

The latter result arises because, apart from its average value $V_0/2$, the time-base is an odd function of t. Again in the notation of section 11.1

$$b_n = \frac{2}{T}\int_0^T \left(\frac{V_0 t}{T}\right)\sin n\omega t\,dt$$

$$= \frac{2V_0}{(n\omega T)^2}\int_0^{2n\pi} x \sin x\,dx$$

$$= \frac{2V_0}{(2n\pi)^2}\left[(-x\cos x)_0^{2n\pi} + \int_0^{2n\pi}\cos x\,dx\right]$$

$$= -\frac{V_0}{n\pi}$$

Hence the time-base is equivalent to the Fourier spectrum

$$V_0\left[\frac{1}{2} - \frac{1}{n\pi}\sin(2\pi nt/T)\right]$$

11.5 Applying Kirchhoff's voltage law to the series circuit shows that, in terms of the resistance R and capacitance C, the potential difference V_C

across the capacitor is related to the time-dependent e.m.f. $\mathscr{E}$ by

$$RC\frac{dV_C}{dt}+V_C=\mathscr{E}$$

Since

$$\mathscr{E}=h(1-t/T)U(t)+h(t/T-1)U(t-T)$$

taking the Laplace transform of the differential equation yields

$$(RCs+1)\mathscr{L}V_C=h\left[\frac{1}{s}-\frac{1}{Ts^2}(1-\exp-Ts)\right]$$

or

$$V_C=\mathscr{L}^{-1}\left\{\frac{h}{RC}\left[\frac{1}{s(s+1/RC)}-\frac{1}{Ts^2(s+1/RC)}\right.\right.$$
$$\left.\left.+\frac{\exp-Ts}{Ts^2(s+1/RC)}\right]\right\}$$

Now

$$\mathscr{L}^{-1}\frac{1}{s(s+1/RC)}=\mathscr{L}^{-1}\left(\frac{1}{s}-\frac{1}{s+1/RC}\right)RC$$
$$=RC[1-\exp-t/RC]U(t)$$

$$\mathscr{L}^{-1}\frac{1}{s^2(s+1/RC)}=\mathscr{L}^{-1}\left(\frac{1}{s^2}-\frac{RC}{s}+\frac{RC}{s+1/RC}\right)RC$$
$$=RC[t-RC+RC\exp-t/RC]U(t)$$

and from the latter

$$\mathscr{L}^{-1}\frac{\exp-Ts}{s^2(s+1/RC)}=RC\left[(t-T)-RC+RC\exp-\frac{(t-T)}{RC}\right]U(t-T)$$

Thus

$$V_C=h\left\{\left[\left(1+\frac{RC}{T}\right)\left(1-\exp-\frac{t}{RC}\right)-\frac{t}{T}\right]U(t)\right.$$
$$\left.+\left[\left(\frac{t-T}{T}\right)-\frac{RC}{T}\left(1-\exp-\frac{(t-T)}{RC}\right)\right]U(t-T)\right\}$$

If $RC=T$, then when $t=2T$

$$V_C=h(-2\exp-2+\exp-1)\approx 0.097h$$

12.1 The filter must be third order to meet the roll-off requirement and, in view of the constant-current nature of the source, its form should be as indicated in the figure. Analysis shows that the transfer function of this network is

$$\mathscr{T}=\frac{\mathbf{I}_o}{\mathbf{I}_i}=[1+(C_1+C_2)Rs+LC_1s^2+LC_1C_2Rs^3]^{-1}$$

Consequently to synthesise the third-order Butterworth response represented by equation (12.17), it is necessary to arrange that

$$(C_1+C_2)R=2/\omega_c$$

$$LC_1 = 2/\omega_c^2$$
$$LC_1C_2R = 1/\omega_c^3$$

Inserting $\omega_c = 6\pi \times 10^3$ rad s^{-1}, $R = 500\,\Omega$ into these relations, the last two combine to give $C_2 = 0.053\,\mu F$, which means that the first two give $C_1 = 0.159\,\mu F$, $L = 35.4$ mH.

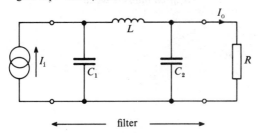

filter

12.2 The required form of second-order, high-pass filter is shown in the figure and analysis shows that its transfer function is

$$\mathcal{T} = \frac{V_o}{V_i} = \left[1 + \frac{1}{RC}\left(\frac{1}{s}\right) + \frac{1}{LC}\left(\frac{1}{s}\right)^2\right]^{-1}$$

Second-order, high-pass, Chebyshev response is, however, that given by equation (12.25) when ω/ω_c is replaced by ω_c/ω. The poles of $|\mathcal{T}|^2$ must therefore be given by

$$(\omega_c/s)^2 = -0.5 \pm j/2\xi$$

where, because the ripple in the pass band is to be 2 dB,

$$10\log_{10}(1+\xi^2) = 2$$

or

$$\xi^2 = 0.585$$

Imposing $\xi = 0.585^{\frac{1}{2}} = 0.765$ leads to poles given by

$$(\omega_c/s)^2 = -0.5 \pm 0.654j = 0.8232\exp j[(127.40° \text{ or } 232.60°) + 2p\pi]$$

or

$$\omega_c/s = \pm 0.9073 \exp j[(63.70° \text{ or } 116.30°) + p\pi] = \pm 0.4020 \pm 0.8134j$$

The appropriate transfer function is consequently

$$\mathcal{T} = [(\omega_c/s) + 0.402 + 0.8134j]^{-1}[(\omega_c/s) + 0.402 - 0.8134j]^{-1}$$
$$= [(\omega_c/s)^2 + 0.804(\omega_c/s) + 0.823]^{-1}$$

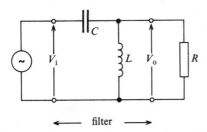

filter

and comparison with the transfer function of the network shown in the figure reveals that it is necessary for

$$1/RC = 0.804\omega_c/0.823$$
$$1/LC = \omega_c^2/0.823$$

In particular, to give the required filtering when $R = 1\,\text{k}\Omega$ and $\omega_c = 600\pi$

$$C = 0.54\,\mu\text{F}$$
$$L = 0.43\,\text{H}$$

INDEX

accumulator, 24, 27
active component or network, 29, 224–5, 230, 233
adder, operational amplifier, 242–3
admittance, complex
  general, 109–10
  small-signal, 222
alternating current
  bridges, 151–66
  comparator, 163, 164, 165–6
  mains, 99
  meters, 144–7
ammeters
  alternating current (a.c.), 144–7
  direct current (d.c.), 54–5
  measurement problems caused by finite resistance, 55, 56
ampere unit, 12, 54, 72
amplification, definition, 233
amplifier
  buffer, 243
  common-emitter, junction transistor, 224–6, 232–4
  feedback, general effects of, 235–40
  gain – separately listed
  general definition, 223
  instability, due to positive feedback, 236, 244–6, 249
  Miller effect in, 234–5
  operational – *see* operational amplifier
  stability under feedback, 244–6, 249–51
amplitude
  definition, 97
  demodulation, 176
  modulation, 126
  resonance, 111–6, 118–9
analogue-to-digital conversion, 55–6
analysis
  branch current, 31–3
  load-line, 62–4, 220–2, 225–6
  mesh current, 37–8, 41, 110

analysis (*continued*)
  node-pair potential, 39–41, 110
  small-signal, 21, 26–7, 220, 222–3, 226–35
angular frequency, 97
apparent power, 120, 123
Argand diagram, 295
astable multivibrator, 248
attenuation
  cascaded filter sections, 189, 202, 206
  definition, 167
  potential difference, 167–72
  power 173
attenuation constant
  definition, 191, 215
  ladder filter section, symmetric: band-pass, constant-k, 199–201; high-pass, constant-k, 197–8; low-pass, constant-k, 193–4; low-pass, m-derived, 206; purely reactive, 191–3, 193–4, 197–8, 199–201, 205–6
  transmission line, 215, 216, 219
attenuator
  loading 49–50, 167, 169–72, 173
  nature, 167
  pole-zero coincidence for, 277
  types, 167–73
auxiliary equation, of differential equation, 297, 299
average power, 120–2, 123–4
averager, operational amplifier, 242–3

back e.m.f., 73–4, 75, 127–8, 131
band-pass and stop filters
  constant-k, ladder L–C, 199–202
  definitions, 173–4
  single-section C–R, 179–85
  single-section L–C–R, 177–8, 179, 185
  synthesised, 286–8
bandwidth for rectangular pulse transmission, 257–8

base terminal, junction transistor, 224, 225–6
battery, 22–4, 25, 27
bias
  forward, 144
  general, 63
  junction transistor, common-emitter configuration, 225–6
  reverse, 144
  small-signal parameter dependence on, 222, 228
  two-terminal nonlinear network, 221
bistable multivibrator, 247, 248–9
Bode diagram or plot, 249–51
boundary conditions in differential equation solution, 298
branch current analysis, 31–3
bridge
  comparator, a.c., 163, 164, 165–6
  Hay, 156, 157, 160
  Heydweiller, 157, 158–9
  Maxwell, 155–6
  Owen's, 155, 156–7
  rectifier, 144–5
  resonance, 160, 161
  Robinson, 161
  Schering, 157–8
  transformer ratio-arm, 161–5, 186
  tuned-arm, 160, 161
  Wheatstone type, 59–62, 151–8, 160–1
  Wien, 160–1, 181
bridged-T attenuator, 170, 171–2
bridged-T rejection filters, 182–3, 185
buffer amplifier, 243
Butterworth filters, 279–82, 285–8

capacitance
  circuit symbols, 68–9
  colour code, 68
  combined, including series and parallel, 69–71
  definition, 65
  measurement, 148–50, 151–5, 157–8, 161–6
  nature, 65
  parallel plate, 66
  relations, basic, 65, 66, 69–71
  stray, 69, 102, 153–5, 164
  transmission line, 213
  unit, 66
capacitance–resistance (C–R) circuit
  attenuator or potential divider, 168–9
  band-pass and stop filters, 179–85
  coupling, 84–5, 86–8, 176–7, 221–2, 225–6, 229
  differentiators, 85–6, 87–8, 176–7, 243, 244, 250–1
  high-pass filters, 85–8, 174, 176–7, 243, 244, 278

capacitance–resistance (C–R) circuit (continued)
  impedance, 109
  integrators, 85, 88, 89, 174–6, 243–4
  low-pass filters, 85, 88–9, 174–6, 243, 244, 258–9, 278
  phase shifter, 186–8
  rectangular pulse response, 258–9
  rejection filters, 179–86
  square-wave response, 86–8, 89
  steady-state sinusoidal response, 109, 174–5, 176, 179–85, 186–8
  time constant, 82, 83, 87, 89, 93, 175, 181
  transient response, 80–4, 86–8, 89, 269–70
capacitive attenuator, 168, 172
capacitive reactance or impedance, 100–1, 102, 108
capacitor
  circuit symbols, 68–9
  leakage resistance, 65, 66–7
  linear, 65
  nature, 65
  nonlinear, 65–6, 68
  polarising voltage, 67
  potential energy, when charged, 71, 121–2
  power dissipation, 122–3
  types, 66–8
carbon resistor, 18
carrier wave, 126
cathode-ray oscilloscope
  input impedance, 102
  potential difference measurement, 102, 148
ceramic chip capacitor, 67
characteristic impedance
  definition and nature, 189, 210, 215, 278–9
  ladder filter section, symmetric: band-pass, L–C, constant-k, 201; high-pass, L–C, constant-k, 198; lattice type, 202; low-pass, L–C, constant-k, 195, 196, 202; m-derived, 203–5; Π-type, 189–90, 191, 195, 198, 202, 204–5, 211–2; purely reactive, 202; T-type, 189–90, 191, 195, 196, 198, 201, 202, 203, 205, 211–2
  transmission line: general, 215; lossless, 216, 217–8
characteristics – see static characteristics
charge
  concept, 1–2
  conservation, 2
  electronic, 2, 7
  ionic, 11
  mobile, 11–2
  mobility, 15
  sign, 1–2
  unit, 7, 12, 72

Index

Chebyshev filters, 282–8
choke, 122
circuit symbol – *see* symbol, circuit
closed-loop frequency response, 236
closed-loop gain, 235–6, 237, 238, 239–40, 241–2, 243, 244–5
closed-loop terminal (input and output) impedance, 236, 237–40, 242, 243
coaxial cable; construction, nonradiating and screening aspects, 212–3
coefficient of coupling, magnetic, 130
collector terminal, junction transistor, 224, 225–6
colour code, 19, 68, 77
common-base configuration, 232
common-collector configuration, 232
common-emitter capacitor-coupled amplifier, 225–6
common-emitter configuration
  nature, 224
  small-signal analysis; current, voltage and power gain and input and output resistance, 232–4
  small-signal *h*-parameter measurement, 228–9
common-mode rejection ratio, 240
common-mode signal, 240
comparator
  operational amplifier, potential difference, 247
  transformer, alternating current, 163, 164, 165–6
complementary function, in differential equation solution, 297–8, 299, 300–1
complex admittance – *see* admittance, complex
complex algebraic representation of Kirchhoff's laws, 107–10, 295–6
complex impedance – *see* impedance, complex
complex quantity, nature, 295
composition resistor, 18
conductance, electrical
  concept, 17, 109
  relations, basic, 17, 109, 222
  small-signal, 222
  transmission line, 213
  unit, 17
conduction, electrical, 11–2
conductivity, electrical
  concept, 14
  magnitude in solids, 17
  mobile electronic in solids, 14–5
  relations, basic, 14, 15, 17
  temperature dependence, 17
  unit, 17
constant-current source, direct, 27
constant-*k* filters and sections, 193–203
constant-voltage source, direct, 27
continuous frequency spectrum
  nonperiodic signal, 256–7
  rectangular pulse, 257–8
copper loss
  inductor, 77
  transformer, 134, 135
core loss
  inductor, 77
  transformer, 134, 135
core, magnetic, 77–8, 129, 133–5
correct termination
  ladder filter, symmetric L–C, constant-*k*: band-pass, 201, 202; high-pass, 197, 198, 202; low-pass, 194, 195, 196, 202–3
  lattice filter, symmetric L–C, constant-*k*, low-pass, 202–3
  meaning, 189, 278–9
  transmission line, 215, 218, 219
coulomb unit, 7, 12, 72
Coulomb's law, 3, 6–7
Coulomb's torsion balance, 3–4
coupling
  coefficient, magnetic, 130
  C–R, 84–5, 86–8, 176–7, 221–2, 225–6, 229
  critical, 141, 142
  transformer, 133, 135–9
C–R circuit – *see* capacitance–resistance circuit
critical coupling, 141, 142
critical damping, 92, 93, 94, 95, 273, 277
critical frequency or pulsatance
  ideal filter, 278, 279
  ladder filter, symmetric L–C, constant-*k*: band-pass, 200–1; band-stop, 202; high-pass, 197–8; low-pass, 193–4
  m-derived filter section, 203, 205–6
  synthesised filters, 279, 282, 283, 285, 286–7, 288
  transmission line, lossless, 216
current density, electric
  mobile electronic in solids, 14–5
  nature, 13–4
  relations, basic, 13–5
  unit, 14
current, electric
  balance, 54
  continuity, 30–1
  eddy, 77, 134
  measurement, 54–5, 57, 144–7
  mesh, 37
  nature, 11–4
  relations, basic, 11, 14, 16
  unit, 12, 54, 72
current gain, small-signal, common-emitter, 232, 233
cut-off frequency or pulsatance–*see* critical frequency or pulsatance

cut-off (or fall-off) of response, 86, 88, 173–5, 176, 178–9, 194, 198, 200–1, 202–3, 205–6, 234–5, 236, 241, 244, 278–9, 282–4, 285, 286, 288

damped oscillation, 91–2, 94, 95, 272, 277
decibel unit, 173
decrement, 94
delay
 ladder filter, low-pass L–C, 195–6
 lattice filter, low-pass L–C, 203
 transmission line, lossless, 216
delay line, 195–6, 203, 216
demodulation, amplitude, 176
dielectric constant, 6–7, 67
dielectric loss or resistance, 66–7, 122–3, 219
differential equations, linear with constant coefficients
 auxiliary equations for, 297, 299
 boundary conditions on, 298
 complementary functions of, 297–8, 299, 300–1
 examples of solutions, 300–1
 method of solution, 297–300
 nature, 296–7
 order, 296–7
 particular integrals of, 298, 299–301
differential resistance
 concept, 21
 four-terminal network, 227, 228
 internal, of source, 26–7
 two-terminal network, 222
differentiator
 C–R, basic passive, 85–6, 87–8, 176–7
 operational amplifier, active, 243, 244, 250–1
digital-to-analogue converter (R–2R), 51, 52
diode
 rectifying, 20, 21, 144, 145
 Zener, 20, 21
direct-current meters, 54–5
discrete component, assumption of, 80–1, 213
discrimination, frequency, 112, 125
discriminator, signal level, 247–8
dispersion, transmission line, 216
distortion, 168, 185–6, 196, 216, 258
double balance, a.c. bridge, 151–3
drift velocity, electronic, 14–5
dual circuits, 96
dual impedances, 199

earth, 10, 69, 154–5, 164, 212–3
eddy current, 77, 134
effective mass, electronic, 14, 15

electric charge, current, current density, field, force – see under charge, etc.
electrical conductance, conduction, conductivity, energy, potential difference, power, resistance, resistivity – see under conductance, etc.
electricity, nature, 1–2
electrodynamometer, 146–8
electrolyte, 11, 12
electrolytic capacitor, 67
electromagnetic induction, phenomenon and laws 73–5
electromotive force (e.m.f.)
 back, 73–4, 75, 127–8, 131
 circuit symbols, 24, 100
 concept, 22, 24–5
 induced, 73–5, 127–8, 131
 power delivered by, 24
 relations, basic, 24–8
 sources of direct, 22–8
 sources of sinusoidal, 97–9
 unit, 24
electron
 drift velocity, 14–5
 effective mass, 14, 15
 fundamental particle, 2
 mean free time, 15
 mobility, 15
 scattering, 14–5
electronic charge, 2, 7
emitter terminal, junction transistor, 224, 225–6
energy, electrical
 involved in charging capacitor, 71, 84, 121–2
 involved in establishing current in inductor, 75–6, 121–2
 nature, 8–9, 18, 22, 24, 71, 75–6, 128
 unit, 10
equivalent circuit
 four-terminal small-signal, 230–2
 hybrid, 231–2
 Norton, 51–3
 source, 25, 27–8
 Thévenin, 45–6, 48, 51–3
 transformer, 134–5
 transmission line, 213–4
 Z-parameter, 230–2
Euler's identity, 293

farad unit, 66
Faraday's law of electromagnetic induction, 73–5
feedback
 concept, 235
 fraction, 235, 244–5
 instability due to positive, 236, 244–6, 249
 negative, definition, 235

Index

feedback (*continued*)
  parallel-inserted, current-derived (shunt-current), 236, 237, 239–240
  parallel-inserted, voltage-derived (shunt-voltage), 236, 237, 238–40, 242
  positive, definition, 235, 245
  series-inserted, current-derived (series-current), 236, 237, 238, 239–40
  series-inserted, voltage-derived (series-voltage), 236–8, 239–40, 242
  stability of amplifier under, 244–6, 249–51
ferrite, inductor core, 77
ferromagnetic material, 73
field, electric
  nature, 7–8
  relations, basic, 7–9, 14–5
  unit, 10
filtering applications, 173, 175–6, 177, 179, 181, 185–6
filtering, definition and nature, 173
filters
  asymmetric section, 209–12
  band-pass and band-stop, 173–4, 177–8, 179–86, 199–202, 286–8
  Butterworth, 279–82, 285–8
  Chebyshev, 282–8
  constant-k, 193–203
  C–R, 84–9, 174–7, 179–85, 258–9, 278
  high-pass, 85–8, 174, 176–7, 197–9, 202, 206, 243, 244, 278, 285–6
  ideal, 173–4, 278, 279
  ladder – *see* ladder filters
  lattice, 202–3
  L–C, 178–9, 193–203, 205–6, 207, 208, 209, 216, 278, 280–2, 284–8
  L–C–R, 177–8, 179, 182, 185
  low-pass, 85, 88–9, 174–6, 177, 178–9, 193–6, 199, 202–3, 205–6, 207, 208, 209, 243, 244, 258–9, 278–85
  L–R, 177, 279–80
  m-derived, 203–9
  operational amplifier, active, 243, 244
  orders of, 278, 279, 282
  rejection, 173–4, 177–8, 179–81, 182–6
  single-section, 84–9, 173–86
  synthesised, 279–88
flip-flops, 247, 248–9
force
  electric, fundamental law and nature, 1, 2–3, 6–7
  magnetic, fundamental law and nature, 71–2
  unit, 7
forward bias, 144
forward transfer characteristic, 224–5
Fourier analysis, 86, 99, 167, 173, 252–9
Fourier coefficients or harmonic amplitudes, formulae for, 253–4, 256

Fourier spectra
  continuous, 256–8
  harmonic, 252–6
Fourier transform, definition, 257
four-terminal network
  concept, 223
  C–R basic types of, 84–9, 174–7
  current sign convention, 224
  h-parameters, 227–32
  hybrid equivalent circuit, 231–2
  load-line graphical analysis, 225–6
  pulse response by Fourier transformation, 258–9
  small-signal (linear approx.) analysis, 226–35
  analysis, 226–35
  static characteristics, 223–5
  Z-parameter equivalent circuit, 230–2
  Z-parameters, 226–7, 228–30
four-terminal resistance measurement technique, 57–8
frequency
  definition and unit, 97
  discrimination, 112, 125
  fundamental, of harmonic spectrum, 252
  measurement by bridge, 160–1
  response – *see* particular circuit
frequency spectrum
  half-wave rectified sinewave, 254–6
  nonperiodic signal, 256–7
  periodic signal, even or odd, 252–4, 256
  rectangular pulse, 257–8
  square wave, 252, 253
  unit-step function, 259–61
full-wave rectification, 144, 145
functions, mathematical
  Butterworth, 279
  Chebyshev, 282
  complex exponential, 293
  cosh, 293
  cosine, 289–91, 293
  exponential, 291–3
  harmonic, 289–91, 293
  sine, 289–91, 293
  sinh, 293
fundamental frequency, of harmonic spectrum, 252

gain
  closed-loop, definition and nature, 235, 245
  junction transistor, small-signal, common-emitter configuration, 232–4
  loop, definition and nature, 235, 245
  margin, for amplifier stability, 249, 250
  open-loop, definition and nature, 235, 244–5
  operational amplifier, inverting and noninverting configurations, 241–2

gain (*continued*)
  under feedback (closed-loop in terms of open-loop), 235–6, 237, 238, 239–40, 241–2, 243, 244–5

half-power points, 113–4
half-sections, filter, 206–9, 211–2
half-wave rectification, 144
harmonic frequencies, 252
harmonic frequency spectrum of periodic signal, 252–4, 256
Hay bridge, 156, 157, 160
Heavyside expansion theorem, 267–9
henry unit, 75, 128–9
hertz unit, 97, 157
Heydweiller bridge, 157, 158–9
high-pass filters
  basic C–R, 85–8, 174, 176–7, 278
  basic L–R, 177
  constant-k, 197–9, 202
  definition, 173
  ladder L–C, 197–9, 202
  m-derived, 206
  operational amplifier, 243, 244
  synthesised, 285–6
high-stability resistor, 18
hot-wire meter, 146
hybrid equivalent circuit
  general, high-frequency and common-emitter, 231–2
  small-signal analysis using, 232–4
hybrid or h-parameters
  active network case, 230
  bias dependence, 228
  definitions, 228, 231, 232
  from small-signal measurements, 228–9
  from static characteristics, 228
  passive network case, 230
  relation to Z-parameters, 229–30
  symmetric passive network case, 230
  transistor configurations, 232
hysteresis
  magnetic, 77, 134
  Schmitt trigger, 247–8

image impedance, 210–2
imaginary part of complex quantity, 295
impedance
  capacitive, 100–1, 102, 108
  characteristic – *see* characteristic impedance
  combined, series and parallel, 109–10
  complex: definition, 107–8; small-signal, 222–3
  C–R circuit, 109
  dual, 199
  image, 210–2
  inductive, 101, 108

impedance (*continued*)
  input, miscellaneous cases – *see* input impedance
  iterative, 209–10, 211
  L–C–R circuit, 110–2, 114–5, 118–9
  L–R circuit, 104, 107, 108
  magnitude, definition, 104
  matching, 123–4, 133, 138, 173, 210, 217
  measurement, 148–60, 161–6
  output, miscellaneous cases – *see* output impedance
  reflected through transformer, 136–7, 138
  resistive, 100, 108
  series circuit, general, 108–9
  unit, 104
incremental resistance, 21, 26–7, 222, 227, 228
induced e.m.f., 73–5, 127–8, 131
inductance, self
  circuit symbols, 77–8
  combined, including series and parallel, 78–80
  definitions, alternative, 75–6
  measurement, 148–50, 151–7, 161–6
  nature, 71–6
  relations, basic, 75–6, 78–9
  stray, 78, 153–4
  transmission line, 213
  unit, 75
induction, electromagnetic – *see* electromagnetic induction
inductive reactance or impedance, 101, 102, 108
inductive voltage surge, 84
inductor
  circuit symbols, 77–8
  core, laminated or otherwise, 77
  linear or nonlinear, 75–6, 77
  losses; copper, eddy current and hysteresis, 77
  potential energy, due to current, 75–6, 121–2
  power dissipation, 122
  types, 77
input characteristics, 224–5
input impedance
  amplifier, under feedback (closed-loop in terms of open-loop), 236, 237, 238, 239, 240, 242, 243
  cathode-ray oscilloscope, 102
  low-pass, symmetric, L–C, filter section terminated in resistance $(L/C)^{\frac{1}{2}}$, 195, 196
  m-derived terminating half-sections, 206–9
  small-signal, of four-terminal network, 228
  transformer, 136, 137, 138

input impedance (*continued*)
  transmission line, lossless, 216–7, 218–9
input resistance, common-emitter
  configuration, 232, 234
instability
  amplifier, due to positive feedback, 236,
    244–6, 249
  operational amplifier differentiator, at
    high frequency, 244, 251
integrated-circuit capacitor, 67–8
integrator
  C–R, basic passive, 85, 88, 89, 174–6
  operational amplifier, active, 243–4
intermediate frequency, 125–6, 142–3
internal resistance, source of e.m.f.
  concept, 25
  determination, 26
  magnitudes, typical, 27
inverse Fourier transform, definition, 257
inverse Laplace transform
  for polynomial ratios, 265–9
  for some common functions, 265
inverse square law of force between point
    charges
  direct verification, 3–4
  indirect verification, 4–6
inverting configuration of operational
    amplifier, 241, 242–4
inverting input terminal of operational
    amplifier, 240
ionic charge, 11
iterative impedance, 209–10, 211

j-operator, 107, 293, 295
joule unit, 10
junction transistor
  common-emitter amplifier, 224–6, 231,
    232–4
  configurations, 224, 232
  static characteristics, 224–5, 228
  symbol, circuit, N–P–N type, 225–6
  terminals; base, collector and emitter,
    224, 225–6

Kirchhoff's laws, 30–1, 36–7, 80–1

ladder attenuators, 170, 172
ladder filters
  asymmetric section, 209–12
  attenuation constant – separately listed
  band-pass, L–C, symmetric, 199–201
  band-stop, L–C, symmetric, 201–2
  constant-k, 193–203
  correct termination – separately listed
  high-pass, L–C, symmetric, 197–9
  lattice, 202–3

ladder filters (*continued*)
  low-pass, L–C, symmetric, 193–6, 202–3
  low-pass, L–C, symmetric, infinitesimal
    section, 216
  m-derived, 203–9
  phase shift – separately listed
  propagation and propagation constant,
    191
  purely reactive, 191–203
  symmetric section, 189–209
laminated core, 77, 134
lamp filament, 20–1
Laplace transform
  definition, 261
  of some common functions, 261–5
  use in deducing transient response,
    269–77
lattice filter, 202–3
L–C filter
  basic single-section, low-pass, 178–9, 278
  ladder types of, 193–203, 205–6, 207,
    208, 209, 216
  synthesised types of, 280–2, 284–8
L–C oscillator, 246
L–C–R circuit
  pair, inductively coupled, 139–43
  steady-state sinusoidal response, 110–9
  transient sinusoidal response, 273–5
  transient step response, 89–96, 272–3,
    277
L–C–R filter
  band-pass, 177–8, 179
  band-stop or rejection, 177–8, 179, 182,
    185
leakage
  magnetic flux, 134–5
  resistance, 65, 66–7
Lenz's law, 73–4
linear capacitor, 65
linear inductor, 75–6, 77
linear resistor, 18
load, 25
load line
  direct or bias, 62–3, 221–2, 225–6
  input, 225–6
  output, 225–6
  signal, 221, 222, 225, 226
load resistor, 25–6
load-line analysis, 62–3, 64, 220–2, 225–6
local oscillator, 125–6
logarithmic decrement, 94
loop gain, 235, 245
losses
  capacitor, 65–7, 122–3, 219
  inductor, 77
  transformer, 134, 135
  transmission line, 212, 213, 219
low-pass filters
  basic C–R, 85, 88–9, 174–6, 258–9

low-pass filters (*continued*)
  basic L–C, 178–9, 278
  basic L–R, 177, 278
  Butterworth, 279–82
  Chebyshev, 282–5
  constant-k, 193–6, 199, 202–3
  definition, 173
  ladder L–C, 193–6, 202–3, 205–6, 207, 208, 209
  m-derived, 205–6, 207, 208, 209
  operational amplifier, active, 243, 244
  synthesised, 279–85
L–R filter, basic low or high-pass, 177, 278
L–R series circuit
  impedance, 104, 107, 108
  steady-state sinusoidal response, 103–4, 106–8
  time constant, 82, 83, 101, 103, 104
  transient sinusoidal response, 101, 103, 104–5
  transient step response, 80–4, 269

magnetic flux
  definition and nature, 73
  leakage, 130, 134–5
  relations basic, 73, 74, 75, 127, 130, 131
  unit, 73
magnetic force – *see* force, magnetic
magnetic induction
  definition and nature, 72–3
  relations, basic, 72, 73
  unit, 73
magnetisation, 73
mains, a.c., 99
matching, 28–9, 123–4, 133, 138, 173, 210, 217
Maxwell bridge, 155–6
m-derived sections
  half, terminating, 206–9
  Π-type, 204–5
  purely reactive, 205–6, 207, 208, 209
  resonance in, 205–6
  T-type, 203–4
mean free time, electron, 15
measurement
  capacitance, 148–50, 151–5, 157–8, 161–6
  current, 54–5, 57, 144–7
  frequency, 160–1
  impedance, 148–60, 161–6
  inductance, 148–50, 151–7, 161–6
  mutual inductance, 158–60
  potential difference, 54, 55–6, 56–7, 58–9, 102, 144–6, 148
  power, 147–8
  power factor, 151, 158
  Q-factor, 150, 151
  reactance, 148–60, 161–6
  resistance, 54, 56, 57–8, 148–50, 151–9, 161–6

mesh current, 37
mesh current analysis, 37–8, 41, 110
mesh, in network, 30
mesh pair, inductively coupled, 127–43
metal-film resistor, 18
meters
  alternating current, 144–7
  capacitance or inductance measurement by, 148–50, 151
  direct current, 54–5
  impedance or reactance measurement by, 148–51
  power measurement by, 147–8
  Q-factor measurement by, 150, 151
  resistance measurement by, 54, 56, 148–50, 151
Miller effect, 234–5
mixing, 125–6
mobility, charge, 15
modulation, amplitude, 126
monostable multivibrator, 248
moving-iron meter, 146
multivibrators, 246–7, 248–9
mutual characteristics, 224–5
mutual inductance
  coefficient of coupling, 130
  definitions, alternative, 127–8
  dependence on current, 134
  measurement, 158–60
  nature, 127–8
  relations, basic, 127–8, 130
  symbol, circuit, 129
  unit, 128—9

negative differential resistance, 21
negative feedback
  definition, 235
  effects of, virtues of, 235–40
newton unit, 7
node
  in network, 30
  of transmission line, 218–9
node-pair potential analysis, 39–41, 110
noninverting input terminal, 240
noninverting operational amplifier configuration, 241–3
nonlinear capacitor, 65–6, 68
nonlinear devices, miscellaneous, 20, 21, 144, 145, 223–5, 228
nonlinear inductor, 76, 77
nonlinear network or circuit analysis
  algebraic, 63–4, 124–6
  four-terminal, 223–35
  load-line graphical, 62–3, 64, 220–2, 225–6
  small-signal (linear approx.), 21, 26–7, 220, 222–3, 226–35
  two-terminal, 21, 26–7, 62–4, 124–6, 220–3

Index

nonlinear resistive circuit
  differential resistance, 21, 26–7
  mixing, 125–6
  sinusoidal response, 124–6
nonlinear resistor, at excessive current, 18
nonlinear transformer, 134
Norton equivalent circuit, 51–3
Norton's theorem
  correspondence with Thévenin's theorem, 51–3
  examples of application, 53
  relevance to hybrid equivalent circuit, 231
  statement, 51, 110
null, balance, 56–7, 58, 59–60, 151–3, 154–5, 186
Nyquist criterion or diagram, 245–6

ohm unit, 17, 54, 102, 104
Ohm's law
  derivation, 14–6
  statement, 15
open-circuit, 25
open-loop gain, 235, 244–5
operating point or bias, 63, 220–2, 225–6
operational amplifier
  adder, 242–3
  averager, 242–3
  comparator, potential difference, 247
  differentiator, 243, 244, 250–1
  filters, 243, 244
  instability at high frequencies, 244, 251
  integrator, 243, 244
  inverting configuration, 241, 242–4
  multivibrators, 246–7, 248–9
  nature, 240–1
  noninverting configuration, 241–3
  scaler, 241–2
  scaling adder, 242–3
  Schmitt trigger, 247–8
  virtual earth, 242
  voltage follower, 242, 243
optical lever, 54
order
  of differential equation, 296–7
  of filter, 278, 279, 282
oscillation
  damped, 91–2, 94, 95, 272, 277
  continuous: due to positive feedback, 99, 236, 245–7, 248–9; parasitic, 245–6; relaxation, nonsinusoidal, 245–7, 248–9; sinusoidal, 99, 245–6
oscillators
  L–C, 246
  local, in receivers, 125–6
  nature, 245–7
  relaxation, 245–7, 248–9
  sinusoidal, 99, 245–6
  Wien-bridge, 181, 246

output characteristics, 224–5, 228
output impedance
  amplifier, under feedback (closed-loop in terms of open-loop), 236, 237–8, 239, 240, 242, 243
  four-terminal, small-signal, 227
output resistnace
  concept, 50
  junction transistor, common-emitter configuration, 233, 234
overdamped behaviour or response, 92, 93, 94, 95, 273, 277
Owens' bridge, 155, 156–7

Π-network, form, 171
Π-section, symmetric, ladder filter, 189–99, 202, 203, 204–6, 209, 211–2
parallel combinations of
  capacitance, 69, 70
  impedance, 110
  inductance, 78–9
  resistance, 33, 34–5
parallel resonance, 114–9
parallel resonant circuit
  entirely: admittance, 114–5; form, 114, 115; phasor diagram, 115; quality or Q-factor, 115–6; resonant frequency, 115; sinusoidal response, amplitude and phase versus frequency, 114–5, 116
  practical: amplitude resonance, 118–9; equivalent entirely parallel form, 116–7; form, 116, 117; impedance, 118–9; phase resonance, 117–8; quality or Q-factor, 117–8; resonant frequency, 117, 118, 119
parallel-plate capacitance, 66
parasitic oscillation, 245–6
particular integral, in differential equation solution, 298, 299–301
passive component or network, 29, 230
period, definition, 97
permeability
  of free space, 72
  of medium, 72
  unit, 72, 75
permittivity
  of free space, 6, 7, 72
  of medium, 6
  unit, 66
phase
  constant, transmission line, 215, 216, 217
  difference, lag or lead, 97–8
  margin, for amplifier stability, 249
  resonance, 117–8
  response of circuit – see particular circuit

phase (*continued*)
  shift in ladder filters, 191, 192–5, 196, 197–9, 200, 201
  shifter, constant amplitude, 186–8
  shifting, ideal, 186
  splitting, 186, 187
  velocity, transmission line, 215, 216
phasor, 105, 295
phasor diagram
  impedance determination from, 148–50
  nature and circuit solution by, 105–7, 294–5
  resonant circuit, 112–3, 115
plastic film capacitor, 66–7
P–N junction
  diode, 20, 21, 144, 145
  capacitor, 67, 68
poles, of functions, 275
pole-zero plots, 276–7
positive feedback
  definition, 235, 245
  effect on amplifier gain and terminal impedance, 235, 236–40
  instability through accidental, 236, 244–6, 249
  oscillation through, 236, 246–9
  switching through, 99, 236, 244–7, 248–9
potential concept and origin, 10
potential difference
  comparator, 247
  concept and definition, 8–9
  due to point charge, 9–10
  measurement, 54, 55–6, 56–7, 58–9, 102, 144–6, 148
  relations, basic, 8, 9, 10, 16
  sign convention, 25
  unit, 10, 54
potential dividers
  capacitive, 168, 172
  circuit function, 167
  ladder, 170, 172
  loading aspect, 49–50, 167, 168–72
  R–C, 168–9
  resistive, 49–50, 167–8, 169–72
  single-section, 49–50, 167, 168–72
  transformer, 172
potential energy
  capacitor when charged, 71, 121–2
  inductor carrying current, 75–6, 121–2
  magnetically coupled meshes, 128
potentiometer instrument
  accuracy and discrimination, 58–9
  circuit operation, potential difference measurement by, 56–7
  current and resistance measurement by, 57–8
potentiometer component, 19–20, 167–8
power, electrical
  apparent, 120, 123

power, electrical (*continued*)
  attenuation, 173
  average, 120–2, 123–4
  concept, 18
  delivered by e.m.f., 24
  dissipation: absence of reactive, 121–2; capacitor, 122–3; inductor, 122; resistive, 18, 24, 120–1
  factor, 120–3, 151, 158
  instantaneous, 119–20, 120–2, 132
  meter, 147–8
  reactive, 123
  relations, basic, 18, 120, 122
  supply, direct mains-derived, 176, 179
  transfer, in lossless, unity-coupled transformer, 137–8, 138–9
  transmission grid, mains, 132–3
  unit, 18, 123
power factor
  capacitor, 122–3
  definition, 120
  general impedance, 122
  inductor, 122
  measurement, 151, 158
power gain, junction transistor, common-emitter configuration, 233–4
preferred range
  capacitor, 68
  inductor, 77
  resistor, 19
primary, winding of transformer, 127
propagation and propagation constant
  ladder filter, 191
  transmission line, 214–5, 216
prototype filter section, 203
pulsatance, definition, 97
pulse response by Fourier transformation
  four-terminal network, 258
  low-pass C–R filter, 258–9

quality or Q-factor meter, 150, 151
quality or Q-factor of resonant circuit
  entirely parallel form, 115–6
  entirely series form, 111–4
  practical parallel form, 117–8
quarter-wavelength lossless transmission line, 217

rationalised units, 7, 72
R–C attenuators, 168–9
reactance
  capacitive, 100–1, 102
  definition, 101
  inductive, 101, 102
  measurement, 148–60, 161–6
  mutually inductive, 128
  unit, 102
reactive power, 123
real part of complex quantity, 295

Index

reciprocity theorem, 41–3, 44–5, 110
rectification, half and full-wave, 144, 145
rectifier, bridge, 144–5
reflected impedance, 136–7, 138
reflection, transmission line, 215, 217–8, 219
rejection filters, 173–4, 177–8, 179–81, 182–6
relaxation oscillators, 245–7, 248–9
resistance–capacitance (R–C) circuit – *see* capacitance–resistance (C–R) circuit
resistance, electrical
  colour code, 19
  combined, series and parallel, 33–5
  concept, 16–7
  differential, incremental or small-signal, 21, 26–7, 222, 227, 228
  input, junction transistor, common-emitter configuration, 232, 234
  leakage, 65, 66–7
  measurement, 54, 56, 57–8, 148–50, 151–9, 161–6
  output, concept, 50
  output, junction transistor, common-emitter configuration, 233–4
  relations, basic, 16–7, 18
  symbols, 19–20
  transmission line, 212, 213, 219
  unit, 17, 54, 102, 104
resistive attenuators, 167–8, 169–72
resistive power dissipation, 18, 24, 120–1
resistivity, electrical, concept and unit, 17
resistor
  nature, 18
  symbols, 19–20
  types, 18–9
resonance
  amplitude, 110–6, 118–9
  bridge, 160, 161
  parallel, 114–9
  phase, 110–8, 119
  series, 110–4
resonant frequency, 111, 114, 115, 117, 118, 119, 272
reverse bias, 144
reverse-transfer characteristics, 224–5
ringing, 91–2, 272, 277
Robinson bridge, 161
root-mean-square (r.m.s.) reading meter 145, 146
root-mean-square (r.m.s.) value, 98–9

scaler, operational amplifier, 241–2
scaling adder, operational amplifier, 242–3
scattering, electron, 14–5
Schering bridge, 157–8
Schmitt trigger, 247–8
screening
  electrostatic, 69, 153–4, 212–3
  inductive, 78, 153–4

secondary, winding of transformer, 127
semiconductor, 11–2, 17
series combinations of
  capacitance, 69–70
  impedance, 109
  inductance, 78, 79
  resistance, 33–5
series resonance, 110–4
series resonant circuit
  form, 110
  half-power points, 113–4
  impedance, 110–2
  phasor diagram, 112–3
  practical, 112
  quality or Q-factor, 111–4
  resonant frequency, 111, 114
  sinusoidal response, amplitude and phase versus frequency, 110–3
  tuning, 112
short-circuit, 27
shunt, ammeter, 55
siemen unit, 17
sign convention, four-terminal network current, 224
sinusoidal oscillators, 99, 245–6
sinusoidal sources, 97–9, 245–6
sinusoidal response
  attenuators, 167–73
  by complex algebraic solution, 107–10
  by differential equation solution, 102–5
  by phasor diagram solution, 105–7
  capacitive, purely, 100–1, 102
  C–R, 109, 174–5, 176, 179–85, 186–8
  fedback or closed-loop, 235–40, 241–4
  inductive, purely, 100, 101, 102
  ladder filters, 190–203, 205–6
  L–C, 178–9, 193–203, 205–6, 280–2, 284–5, 288
  L–C–R, 110–9, 177–8, 185, 273–5
  L–R, 103–5, 106–8, 177
  nonlinear network, 124–6, 222–3, 226–7, 227–8, 232–4
  operational amplifier, 241–4
  phase shifter, 186–8
  resistive, purely, 100, 102
  single-section filters, 173–5, 176, 177–9, 180–5
  steady-state, 99–104, 105–26, 128, 135–42, 167–73, 173–5, 176, 177–9, 180–5, 186–8, 190–203, 205–6, 213–9, 222–3, 226–7, 227–8, 232–5, 235–40, 241–4, 273–5, 279, 280–4, 285, 286, 288
  synthesised, 279, 282–4, 285, 286, 288
  transformer, loaded, 135–9
  transformer-coupled L–C–R circuits, 139–42
  transient, 101, 103, 104–5, 273–5
  transistor amplifier, 232–5
  transmission line, 213–9

Index

skin effect, transmission line, 219
small-signal admittance and impedance, 222, 227, 228
small-signal analysis
  four-terminal, 226–35
  h-parameter, 227–9, 231–4
  junction transistor, common-emitter configuration, 232–4
  two-terminal, 21, 26–7, 220, 222–3
  Z-parameter, 226–7, 228, 230–1
small-signal conductance and resistance, 21, 26–7, 222, 227, 228
small-signal equivalent circuits, 230–2
small-signal linearity of nonlinear network, 220, 222, 226, 228–9
small-signal parameters, h and Z, 226–32
source
  direct, constant-current or constant-voltage, 27
  equivalent circuits, 25, 27–8
  matching, 28–9, 123–4, 133, 138, 173, 210, 217
  of e.m.f., direct, 22–8
  sinusoidal, 97–9, 245–6
  terminal behaviour, 25–7
  Van-de-Graaff, 27
spectra, signal equivalent, 252–8, 259–65, 276–7
s-plane diagrams, 276–7
square-wave response, C–R, 86–8, 89
stability, amplifier, 244–6, 249–51
standard e.m.f. and resistance, 54
standard impedances, for matching, 173
standing wave, transmission line, 217–9
star-delta transformation, 35
static characteristics
  current sign convention, 224
  four-terminal, 223–5
  Gunn device, 20, 21
  h-parameters from, 228
  input, output, transfer and mutual, general nature, 224
  junction diode, rectifying, 20, 21
  junction transistor, common-emitter configuration, 224–5, 228
  lamp filament, 20–1
  nonlinear devices, miscellaneous, 20–1, 224–5, 228
  two-terminal, 20–1, 220–1
  Zener diode, 20, 21
  Z-parameters from, 227
stray capacitance, 69, 102, 153–5, 164
stray inductance, 78, 153–4
stub, transmission line, 219
superconductivity, 17
superposition theorem
  examples of application, 44, 51
  proof, 41–3
  statement, 43–4, 110, 126

susceptance, 109–10
switching, through positive feedback, 236, 246–9
symbol, circuit
  capacitance or capacitor, 68–9
  diode, 144, 145
  electromotive force (e.m.f.), 24, 100
  inductance or inductor, 77–8
  junction transistor, N–P–N bipolar, 225–6
  mutual inductance, 129
  node, 30
  resistance or resistor, 19–20
synthesis, filter
  band-pass, 286–8
  band-stop, 288
  high-pass, 285–6
  low-pass, 279–85
  scaling in, 282

tesla unit, 73
theorem
  Norton, 51–3, 110, 231
  reciprocity, 41–3, 44–5, 110
  superposition, 41–4, 51, 110, 126
  Thévenin, 45–53, 110, 231
thermocouple meter, 146
Thévenin equivalent circuit, 45–6, 48, 51–3
Thévenin's theorem
  correspondence with Norton's theorem, 51–3
  examples of application, 49–51
  relevance to Z and h-parameter equivalent circuits, 231
  statement, 45–6, 110
  verification, 46–8
threshold detector, 247
time constant, 82, 83, 87, 89, 91, 93, 101, 103, 104–5, 175, 177, 181
T-network, form, 171
transfer characteristics, 224–5, 228
transfer function
  definition, 174
  see also particular circuit
transform
  Fourier, definition, 257
  Laplace, definition, 261
transformer
  audio-frequency, 133
  coefficient of magnetic coupling, 130
  comparator, 163, 164, 165–6
  core, 129, 133–4
  coupled L–C–R circuits, 139–43
  coupling, of circuits, 133, 135–9
  equivalent circuits, 134–5
  ideal, 132, 133, 136, 138–9
  imperfectly coupled magnetically, 135–7
  input or terminal impedance, 136, 137, 138

Index

transformer (*continued*)
  isolation, 133
  lamination, 134
  linearity, 134
  losses, 134, 135
  lossless, behaviour, 135–9
  magnetic flux leakage, 134–5
  mains-frequency, 133
  matching, 133, 138
  nature, 131–2
  perfectly coupled magnetically (unity-coupled), 129–30, 131–2, 137–9
  power transfer in unity-coupled, lossless, 137–8, 138–9
  primary, 127
  radio-frequency, 133
  reflected impedance, 136–7, 138
  relations, basic, 131–2
  secondary, 127
  turns ratio, 131, 137, 138
  unity-coupled, 129–30, 131–2, 137–9
transformer ratio-arm bridge
  current comparator in, 163, 164
  double, 163–5
  phase-splitting in, 161–2, 186, 187
  single, 161–3
  three-decade divider in, 162–3
transient response, 80–4, 86–8, 89–96, 101, 103, 104–5, 168, 258–9, 269–75, 277
transmission line, general
  attenuation constant, 215, 216, 219
  basic nature, 212–3
  capacitance, 213
  characteristic impedance, 215
  coaxial cable type, 212–3
  conductance, parallel, 213
  current and potential distribution along, 213–5
  dielectric loss, 219
  distributed properties, 213
  equivalent circuit, 213–4
  inductance, 213
  infinite length case, 215
  phase constant and velocity, 215
  propagation constant, 214–5
  propagation of signal along, 215
  skin effect loss, 219
  reflection at irregularities, 219
  reflection at termination, 215
  resistance, series, 212, 213, 219
  wavelength on, 213
transmission line, lossless approximation
  attenuation constant, 216
  characteristic impedance, 216, 217–8
  cut-off pulsatance, 216
  delay along, 216
  dispersion, 216
  input impedance, dependence on position, 218–9; dependence on

transmission line, lossless approximation (*continued*)
  terminating load, 216–7
  low-pass, L–C, ladder-filter treatment, 216
  matching by, 217
  nodes, 218–9
  parameter determination, 217
  phase constant and velocity, 216, 217
  propagation constant, 216
  quarter wavelength, 217
  reflection coefficient at termination, 217–8
  standing waves on, 217–9
  stub for correct termination, 219
  voltage standing wave ratio, 218
T-section, asymmetric, 209–11
T-section, symmetric, ladder filter, 189–202, 203–4, 205–6, 207, 208–11
tunable components – *see* variable
tuned-arm bridge, 160, 161
tuning, resonant circuit, 112, 125–6
twin-T rejection filter, 183–5
two-terminal network
  algebraic analysis of nonlinear, 63–4, 124–6
  load-line graphical analysis, 62–3, 64, 220–2
  small-signal (linear approx.) analysis, 21, 26–7, 220, 222–3

underdamped behaviour or response, 91–2, 94, 95, 272, 275, 277
unit, fundamental (S.I.) for
  apparent power, 123
  capacitance, 66
  charge, 7, 12, 72
  conductance, 17
  conductivity, 17
  current, 12, 54, 72
  current density, 14
  electromotive force (e.m.f.), 24
  energy, 10
  field, 10
  force, 7
  frequency, 97
  impedance, 104
  inductance, 75
  length, 7
  magnetic flux, 73
  magnetic induction, 73
  mass, 7
  mutual inductance, 128–9
  permeability, 72, 75
  permittivity, 66
  potential difference, 10, 54
  power, 18, 123
  power attenuation, 173
  reactance, 102

unit, fundamental (S.I.) for (*continued*)
   reactive power, 123
   resistance, 17, 54, 102, 104
   resistivity, 17
   time, 7
   work, 10

Van-de-Graaff source, 27
var unit, 123
variable capacitors, 68, 69, 112
variable inductors, 77, 78, 112
variable resistors, 19–20
virtual earth, 242
volt unit, 10, 54
volt amp unit, 123
voltage
   follower, 242, 243
   gain, junction transistor, common-emitter configuration, 232–3
   standing wave ratio, 218
voltmeters
   alternating signal, 144–6, 148
   cathode-ray oscilloscopes as, 102, 148
   direct, 55–6
   errors caused by noninfinite impedance, 55, 56, 102, 146, 148

Wagner earth, 154–5
watt unit, 18, 123
wavelength, transmission line, 213
weber unit, 73
Wheatstone bridge, direct current
   accuracy, 60
   balance condition, 59–60, 61
   circuit and operation, 59–60
   detector, 59, 60, 61–2
   sensitivity optimisation, 60–2
Wheatstone bridge rectifier, 144–5
Wheatstone form of alternating current bridge
   comparator, 165–6

Wheatstone form of alternating current bridge (*continued*)
   components, 151, 153
   detection system, 151, 153
   difference technique, 153
   double balance, 151–3
   general form, 151, 152
   Hay type, 156, 157, 160
   Maxwell type, 155–6
   Owen type, 155, 156–7
   resonance type, 160, 161
   Robinson type, 161
   Schering type, 157–8
   screening, 153–4
   sensitivity, 153
   source, 151, 153
   substitution technique, 153
   tuned-arm type, 160, 161
   Wagner earth, 154–5
   Wien type, 160–1, 181
Wien bridge, 160–1, 181
Wien bridge oscillator, 181, 246
Wien filters, 179–82
wire-wound resistors, 18
work unit, 10

Zener diode, 20, 21
zero-crossing detector, 247
zeros of functions, 275–6
Z-parameter equivalent circuit, 230–2
Z-parameters
   active network case, 230
   bias dependence, 228
   definitions, 226–7
   from small-signal measurements, 227, 228–9
   from static characteristics, 227
   passive network case, 230
   relation to h-parameters, 229–30
   symmetric passive network case, 230